ISAAC TODHUNTER

A TREATISE ON

PLANE CO-ORDINATE GEOMETRY

AS APPLIED TO THE STRAIGHT LINE

AND THE

CONIC SECTIONS

Elibron Classics
www.elibron.com

Elibron Classics series.

ISBN 1-4021-0021-3 (paperback)
ISBN 1-4021-3142-9 (hardcover)

This Elibron Classics Replica Edition is an unabridged facsimile of the edition published in 1888 by Macmillan and Co., London and New York.

A TREATISE ON

PLANE CO-ORDINATE GEOMETRY.

A TREATISE ON

PLANE CO-ORDINATE GEOMETRY

AS APPLIED TO THE STRAIGHT LINE

AND THE

CONIC SECTIONS.

With Numerous Examples.

BY I. TODHUNTER, Sc.D., F.R.S.,

LATE HONORARY FELLOW OF ST JOHN'S COLLEGE, CAMBRIDGE.

London:

MACMILLAN AND CO.

AND NEW YORK.

1888

PREFACE.

I HAVE endeavoured in the following Treatise to exhibit the subject in a simple manner for the benefit of beginners, and at the same time to include in one volume all that students usually require. In addition, therefore, to the propositions which have always appeared in such treatises, I have introduced the methods of *abridged notation*, which are of more recent origin; these methods which are of a less elementary character than the rest of the work, are placed in separate Chapters, and may be omitted by the student at first.

The Examples at the end of each Chapter, will, it is hoped, furnish sufficient exercise on the principles of the subject, as they have been carefully selected with the view of illustrating the most important points, and have been tested by repeated experience with pupils. At the end of the volume will be found the results of the Examples, together with hints for the solution of some which appear difficult.

The properties of the parabola, ellipse, and hyperbola, have been separately considered before the discussion of the general equation of the second degree, from the belief that the subject is thus presented in its most accessible form to students in the early stages of their progress.

I. TODHUNTER.

ST JOHN'S COLLEGE,
July, 1855.

In the fourth edition the work has been carefully revised, and a large amount of new matter has been introduced, chiefly relating to the more recent methods of investigating the properties of the conic sections. The work was originally designed for early students, and in the additions which have been made this object has been constantly regarded. Accordingly great attention has been given to the explanation and illustration of the principles of the methods which are employed; so that it will be easy for a student hereafter to develope these principles to any required extent.

The favour with which the work has been received indicates that it has been found adapted for the purpose of elementary instruction; and the hope may be expressed that the improvements now effected will increase its utility.

May, 1867.

CONTENTS.

PLANE CO-ORDINATE GEOMETRY.

Students reading this work for the first time may omit Chapters IV, VII, XIV, XV, XVI.

PLANE CO-ORDINATE GEOMETRY.

CHAPTER I.

CO-ORDINATES OF A POINT.

1. In Plane Co-ordinate Geometry we investigate the properties of straight lines and curves lying in one plane by means of *co-ordinates;* we commence by explaining what we mean by the *co-ordinates of a point.*

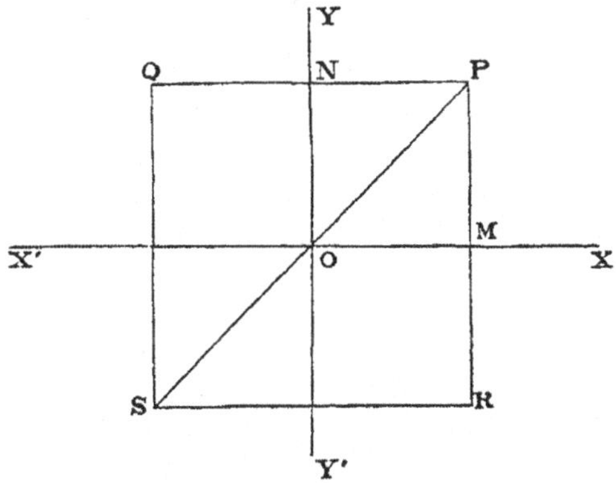

Let O be a fixed point in a plane through which the straight lines $X'OX$, $Y'OY$, are drawn at right angles. Let P be any other point in the plane; draw PM parallel to OY meeting OX at M, and PN parallel to OX meeting OY at N. The position of P is evidently known if OM and ON are known; for if through N and M straight lines be drawn parallel to OX and OY respectively, they will intersect at P.

The point O is called the *origin;* the straight lines OX and OY are called *axes;* OM is called the *abscissa* of the point P; and ON, or its equal MP, is called the *ordinate* of the point P. Also OM and MP are together called *co-ordinates* of the point P.

2. Let $OM = a$, and $ON = b$. then according to our definitions we may say that the point P has *its abscissa equal to a, and its ordinate equal to b*; or, more briefly, *the co-ordinates*

of the point P are a and b. We shall often speak of the point which has a for its abscissa and b for its ordinate, as *the point* (a, b).

3. A distance measured along the axis OX is however most frequently denoted by the symbol x, and a distance measured along the axis OY by the symbol y. Hence OX is called the axis of x, and OY the axis of y. Thus x and y are symbols to which we may ascribe different numerical values corresponding to the different points we consider, and we may express the statement that the co-ordinates of the point P are a and b, thus: for the point P, $x = a$ and $y = b$.

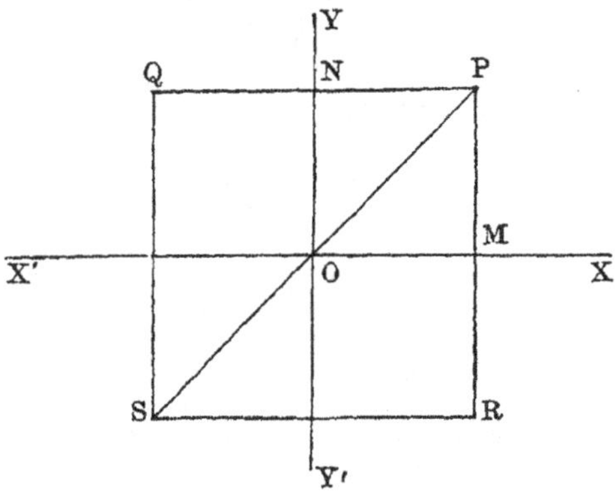

4. The straight lines $X'OX$, $Y'OY$, being indefinitely produced divide the plane in which they lie into four compartments. It becomes therefore necessary to distinguish points in one compartment from points in the others. For this purpose the following convention is adopted, which the reader has already seen in works on Trigonometry; straight lines measured along OX are considered positive and along OX' negative; straight lines measured along OY are considered positive, and along OY' negative. (See *Trigonometry*, Chap. IV.) If then we produce PN to a point Q such that $NQ = NP$, we have for the point Q, $x = -a$, $y = b$. If we produce PM to R so that $MR = MP$, we have for the point R, $x = a$, $y = -b$. Finally, if we produce PO to S so that $OS = OP$, we have for the point S, $x = -a$, $y = -b$.

5. In the figure in Art. 1 we have taken the angle YOX a right angle; the axes are then called *rectangular*. If the angle YOX be not a right angle, the axes are called *oblique*. All that has been hitherto said applies whether the axes are rectangular or oblique. We shall always suppose the axes rectangular unless the contrary be stated; *this remark applies both to our investigations and to the examples which are given for the exercise of the student.*

6. Another method of determining the position of a point in a plane is by means of *polar co-ordinates.*

Let O be a fixed point, and OX a fixed straight line. Let P be any other point; join OP; then the position of P is determined if we know the angle XOP and the distance OP. The angle is usually denoted by θ, and the distance by r.

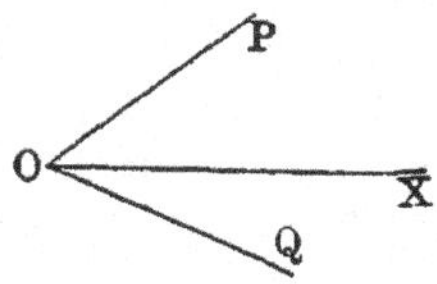

O is called the *pole*, OX the *initial line*; OP the *radius vector* of the point P, and POX the *vectorial angle*.

7. The position of *any* point *might* be expressed by *positive* values of the polar co-ordinates θ and r, since there is here no ambiguity corresponding to that arising from the four compartments of the figure in Art. 4. It is however found convenient to use a similar convention to that in Art. 4; angles measured in one direction from OX are considered *positive* and in the other *negative*. Thus if in the figure XOP be a *positive* angle, XOQ will be a *negative* angle; if the angle XOQ be a quarter of a right angle, we may say that for XOQ, $\theta = -\frac{\pi}{8}$. It is, as we have stated, not absolutely *necessary* to introduce *negative* angles, but *convenient;* the position of the straight line OQ, for instance, might be determined by measuring from OX in the positive direction an angle $= 2\pi - \frac{\pi}{8}$, as well as by measuring in the negative direction an angle $= \frac{\pi}{8}$.

Also positive and negative values of the radius vector are admitted. Thus, suppose the co-ordinates of P to be $\frac{\pi}{4}$ and a, that is, let $XOP=\frac{\pi}{4}$ and $OP=a$; produce PO to P', so that $OP'=OP$, then P' *may* be determined by saying its co-ordinates are $\frac{\pi}{4}$ and $-a$. Thus when the radius vector is a negative quantity, we measure it on the same straight line as if it had been a positive quantity but in the *opposite* direction from O.

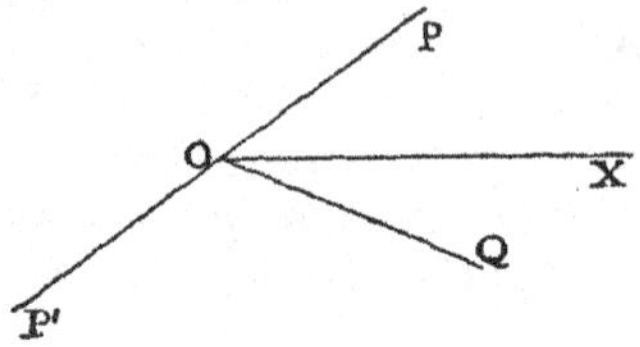

Hence if β represent any angle and c any length, the *same* point is determined by the polar co-ordinates β and $-c$ as by the polar co-ordinates $\pi+\beta$ and c.

8. Let x, y denote the co-ordinates of P referred to OX as the axis of x, and to a straight line through O at right angles to OX as the axis of y. Also let θ and r be the polar co-ordinates of P. If we draw from P a perpendicular on OX, we see that

$$x=r\cos\theta, \text{ and } y=r\sin\theta.$$

These equations connect the rectangular and polar co-ordinates of a point. From them, or from the figure, we may deduce

$$x^2+y^2=r^2,\quad \frac{y}{x}=\tan\theta.$$

We proceed to investigate expressions for some geometrical quantities in terms of co-ordinates.

9. *To find an expression for the length of the straight line joining two points.*

Let P and Q be the two points; ω the incliuation of the axes OX, OY. Draw PM, QN parallel to OY; let x_1, y_1 be the co-ordinates of P, and x_2, y_2 those of Q. Draw PR parallel to OX. Then, by Trigonometry,

$$\begin{aligned} PQ^2 &= PR^2+QR^2-2PR\,.\,QR\cos PRQ \\ &= PR^2+QR^2+2PR\,.\,QR\cos\omega. \end{aligned}$$

But $PR = x_2 - x_1$, and $QR = y_2 - y_1$; therefore

$$PQ^2 = (x_2 - x_1)^2 + (y_2 - y_1)^2 + 2(x_2 - x_1)(y_2 - y_1)\cos\omega \ldots (1),$$

and thus the distance PQ is determined.

If the axes are rectangular, we have

$$PQ^2 = (x_2 - x_1)^2 + (y_2 - y_1)^2 \ldots\ldots\ldots\ldots\ldots\ldots (2).$$

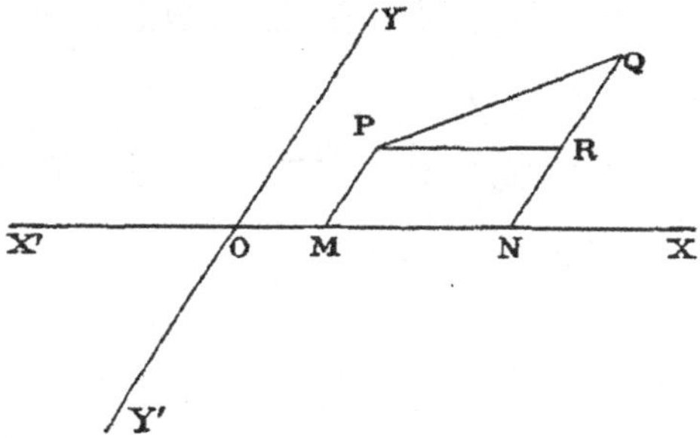

The student should draw figures placing P and Q in the different compartments and in different positions; the equations (1) and (2) will be found universally true.

From the equation (2) we have

$$PQ^2 = x_1^2 + y_1^2 + x_2^2 + y_2^2 - 2(x_1x_2 + y_1y_2) \ldots\ldots (3).$$

The following particular cases may be noted.

If P be at O then $x_1 = 0$ and $y_1 = 0$; thus $PQ^2 = x_2^2 + y_2^2$.

If P be on the axis of x and Q on the axis of y, then $y_1 = 0$ and $x_2 = 0$; thus $PQ^2 = x_1^2 + y_2^2$.

Let θ_1, r_1 be the polar co-ordinates of P, and θ_2, r_2 those of Q; then, by Art. 8,

$$x_1 = r_1\cos\theta_1, \quad y_1 = r_1\sin\theta_1, \quad x_2 = r_2\cos\theta_2, \quad y_2 = r_2\sin\theta_2.$$

Substitute these values in (3) and we have

$$PQ^2 = r_1^2 + r_2^2 - 2r_1r_2\cos(\theta_2 - \theta_1).$$

This result can also be obtained immediately from the triangle POQ formed by drawing straight lines from P and Q to the origin.

10. *To find the co-ordinates of the point which divides in a given ratio the straight line joining two given points.*

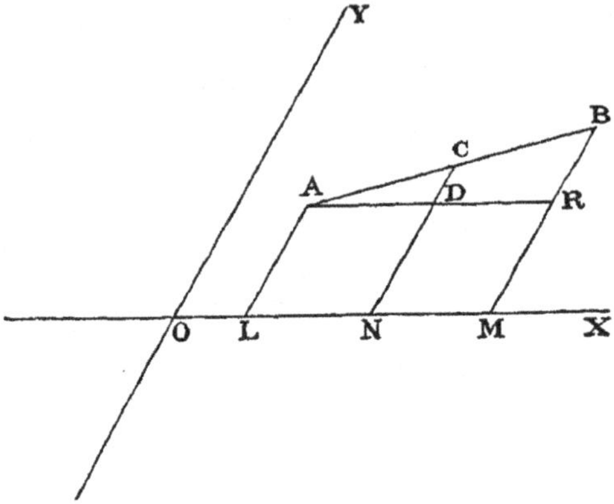

Let A and B be the given points, x_1, y_1 the co-ordinates of A, and x_2, y_2 those of B; and let the given ratio be that of n_1 to n_2. Suppose C the required point, so that AC is to CB as n_1 is to n_2. Draw the ordinates AL, BM, CN; and draw AR parallel to OX meeting CN at D. Let x, y be the co-ordinates of C.

It is obvious from the figure that

$$\frac{LN}{NM} = \frac{AD}{DR} = \frac{AC}{CB};$$

that is,

$$\frac{x - x_1}{x_2 - x} = \frac{n_1}{n_2};$$

therefore

$$x = \frac{n_1 x_2 + n_2 x_1}{n_1 + n_2}.$$

Similarly

$$y = \frac{n_1 y_2 + n_2 y_1}{n_1 + n_2}.$$

In this Article the axes may be oblique or rectangular. A simple case is that in which we require the co-ordinates of the point midway between two given points; then $n_1 = n_2$, and

$$x = \tfrac{1}{2}(x_1 + x_2), \qquad y = \tfrac{1}{2}(y_1 + y_2).$$

11. *To express the area of a triangle in terms of the co-ordinates of its angular points.*

Let ABC be a triangle; let x_1, y_1 be the co-ordinates of A; x_2, y_2 those of B; x_3, y_3 those of C. Draw the ordinates AL, BM, CN. The area of the triangle is equal to the trapezium $ABML$ + trapezium $BCNM$ − trapezium $ACNL$.

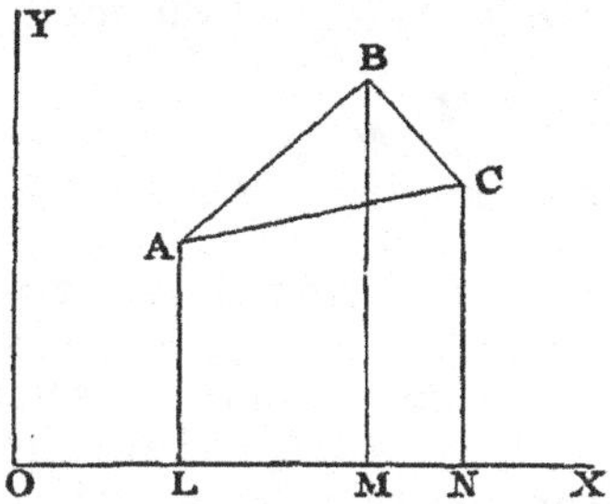

The area of the trapezium $ABML$ is $\frac{1}{2}LM(AL+BM)$. This is obvious, because if we join BL we divide the trapezium into two triangles, one having AL for its base, and the other BM, and each having LM for its height;

thus trapezium $ABML = \frac{1}{2}(x_2 - x_1)(y_1 + y_2)$;

also trapezium $BCNM = \frac{1}{2}(x_3 - x_2)(y_2 + y_3)$;

and trapezium $ACNL = \frac{1}{2}(x_3 - x_1)(y_1 + y_3)$;

therefore the triangle ABC

$$= \tfrac{1}{2}\{(x_2 - x_1)(y_1 + y_2) + (x_3 - x_2)(y_2 + y_3) - (x_3 - x_1)(y_1 + y_3)\}.$$

This expression may be written more symmetrically thus:

$$\tfrac{1}{2}\{(x_2 - x_1)(y_2 + y_1) + (x_3 - x_2)(y_3 + y_2) + (x_1 - x_3)(y_1 + y_3)\} \ldots (1).$$

By reducing it, we shall find the area of the triangle

$$= \tfrac{1}{2}\{x_2y_1 - x_1y_2 + x_3y_2 - x_2y_3 + x_1y_3 - x_3y_1\} \ldots\ldots (2).$$

If the axes be oblique and inclined at an angle ω, the area of the trapezium $ABML = \frac{1}{2}LM(AL + BM)\sin\omega$, and similarly for the other trapeziums. Thus the area of the triangle will be found by multiplying the expressions given above by sin ω.

However the relative situations of A, B, C may be changed, the student will always find for the area of the triangle the expression (2), or that expression with the *sign of every term changed.* Hence we conclude, that we shall always obtain the area of the triangle by calculating the value of the expression (2), and changing the sign of the result if it should prove negative.

Locus of an equation. Equation to a curve.

12. Suppose an equation to be given between two unknown quantities, for example, $y - x - 2 = 0$. We see that this equation has an indefinite number of solutions, for we may assign to x any value we please, and from the equation determine the corresponding value of y. Thus corresponding to the values 1, 2, 3,... of x, we have the values 3, 4, 5,... of y. Now suppose a line, straight or curved, such that it passes through *every point* determined by giving to x and y values that satisfy the equation $y - x - 2 = 0$; such a line is called the *locus of the equation.* It will be shewn in the next Chapter that the locus of the equation in question is a straight line. We shall see as we proceed that generally every equation between the quantities x and y has a corresponding locus.

But instead of starting with an equation and investigating what locus it represents, we may give a geometrical definition of a curve and deduce from that definition an appropriate equation; this will likewise appear as we proceed: we shall take successively different curves, define them, deduce their equations, and then investigate the properties of these curves by means of their equations. We shall in the next Chapter begin with the *equation to a straight line.*

The connexion between a *locus* and an *equation* is the fundamental idea of the subject and must therefore be carefully considered; we shall place here a formal definition which we shall illustrate in the next Chapter by applying it to a straight line.

The equation which expresses the invariable relation which exists between the co-ordinates of every point of a curve is called the equation to the curve; and the curve, the co-ordinates of every point of which satisfy a given equation, is called the locus of that equation.

13. The student has probably already become familiar with the division of algebraical equations into equations of the first, second, third,... degree. When we speak of an equation of the nth degree between two variables we mean that every term is of the form $Ax^\alpha y^\beta$ where α and β are zero or positive integers such that $\alpha + \beta$ is equal to n for one or more of the terms but not greater than n for any term, and A is a constant numerical quantity; and the equation is formed by connecting a series of such terms by the signs $+$ and $-$, and putting the result $= 0$.

EXAMPLES.

1. Find the polar co-ordinates of the points whose rectangular co-ordinates are

(1) $x = 1,\ y = 1$; (2) $x = -1,\ y = 2$;

(3) $x = -1,\ y = 1$; (4) $x = -1,\ y = -1$;

and indicate the points in a figure.

2. Find the rectangular co-ordinates of the points whose polar co-ordinates are

(1) $\theta = \frac{\pi}{3},\ r = 3$; (2) $\theta = -\frac{\pi}{3},\ r = 3$;

(3) $\theta = \frac{\pi}{3},\ r = -3$; (4) $\theta = -\frac{\pi}{3},\ r = -3$;

and indicate the points in a figure.

3. The co-ordinates of P are -1 and 4, and those of Q are 3 and 7: find the length of PQ.

4. Find the area of the triangle formed by joining the first three points in Example 1.

5. A is a point on the axis of x, and B a point on the axis of y: express the co-ordinates of the middle point of AB in terms of the abscissa of A and the ordinate of B; shew also that the distance of this point from the origin $= \frac{1}{2} AB$.

6. Transform equation (2) of Art. 11 so as to give an expression for the area of a triangle in terms of the *polar* co-ordinates of its angular points. Also obtain the result directly from the figure.

7. A and B are two points and O is the origin: express the area of the triangle AOB in terms of the co-ordinates of A and B, and also in terms of the polar co-ordinates of A and B.

8. A, B, C are three points the co-ordinates of which are expressed as in Art. 11; suppose D the middle point of AB; join CD and divide it at G so that $CG = 2GD$: find the co-ordinates of G.

9. Shew that each of the triangles GAB, GBC, GAC, formed by joining the point G in the preceding Example to the points A, B, C, is equal in area to one-third of the triangle ABC. See Art. 11.

10. A and B are two points; the polar co-ordinates of A are θ_1, r_1; and those of B are θ_2, r_2. A straight line is drawn from the origin O bisecting the angle AOB; if C be the point where this straight line meets AB shew that the polar co-ordinates of C are $\theta = \frac{1}{2}(\theta_1 + \theta_2)$ and $r = \dfrac{2r_1 r_2 \cos \frac{1}{2}(\theta_2 - \theta_1)}{r_1 + r_2}$.

11. Find the value of CD^2 and AD^2 in Example 8 in terms of the co-ordinates there used; and shew that

$$AC^2 + BC^2 = 2CD^2 + 2AD^2.$$

12. Find the value of GA^2, GB^2, and GC^2, in Example 9 in terms of the co-ordinates there used; and shew that

$$3(GA^2 + GB^2 + GC^2) = AB^2 + BC^2 + CA^2.$$

CHAPTER II.

ON THE STRAIGHT LINE.

14. *To find the equation to a straight line.*

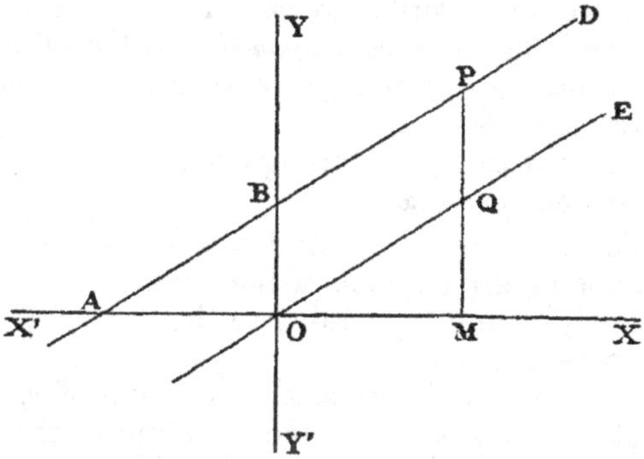

We shall first suppose the straight line not parallel to either axis.

Let ABD be a straight line meeting the axis of y at B. Draw a straight line OE through the origin parallel to ABD. In ABD take any point P; draw PM parallel to OY, meeting OX at M, and OE at Q.

Suppose $OB = c$, and the tangent of $EOX = m$; and let x, y be the co-ordinates of P; then

$$y = PM = PQ + QM = OB + QM$$
$$= c + OM \tan QOM = c + mx.$$

Hence the required equation is

$$y = mx + c.$$

OB is called the *intercept* on the axis of y; if the straight line crosses the axis of y on the negative side of O, then c will be negative.

We denote by m the tangent of the angle QOM or BAO, that is, the tangent of the angle which that part of the straight line which is above the axis of x makes with the axis of x produced in the positive direction. Hence if the straight line through the origin parallel to the given straight line falls between OY and OX, m is the tangent of an acute angle and is positive; if between OY and OX produced to the left, m is the tangent of an obtuse angle and is negative. So long as we consider the same straight line m and c remain unchangeable, they are therefore called *constant quantities* or *constants*. But x and y may have an *indefinite* number of values since we may ascribe to *one* of them, as x, any value we please, and find the *corresponding value* of y from the equation $y = mx + c$; x and y are therefore called *variable quantities* or *variables*.

If the straight line pass through the origin, $c = 0$, and the equation becomes $y = mx$.

15. We have now to consider the cases in which the straight line is parallel to one of the axes.

If the straight line be parallel to the axis of x, $m = 0$, and the equation becomes $y = c$.

If the straight line be parallel to the axis of y, m becomes the tangent of a right angle and is infinite; the preceding investigation is then no longer applicable. We shall now give separate investigations of these two cases.

To investigate the equation to a straight line parallel to one of the axes.

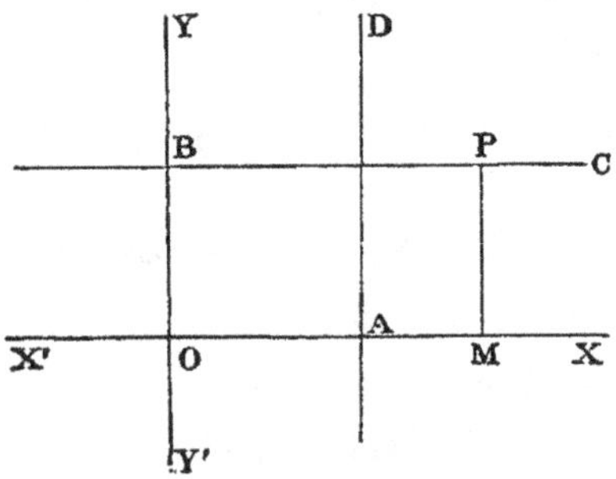

First suppose the straight line parallel to the axis of x. Let BC be the straight line meeting the axis of y at B; suppose $OB = b$.

Since the straight line is parallel to the axis of x, the ordinate PM of *any* point of it is equal to OB. Hence calling y the ordinate of any point P, we have for the equation to the straight line $y = b$.

Next suppose the straight line parallel to the axis of y. Let AD be the straight line meeting the axis of x at A; suppose $OA = a$. Since the straight line is parallel to the axis of y, the abscissa of *any* point of it is OA. Hence calling x the abscissa of any point, we have for the equation to the straight line $x = a$.

16. We have thus shewn that any straight line whatsoever is represented by an equation of the first degree; we shall now shew that any equation of the first degree with two variables represents a straight line.

The general equation of the first degree with two variables is of the form

$$Ax + By + C = 0 \dots\dots\dots\dots\dots\dots (1),$$

A, B, C being finite or zero.

First suppose B not zero; divide by B, then from (1)

$$y = -\frac{C}{B} - \frac{A}{B}x \dots\dots\dots\dots\dots\dots (2).$$

Now we have seen in Art. 14, that if a straight line be drawn meeting the axis of y at a distance $-\frac{C}{B}$ from the origin and making with the axis of x an angle of which the tangent is $-\frac{A}{B}$, then (2) will be the equation to this straight line. Hence (2), and therefore also (1), represents a straight line.

If $A = 0$, then by Art. 15 the straight line represented by (1) is parallel to the axis of x.

If $B = 0$, then (1) becomes $Ax + C = 0$,

or $$x = -\frac{C}{A},$$

and from Art. 15 we know that this equation represents a straight line parallel to the axis of y.

Hence the equation $Ax + By + C = 0$ always represents a straight line.

17. *Equation in terms of the intercepts.* The equation to a straight line may also be expressed in terms of its *intercepts* on the two axes.

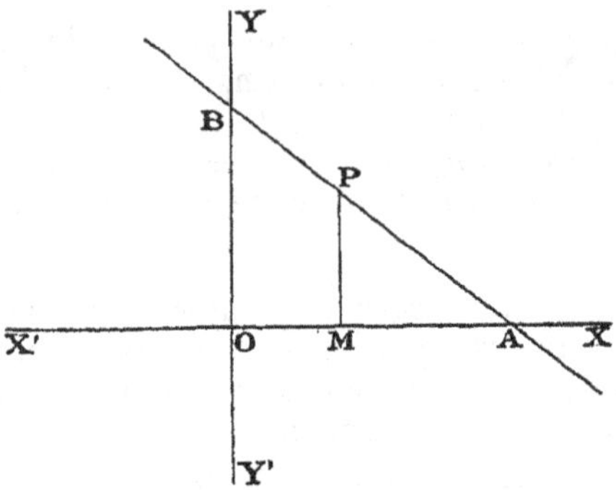

Let A and B be the points where the straight line meets the axes of x and y respectively. Suppose $OA = a$, $OB = b$. Let P be any point in the straight line; x, y its co-ordinates; draw PM parallel to OY. Then by similar triangles,

$$\frac{PM}{OB} = \frac{AM}{AO};$$

that is $$\frac{y}{b} = \frac{a-x}{a};$$

therefore $$\frac{x}{a} + \frac{y}{b} = 1.$$

18. It will be a useful exercise for the student to draw the straight lines corresponding to some given equations. Thus suppose the equation $2y + 3x = 7$ proposed; since a straight line is determined when two of its points are known, we may find in any manner we please two points that lie on the straight line, and by joining them obtain the straight line. Suppose then $x = 1$, it follows from the equation that $y = 2$; hence the point which has its abscissa $= 1$, and its ordinate $= 2$, is on the straight line. Again, suppose $x = 2$, then $y = \frac{1}{2}$; the point which has its abscissa $= 2$ and its ordinate $= \frac{1}{2}$ is therefore on the straight line. Join the two points thus determined, and the straight line so formed, produced indefinitely both ways, is the locus of the given equation. The two points

that will be most easily determined are generally those at which the required straight line *cuts the axes*. Suppose $x=0$ in the given equation, then $y=\frac{7}{2}$, that is, the straight line passes through a point *on the axis of* y at a distance $\frac{7}{2}$ from the origin. Again, suppose $y=0$, then $x=\frac{7}{3}$, that is, the straight line passes through a point *on the axis of* x at a distance $\frac{7}{3}$ from the origin. Join the two points thus determined, and the straight line so formed, produced indefinitely both ways, is the locus of the given equation. What we have here ascertained as to the points where the straight line cuts the axes, may be obtained immediately from the equation; for if we write it in the form

$$\frac{3x}{7}+\frac{2y}{7}=1,$$

and compare it with the equation in Art. 17,

$$\frac{x}{a}+\frac{y}{b}=1,$$

we see that $a=\frac{7}{3}$ and $b=\frac{7}{2}$.

Again, suppose the equation $y=x$ proposed. Since this equation can be satisfied by supposing $x=0$ and $y=0$, the origin is a point of the straight line which the equation represents; therefore we need only determine *one other* point in it. Suppose $x=1$, then $y=1$; here another point is determined and the straight line can be drawn. The straight line may also be constructed by comparing the given equation with the form in Art. 14,

$$y=mx.$$

This we know represents a straight line passing through the origin and making with the axis of x an angle of which the tangent is m. Hence $y=x$ represents a straight line passing through the origin and inclined at an angle of 45° to the axis of x.

Similarly the equation $y=-x$ represents a straight line inclined to the axis of x at an angle of which the tangent is -1; that is, at an angle of 135°. Hence this equation represents a straight line through O bisecting the angle between OY and OX produced to the left in the figure to Art. 14.

19. The student is recommended to make himself thoroughly acquainted with the previous Articles before proceeding with the subject. In Algebra the theory of *indeterminate* equations does not usually attract much attention, and the student is sometimes perplexed on commencing a subject in which he has to consider *one* equation between two unknown quantities, which generally has an infinite number of solutions.

Our principal result up to the present point is, that a straight line corresponds to an equation of the first degree, and the student must accustom himself to perceive the appropriate straight line as soon as any equation is presented to him. The straight line can be determined by ascertaining two points through which it passes, that is, by finding two points such that the co-ordinates of each satisfy the given equation; and the straight line being thus determined, the co-ordinates of *any* point of it will satisfy the given equation.

20. *Equation to a straight line in terms of the perpendicular from the origin, and the inclination of this perpendicular to one of the axes.*

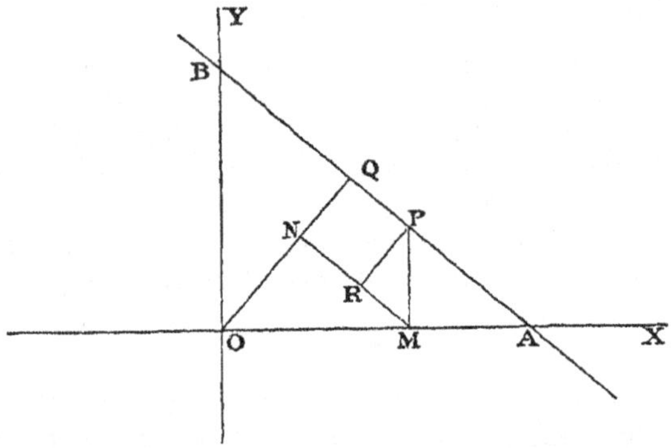

Let OQ be the perpendicular from the origin O on a straight line AB. Take any point P in the straight line; draw PM perpendicular to OA, MN perpendicular to OQ, and PR perpendicular to MN. Suppose $OQ = p$, and the angle $QOA = \alpha$. Let x, y be the co-ordinates of P; then

$$\begin{aligned} OQ = ON + NQ &= ON + PR \\ &= OM \cos QOA + PM \sin PMR \\ &= x \cos \alpha + y \sin \alpha. \end{aligned}$$

Therefore the equation to the straight line is

$$x \cos \alpha + y \sin \alpha = p.$$

21. We have given separate investigations of the different forms of the equation to a straight line in Articles 14, 17, 20; any one of these forms may however be readily deduced from any other by making use of the relations which exist between the constant quantities.

The quantity which we have denoted by b in Art. 17, that is OB, is denoted by c in Art. 14;

therefore $$b = c \quad\text{..........................}(1).$$

In Art. 17,

$$\frac{b}{a} = \tan BAO = \tan (\pi - BAX) = -\tan BAX;$$

in Art. 14 we have denoted the tangent of BAX by m,

therefore $$\frac{b}{a} = -m \quad\text{......................}\ (2).$$

In Art. 20, $OA \cos \alpha = OQ$, and $OB \sin \alpha = OQ$; that is,

$$p = a \cos \alpha = b \sin \alpha \quad\text{................}\ (3);$$

therefore from (2) and (3), $m = -\cot \alpha$ .... (4).

Also if the equation $Ax + By + C = 0$ represent the straight line under consideration, then by Art. 16,

$$-\frac{A}{B} = m, \qquad -\frac{C}{B} = c \quad\text{....................}\ (5);$$

therefore $$\frac{A}{B} = \cot \alpha, \text{ and } \frac{C}{B} = -\frac{p}{\sin \alpha} \quad\text{..............}\ (6).$$

And, by comparing Arts. 16 and 17, we have

$$a = -\frac{C}{A}, \quad b = -\frac{C}{B} \quad\text{..................}\ (7).$$

By means of these relations we may shew the agreement of the equations in Arts. 14, 17, 20, or from one of them deduce the others.

22. The student may exercise himself by varying the figures which we have used in investigating the equations. Thus, for example, in the figure to Art. 17, suppose the point P to be in BA produced, so that it falls *below* the axis of x. We shall still have

$$\frac{PM}{OB} = \frac{AM}{AO}, \quad \text{or } \frac{PM}{b} = \frac{x-a}{a}.$$

Now since P is below the axis of x, its ordinate y is a negative quantity, hence we must not put $PM = y$ but $PM = -y$, because by PM we mean a certain length estimated positively. Thus

$$-\frac{y}{b} = \frac{x-a}{a}, \quad \text{and therefore, as before, } \frac{x}{a} + \frac{y}{b} = 1.$$

Oblique Co-ordinates.

23. *Equation to a straight line.*

We shall denote the inclination of the axes by ω.

Suppose first, that the straight line is not parallel to either axis. Let ABD be a straight line meeting the axis of y at B. Draw a straight line OE through the origin parallel to ABD. In ABD take any point P; draw PM parallel to OY, meeting OX at M, and OE at Q. Suppose $OB = c$, and the angle $QOM = \alpha$.

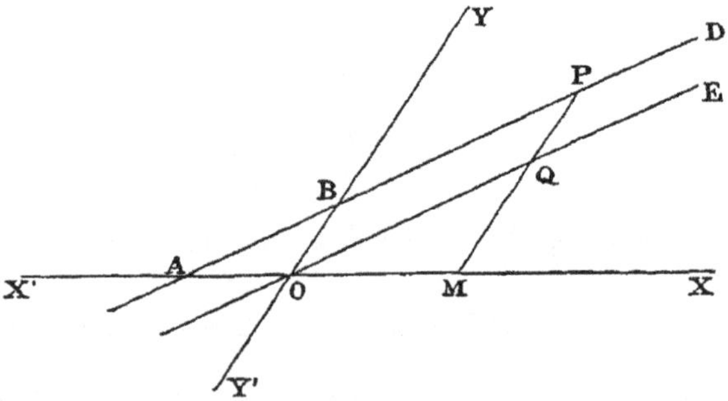

Let x, y be the co-ordinates of P; then

$$y = PM = PQ + QM = OB + QM.$$

But $\dfrac{QM}{OM} = \dfrac{\sin\alpha}{\sin(\omega-\alpha)}$; therefore $QM = \dfrac{x\sin\alpha}{\sin(\omega-\alpha)}$.

Hence the required equation is

$$y = \frac{x\sin\alpha}{\sin(\omega-\alpha)} + c.$$

If we put m for $\dfrac{\sin\alpha}{\sin(\omega-\alpha)}$ the equation becomes

$$y = mx + c,$$

as in Art. 14. The meaning of c is the same as before; m is the ratio of the sine of the inclination of the straight line to the axis of x to the sine of its inclination to the axis of y. Since $\sin\alpha$ is always positive, m will be positive or negative according as $\sin(\omega-\alpha)$ is positive or negative; thus, as before, m will be positive or negative according as the straight line through the origin parallel to the given straight line falls between OY and OX, or between OY and OX'. The meaning of m coincides with that in Art. 14 when $\omega = \dfrac{\pi}{2}$, for then $m = \tan\alpha$.

24. Since $$m = \frac{\sin\alpha}{\sin(\omega-\alpha)};$$

therefore $$m(\sin\omega\cos\alpha - \cos\omega\sin\alpha) = \sin\alpha;$$

therefore $$m(\sin\omega - \cos\omega\tan\alpha) = \tan\alpha;$$

therefore $$\tan\alpha = \frac{m\sin\omega}{1 + m\cos\omega}.$$

Hence $$\sin\alpha = \frac{m\sin\omega}{\pm\sqrt{(1 + 2m\cos\omega + m^2)}},$$

$$\cos\alpha = \frac{1 + m\cos\omega}{\pm\sqrt{(1 + 2m\cos\omega + m^2)}}.$$

Since $\sin\alpha$ is positive, we must take the upper or lower sign according as m is positive or negative.

25. The investigations in Arts. 15 and 17 apply without alteration to the case of oblique axes, and those in Art. 16 with the requisite change in the meaning of the constant m.

26. *To find the equation to a straight line in terms of the perpendicular from the origin, and the inclinations of the perpendicular to the axes.*

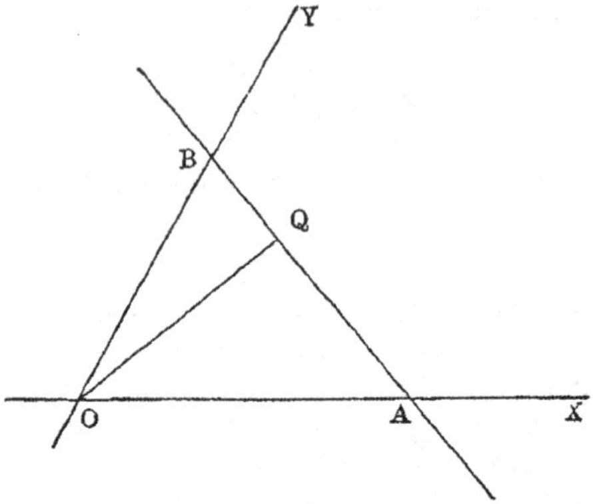

Let OQ be the perpendicular from the origin on a straight line AB; let $OQ=p$, $OA=a$, $OB=b$. If we suppose $QOA=\alpha$, we have $QOB=\omega-\alpha$; denote this by β; then

$$OQ=a\cos\alpha; \quad \text{therefore } a=\frac{p}{\cos\alpha}.$$

$$OQ=b\cos\beta; \quad \text{therefore } b=\frac{p}{\cos\beta}.$$

Substitute in the equation, Art. 17, $\frac{x}{a}+\frac{y}{b}=1$, and we obtain $x\cos\alpha+y\cos\beta=p$.

27. The following form of the equation to a straight line is often useful.

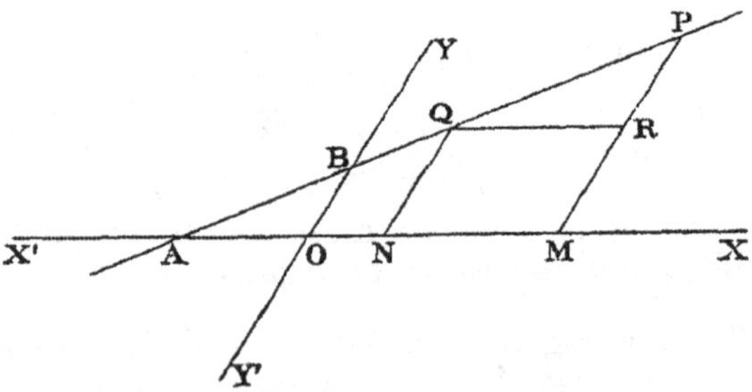

Let Q be a fixed point in any straight line AB; h, k its co-ordinates; let P be any other point in the straight line; x, y its co-ordinates; let $QP=r$, and the angle $BAX=\alpha$. Draw the ordinates PM, QN; and QR parallel to OX; then $\frac{x-h}{r}=\frac{\sin(\omega-\alpha)}{\sin\omega}=l$ suppose, and $\frac{y-k}{r}=\frac{\sin\alpha}{\sin\omega}=n$ suppose,

thus $$\frac{x-h}{l}=\frac{y-k}{n}=r.$$

For the equation to the straight line it is sufficient to put $\frac{x-h}{l}=\frac{y-k}{n}$, but it is useful to remember that each of these quantities is equal to r.

The constants l and n are connected by a certain condition. For, by Art. 9,

$$(x-h)^2+(y-k)^2+2(x-h)(y-k)\cos\omega=r^2;$$

substitute for $x-h$ and $y-k$: thus

$$l^2+n^2+2ln\cos\omega=1.$$

If the axes are rectangular, l and n become respectively $\cos\alpha$ and $\sin\alpha$, that is, the *cosines* of the inclinations of the straight line to the axes of x and y respectively.

In the preceding figure P falls to the right of Q and $x-h$ is positive. If P were to the left of Q then $x-h$ would be negative. Thus since $x-h=lr$, the product lr must be capable of changing its sign; this leads us to consider r as positive or negative according to circumstances. When therefore we write the equation to a straight line under the form

$$\frac{x-h}{l}=\frac{y-k}{n},$$

and ascribe to l and n the values given above, we conclude that each of the expressions $\frac{x-h}{l}$ and $\frac{y-k}{n}$ is *numerically* equal to the distance between the point (h, k) and the point (x, y), but that the sign of each expression will depend upon the relative positions of the two points.

Polar Co-ordinates.

28. *Polar equation to a straight line.*

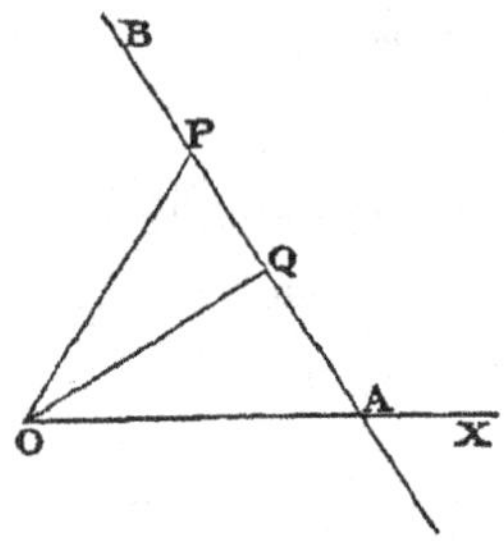

Let AB be a straight line, OQ the perpendicular on it from the origin, OX the initial line, P any point in the straight line. Suppose $OQ = p$, and the angle $QOX = \alpha$. Let r, θ be the polar co-ordinates of P; then

$$OQ = OP \cos POQ;$$

$$\text{that is, } p = r \cos(\theta - \alpha).$$

This is the polar equation to the straight line.

29. The polar equation may also be derived from the equation referred to rectangular co-ordinates. Let

$$Ax + By + C = 0$$

be the equation to a straight line referred to rectangular co-ordinates. Put $r \cos\theta$ for x, and $r \sin\theta$ for y, Art. 8; thus

$$Ar\cos\theta + Br\sin\theta + C = 0 \quad\ldots\ldots\ldots\ldots\ldots(1)$$

is the polar equation. This equation may be shewn to agree with

$$p = r\cos(\theta - \alpha) \quad\ldots\ldots\ldots\ldots\ldots\ldots\ldots(2).$$

For by Art. 21 we have

$$\frac{A}{B} = \cot\alpha \text{ and } \frac{C}{B} = -\frac{p}{\sin\alpha}.$$

Hence (1) becomes

$$\cot\alpha\, r\cos\theta + r\sin\theta - \frac{p}{\sin\alpha} = 0,$$

which agrees with (2).

30. The equation to a straight line passing through the origin is, by Art. 14,

$$y = mx.$$

Put $r\cos\theta$ for x and $r\sin\theta$ for y; the equation then becomes

$$r\sin\theta = mr\cos\theta;$$

therefore $$\tan\theta = m;$$

therefore $$\theta = \text{a constant};$$

this is therefore the polar equation to a straight line passing through the origin.

31. We will collect here the different forms of the equation to a straight line which have been investigated,

$y = mx + c,$	Arts. 14 and 23.
$x = \text{constant, or, } y = \text{constant},$	Arts. 15 and 25.
$\frac{x}{a} + \frac{y}{b} - 1 = 0,$	Arts. 17 and 25.
$x\cos\alpha + y\sin\alpha - p = 0,$	Art. 20.
$y = \frac{\sin\alpha}{\sin(\omega-\alpha)}x + c,$	Art. 23.
$x\cos\alpha + y\cos\beta - p = 0,$	Art. 26.
$\frac{x-h}{l} = \frac{y-k}{n} = r,$	Art. 27.
$p = r\cos(\theta-\alpha),$	Art. 28.
$Ar\cos\theta + Br\sin\theta + C = 0,$	Art. 29.
$\theta = \text{constant},$	Art. 30.

EXAMPLES.

Draw the straight lines represented by the following equations:

(1) $y + 2x = 4$; (2) $2y - x = 2$;

(3) $y + x = -2$; (4) $x - 2y = 4$;

(5) $y + 2x = 0$; (6) $1 = \cos\left(\theta - \frac{\pi}{4}\right)$;

(7) $x = 1$; (8) $\theta = \frac{\pi}{3}$;

(9) $\theta = 0$; (10) $\theta = 1$.

CHAPTER III.

PROBLEMS ON THE STRAIGHT LINE.

32. WE will now apply the results of the preceding Articles to the solution of some problems.

To find the form of the equation to a straight line which passes through a given point.

Let x_1, y_1 be the co-ordinates of the given point, and suppose

$$y = mx + c \quad\text{..........................}(1)$$

to represent the straight line. Since the point (x_1, y_1) is on the straight line, its co-ordinates must satisfy (1); hence

$$y_1 = mx_1 + c \quad\text{........................}(2).$$

By subtraction we have

$$y - y_1 = m(x - x_1) \quad\text{.....................}(3);$$

this is the required equation.

33. The equation (3) of the preceding Article obviously represents what is required, namely, a straight line passing through the point (x_1, y_1). For the equation is of the first degree in the variables x, y, and therefore, by Art. 16, must represent *some* straight line. Also the equation is obviously satisfied by the values $x = x_1$, $y = y_1$; that is, the straight line which the equation represents *does* pass through the given point. The constant m is the tangent of the angle which the straight line makes with the axis of x, and by giving a suitable value to m we may make the equation (3) represent *any* straight line which passes through the assigned point.

The geometrical meaning of equation (3) is obvious. For

let AB be any straight line passing through the given point Q. Let P be any point on the straight line; x, y its co-

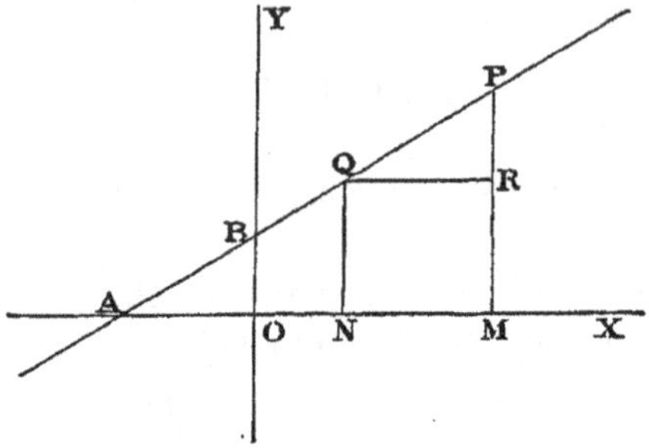

ordinates. Draw the ordinates PM, QN; and QR parallel to OX; then

$$\frac{PR}{QR} = \text{tangent } PQR$$

that is $$\frac{y - y_1}{x - x_1} = \tan BAX = m,$$

which agrees with equation (3).

34. In Art. 32 we eliminated c between the equations (1) and (2) and retained m; we may if we please eliminate m and retain c. From (2)

$$m = \frac{y_1 - c}{x_1}.$$

Substitute in (1), thus $y = \dfrac{y_1 - c}{x_1} x + c$;

therefore $$yx_1 - xy_1 + c(x - x_1) = 0.$$

This equation obviously represents a straight line passing through the given point, because it is of the first degree in the variables x, y; and it is satisfied by the values

$$x = x_1, \quad y = y_1.$$

35. *To find the equation to the straight line which passes through two given points.*

Let x_1, y_1 be the co-ordinates of one given point; x_2, y_2 those of the other; suppose the equation to the straight line to be

$$y = mx + c \quad \text{.........................(1)}.$$

Since the straight line passes through (x_1, y_1) and (x_2, y_2),

$$y_1 = mx_1 + c \text{..........................(2)},$$

$$y_2 = mx_2 + c \text{..........................(3)}.$$

From (1) and (2) by subtraction

$$y - y_1 = m(x - x_1) \text{....................(4)}.$$

From (2) and (3) by subtraction

$$y_2 - y_1 = m(x_2 - x_1),$$

$$\text{hence } m = \frac{y_2 - y_1}{x_2 - x_1}.$$

Substitute the value of m in (4) and we have for the required equation

$$y - y_1 = \frac{y_2 - y_1}{x_2 - x_1}(x - x_1) \text{....................(5)}.$$

We may also write the equation thus,

$$(x_2 - x_1)(y - y_1) = (y_2 - y_1)(x - x_1) \text{.........(6)}.$$

Some particular cases may here be noted. Suppose $y_2 = y_1$, then (6) becomes $(x_2 - x_1)(y - y_1) = 0$, therefore $y = y_1$; the required straight line is thus parallel to the axis of x. Similarly if we suppose $x_2 = x_1$, then (6) becomes $(y_2 - y_1)(x - x_1) = 0$, therefore $x = x_1$; thus the required straight line is parallel to the axis of y. Lastly, suppose the point (x_1, y_1) to be the origin; then $x_1 = 0$ and $y_1 = 0$; thus (6) becomes $x_2 y = y_2 x$. The student should illustrate these particular cases by figures.

36. The equation (6) of Art. 35 becomes by reduction

$$x_1 y - xy_1 + x_2 y_1 - x_1 y_2 + xy_2 - x_2 y = 0.$$

If we compare the expression on the left-hand side of this equation with the expression in brackets in equation (2) of Art. 11, we see the only difference is that we have x and y in the place of x_3 and y_3 respectively. Thus the equation informs us that the area of the triangle formed by joining (x, y), (x_1, y_1), (x_2, y_2) vanishes, as should evidently be the case since the vertex (x, y) falls on the base, that is, on the straight line joining (x_1, y_1) to (x_2, y_2).

37. *To find the equation to the straight line which passes through a given point and divides the straight line joining two other given points in a given ratio.*

Let (h, k) be the first given point; let (x_1, y_1), (x_2, y_2) be the two other given points; let the given ratio in which the straight line joining the last two points is to be divided be that of n_1 to n_2; then, by Art. 10, the co-ordinates of the point of division are

$$\frac{n_1x_2 + n_2x_1}{n_1 + n_2}, \quad \frac{n_1y_2 + n_2y_1}{n_1 + n_2}.$$

Hence by equation (5) of Art. 35 the equation required is

$$y - k = \frac{\dfrac{n_1y_2 + n_2y_1}{n_1 + n_2} - k}{\dfrac{n_1x_2 + n_2x_1}{n_1 + n_2} - h}(x - h);$$

or

$$y - k = \frac{n_1(y_2 - k) + n_2(y_1 - k)}{n_1(x_2 - h) + n_2(x_1 - h)}(x - h).$$

38. *To find the form of the equation to a straight line which is parallel to a given straight line.*

Let the equation to the given straight line be

$$y = m_1x + c_1 \quad \text{.........................(1)},$$

and the equation to the required straight line

$$y = mx + c \quad \text{.........................(2)}.$$

Since the straight lines represented by (1) and (2) are parallel, they must have the same inclination to the axis of x; hence

$$m = m_1.$$

Thus (2) becomes

$$y = m_1x + c.$$

The quantity c remains undetermined, since an indefinite number of straight lines can be drawn parallel to a given straight line.

39. *To determine the co-ordinates of the point of intersection of two given straight lines.*

Let the equation to one straight line be

$$y = m_1x + c_1 \qquad\qquad (1),$$

and the equation to the other

$$y = m_2x + c_2 \qquad\qquad (2).$$

The co-ordinates of the point where the straight lines intersect must satisfy *both* equations; we must therefore find the values of x and y from (1) and (2). Thus

$$x = \frac{c_1 - c_2}{m_2 - m_1}, \quad y = \frac{c_1m_2 - c_2m_1}{m_2 - m_1};$$

these are the required co-ordinates.

40. *To find the condition in order that three straight lines may meet at a point.*

Let the equations to the straight lines be respectively

$$y = m_1x + c_1 \ldots (1), \quad y = m_2x + c_2 \ldots (2), \quad y = m_3x + c_3 \ldots (3).$$

The co-ordinates of the point of intersection of (1) and (2) are

$$x = \frac{c_1 - c_2}{m_2 - m_1}, \quad y = \frac{c_1m_2 - c_2m_1}{m_2 - m_1}.$$

If the third straight line passes through the intersection of the first and second, these values must satisfy (3). Hence the necessary and sufficient condition is

$$\frac{c_1m_2 - c_2m_1}{m_2 - m_1} = \frac{m_3(c_1 - c_2)}{m_2 - m_1} + c_3,$$

that is,

$$c_1m_2 - c_2m_1 + c_2m_3 - c_3m_2 + c_3m_1 - c_1m_3 = 0.$$

41. *To find the angle between two given straight lines.*

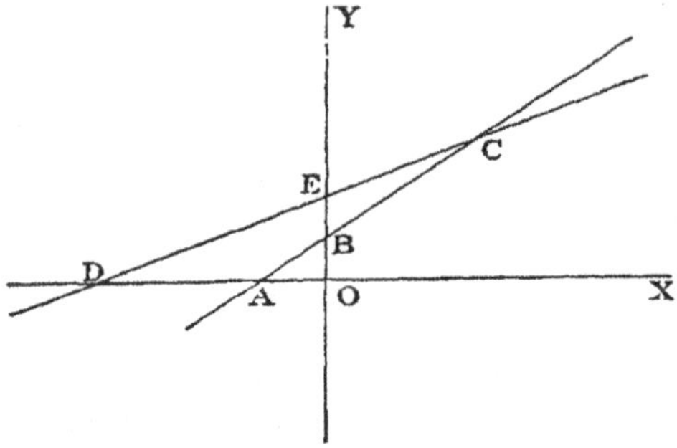

Let ABC be one straight line and DEC the other; let the equation to the former be $y = m_1 x + c_1$, and the equation to the latter $y = m_2 x + c_2$.

Then $\tan ACD = \tan(CAX - CDX)$

$$= \frac{\tan CAX - \tan CDX}{1 + \tan CAX \tan CDX} = \frac{m_1 - m_2}{1 + m_1 m_2}.$$

From this we may deduce

$$\cos ACD = \frac{1 + m_1 m_2}{\sqrt{\{(1 + m_1^2)(1 + m_2^2)\}}};$$

$$\sin ACD = \frac{m_1 - m_2}{\sqrt{\{(1 + m_1^2)(1 + m_2^2)\}}}.$$

42. *To find the form of the equation to a straight line which is perpendicular to a given straight line.*

Let $y = mx + c$ be the equation to the given straight line, and $y = m'x + c'$ the equation to another straight line. Then the tangent of the angle between these straight lines is

$$\frac{m - m'}{1 + mm'}.$$

If these straight lines are perpendicular to each other, $1 + mm' = 0$; therefore $m' = -\frac{1}{m}$.

Hence $$y = -\frac{x}{m} + c'$$

represents a straight line perpendicular to the straight line

$$y = mx + c.$$

43. The result of the last Article may also be obtained thus.

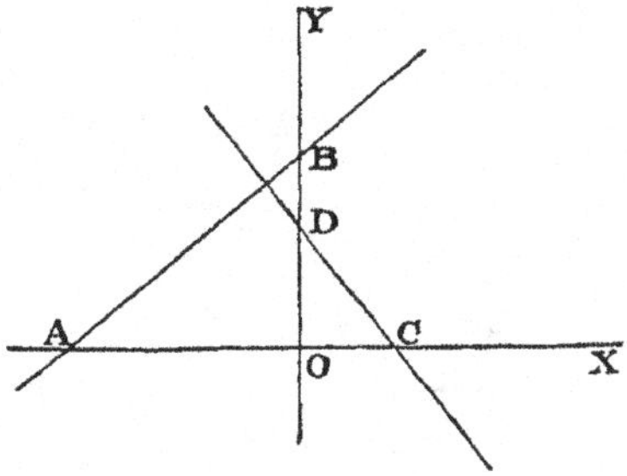

Let AB be the given straight line, so that $\tan BAX = m$. Let CD be a straight line perpendicular to AB; then

$$\tan DCX = -\tan DCO = -\cot BAO = -\frac{1}{m}.$$

Hence the equation to CD is

$$y = -\frac{x}{m} + c',$$

where $$c' = OD.$$

44. *To find the equation to the straight line which passes through a given point, and is perpendicular to a given straight line.*

Let x_1, y_1 be the co-ordinates of the given point, and

$$y = mx + c \text{.......................} (1)$$

the equation to the given straight line. The form of the equation to a straight line through (x_1, y_1) is

$$y - y_1 = m'(x - x_1) \text{..............} (2).$$

If (2) is perpendicular to (1), we have $m'm + 1 = 0$.

Hence the required equation is

$$y - y_1 = -\frac{1}{m}(x - x_1).$$

45. *To find the equations to the straight lines which pass through a given point and make a given angle with a given straight line.*

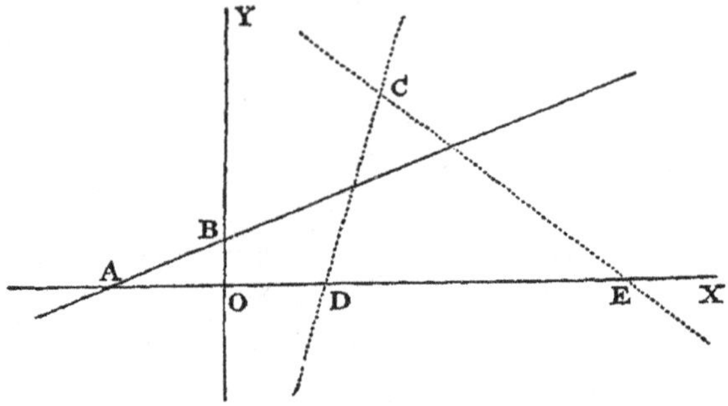

Let AB be the given straight line; C the given point; h, k its co-ordinates; β the given angle. Let the equation to AB be

$$y = mx + c.$$

Suppose CD and CE the two straight lines which can be drawn through C, each making an angle β with AB. Then

$$\tan CDX = \tan (BAX + \beta) = \frac{m + \tan \beta}{1 - m \tan \beta},$$

$$\tan CEX = -\tan CEA = -\tan (\beta - BAX) = \frac{m - \tan \beta}{1 + m \tan \beta}.$$

Hence the equation to CD is

$$y - k = \frac{m + \tan \beta}{1 - m \tan \beta} (x - h);$$

and the equation to CE is

$$y - k = \frac{m - \tan \beta}{1 + m \tan \beta} (x - h).$$

46. The following particular cases of the preceding results may be noted.

(1) Suppose $m = 0$; then the given straight line is parallel to the axis of x. The required equations then are

$$y - k = \tan \beta (x - h), \text{ and } y - k = -\tan \beta (x - h).$$

(2) Suppose $m=\infty$; then the given straight line is parallel to the axis of y. And since

$$\frac{m+\tan\beta}{1-m\tan\beta}=\frac{1+\frac{1}{m}\tan\beta}{\frac{1}{m}-\tan\beta},$$

we have when $m=\infty$, and therefore $\frac{1}{m}=0$, for the equation to CD, $y-k=-\frac{1}{\tan\beta}(x-h)=-\cot\beta\,(x-h)$.

Similarly the equation to CE becomes $y-k=\cot\beta\,(x-h)$.

(3) Suppose $m=\tan\beta$. In this case the equation to CD becomes $y-k=\frac{2\tan\beta}{1-\tan^2\beta}(x-h)$, that is, $y-k=\tan 2\beta(x-h)$.

The equation to CE becomes $y-k=0$, so that CE is parallel to the axis of x.

(4) Suppose $m=\cot\beta$. The equation to CD may be written in the form $(y-k)(1-m\tan\beta)=(m+\tan\beta)(x-h)$, and we see that when $m=\cot\beta$ the left-hand side is zero; thus the required equation is then $x-h=0$.

The equation to CE becomes $y-k=\frac{\cot\beta-\tan\beta}{2}(x-h)$

$$=\frac{\cos^2\beta-\sin^2\beta}{2\cos\beta\sin\beta}(x-h)=\cot 2\beta\,(x-h).$$

(5) Suppose $m=-\tan\beta$. Then the equation to CD becomes $y-k=0$; and the equation to CE becomes

$$y-k=\frac{-2\tan\beta}{1-\tan^2\beta}(x-h)=-\tan 2\beta\,(x-h).$$

(6) Suppose $m=-\cot\beta$. Then the equation to CD becomes $y-k=\frac{\tan\beta-\cot\beta}{2}(x-h)=-\cot 2\beta\,(x-h)$.

The equation to CE may be written in the form

$$(y-k)(1+m\tan\beta)=(m-\tan\beta)(x-h),$$

and we see that when $m = -\cot\beta$ the left-hand member is zero; thus the required equation is then $x - h = 0$.

(7) Suppose $\beta = \frac{\pi}{2}$. The equation to CD may be written

$$y - k = \frac{m\cot\beta + 1}{\cot\beta - m}(x - h).$$

When $\beta = \frac{\pi}{2}$ we have $\cot\beta = 0$; thus the equation becomes

$$y - k = -\frac{1}{m}(x - h).$$

Similarly the equation to CE takes the same form; and thus the result agrees with that of Art. 44.

We have discussed these particular cases as an example of the manner in which the student should test his comprehension of the subject by applying the general formulæ to special examples. He will find it useful to illustrate these cases by figures.

47. *To find the length of the perpendicular drawn from a given point on a given straight line.*

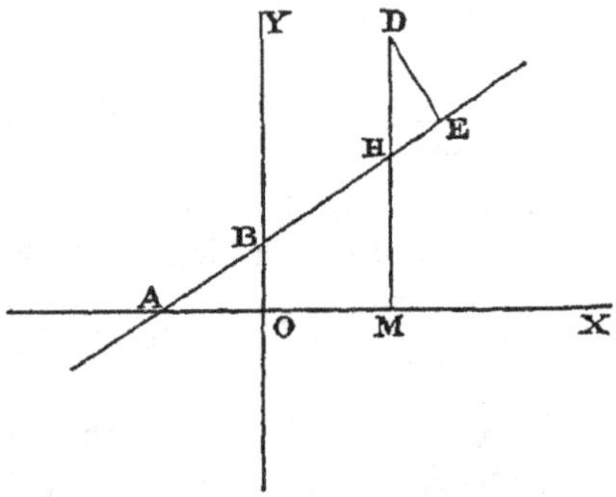

Let AB be the given straight line; D the given point; h, k its co-ordinates. Let the equation to AB be

$$y = mx + c \quad\text{.........................(1).}$$

The equation to the straight line through D perpendicular to AB is, by Art. 44,

$$y-k=-\frac{1}{m}(x-h) \quad \ldots\ldots(2).$$

Let x_1, y_1 be the co-ordinates of E; then, by Art. 9,

$$DE^2=(x_1-h)^2+(y_1-k)^2 \quad \ldots\ldots(3).$$

It remains then to substitute for x_1 and y_1 their values in (3). Now, since x_1, y_1 are the co-ordinates of E, which is the point where (1) and (2) meet, we have

$$y_1=mx_1+c, \quad \text{and} \quad y_1-k=-\frac{1}{m}(x_1-h);$$

therefore $mx_1+c=k-\dfrac{1}{m}(x_1-h)$, and $x_1=\dfrac{mk+h-mc}{1+m^2}$;

thus $$x_1-h=\frac{mk-m^2h-mc}{1+m^2}=\frac{m}{1+m^2}(k-mh-c).$$

Also $$y_1-k=-\frac{1}{m}(x_1-h)=\frac{mh+c-k}{1+m^2};$$

therefore by (3)

$$DE^2=\frac{m^2}{(1+m^2)^2}(k-mh-c)^2+\frac{(k-mh-c)^2}{(1+m^2)^2}=\frac{(k-mh-c)^2}{1+m^2}.$$

Hence $$DE=\frac{k-mh-c}{\sqrt{(1+m^2)}}.$$

The radical in the denominator may be taken with the positive or negative sign, according as the numerator is positive or negative, so as to give for DE a positive value.

We may also obtain the value of DE thus: draw the ordinate DM meeting the straight line AB at H; then

$$DE=DH\sin DHE=DH\cos HAM.$$

Now $OM=h$; therefore $HM=mh+c$, and $DM=k$;

therefore $DH=k-mh-c$.

Also $\tan HAM=m$; therefore $\cos HAM=\dfrac{1}{\sqrt{(1+m^2)}}$;

therefore $$DE=\frac{k-mh-c}{\sqrt{(1+m^2)}}.$$

Hence if on the straight line $y - mx - c = 0$ a perpendicular be drawn from the point (h_1, k_1) and also a perpendicular from the point (h_2, k_2), the ratio of the length of the first perpendicular to the length of the second is equal to the numerical ratio of $k_1 - mh_1 - c$ to $k_2 - mh_2 - c$.

48. *To find the length of the straight line drawn from a given point in a given direction to meet a given straight line.*

Let (h, k) be the given point; and suppose a straight line drawn from this point at an inclination α to the axis of x to meet the straight line

$$Ax + By + C = 0 \qquad (1).$$

Let r be the required length; x_1, y_1 the co-ordinates of the point where the straight line drawn from (h, k) meets (1); then, by Art. 27,

$$x_1 - h = r \cos \alpha, \qquad y_1 - k = r \sin \alpha \qquad (2).$$

But (x_1, y_1) is on (1),

therefore $A(h + r \cos \alpha) + B(k + r \sin \alpha) + C = 0$;

$$\text{therefore } r = -\frac{Ah + Bk + C}{A \cos \alpha + B \sin \alpha}.$$

49. In this Chapter we have used equations of the form $y = mx + c$ to represent straight lines. The student may exercise himself by solving the problems by means of the more symmetrical forms of the equation to a straight line,

$$Ax + By + C = 0, \quad \frac{x}{a} + \frac{y}{b} - 1 = 0, \quad x \cos \alpha + y \sin \alpha - p = 0.$$

The results of course can be easily compared with those we have obtained. For example, if in Art. 47 we represent the given straight line by the equation $Ax + By + C = 0$, the result obtained should coincide with the value of $\frac{k - mh - c}{\sqrt{(1 + m^2)}}$ when we write $-\frac{A}{B}$ for m, and $-\frac{C}{B}$ for c; that is, the result must be

$$\frac{Ah + Bk + C}{\sqrt{(A^2 + B^2)}}.$$

Similarly, if the given straight line be represented by the equation $x \cos\alpha + y \sin\alpha - p = 0$, we shall find for the length of the perpendicular on it from (h, k)

$$\pm (h \cos\alpha + k \sin\alpha - p).$$

Thus if the equation to a straight line be in the form

$$x \cos\alpha + y \sin\alpha - p = 0,$$

the length of the perpendicular drawn from a point on this straight line is the numerical *value of the expression on the left-hand side of this equation, when for* x *and* y *are substituted the co-ordinates of the given point.* This result is of such great importance that we shall proceed to examine it more closely.

50. We may however previously observe that if the equation to a straight line be given in any form, we can immediately transform it so that it may be expressed in terms of the length of the perpendicular from the origin and the inclination of this perpendicular to the axis of x. For example, suppose the equation to be

$$2x + 3y + 4 = 0.$$

Change the sign of every term so that the last term may be negative; thus the equation becomes

$$-2x - 3y - 4 = 0.$$

Divide by $\sqrt{(2^2 + 3^2)}$; thus

$$-\frac{2x}{\sqrt{13}} - \frac{3y}{\sqrt{13}} - \frac{4}{\sqrt{13}} = 0.$$

This is of the form

$$x \cos\alpha + y \sin\alpha - p = 0,$$

$$\text{and } \cos\alpha = -\frac{2}{\sqrt{13}}, \quad \sin\alpha = -\frac{3}{\sqrt{13}}, \quad p = \frac{4}{\sqrt{13}}.$$

In this example α is an angle lying between π and $\frac{3\pi}{2}$.

Any other example may be treated in a similar manner, the rule being the following. Collect the terms on one side, and, if necessary, change the signs so that the equation may be in the form $Ax + By - C = 0$, where C is positive; then divide by $\sqrt{(A^2 + B^2)}$; thus the equation becomes

$$\frac{Ax}{\sqrt{(A^2+B^2)}} + \frac{By}{\sqrt{(A^2+B^2)}} - \frac{C}{\sqrt{(A^2+B^2)}} = 0;$$

this is of the required form, and

$$\cos\alpha = \frac{A}{\sqrt{(A^2+B^2)}}, \quad \sin\alpha = \frac{B}{\sqrt{(A^2+B^2)}}, \quad p = \frac{C}{\sqrt{(A^2+B^2)}}.$$

Thus every equation representing a straight line may be brought to the form $x\cos\alpha + y\sin\alpha - p = 0$, where p is a *positive* quantity, unless the straight line passes through the origin, and then $p = 0$.

When we use the equation $x\cos\alpha + y\sin\alpha - p = 0$ *we shall always suppose* p *positive.*

51. The straight line whose equation is

$$x\cos\alpha + y\sin\alpha - p = 0$$

might be called "*the straight line* (p, α)," since the constants p and α determine the straight line; but when there is no risk of confounding it with another straight line, it may be more shortly called "*the straight line* α," and the equation may be expressed shortly, thus, "$\alpha = 0$."

We shall now give another investigation of the expression for the length of the perpendicular from a given point on the straight line (p, α).

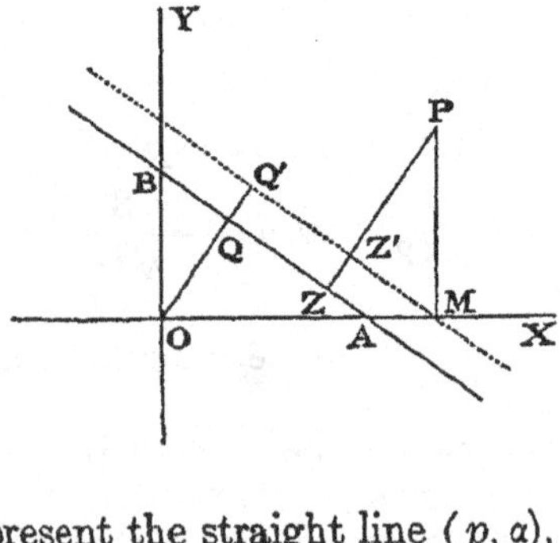

Let AB represent the straight line (p, α), O the origin, P the point (x, y), so that P and O are on *opposite* sides of AB.

Draw OQ, PZ perpendicular to AB, and PM parallel to OY; through M draw a straight line parallel to AB, meeting OQ and PZ, produced if necessary, at Q' and Z' respectively.

Then $OQ' = OM \cos\alpha = x \cos\alpha$; $PZ' = PM \sin\alpha = y \sin\alpha$;

$$PZ = OQ' + PZ' - OQ = x \cos\alpha + y \sin\alpha - p.$$

If P and O be on the *same* side of AB we shall obtain for the length of the perpendicular

$$p - x \cos\alpha - y \sin\alpha.$$

It will be found that these results will hold for all varieties of the figure.

52. Or we may proceed thus.

Let $$x \cos\alpha + y \sin\alpha - p = 0 \quad\quad (1)$$

be the equation to a straight line, and let x', y' be the coordinates of the point from which a perpendicular is drawn on the straight line: it is required to find the length of this perpendicular. The equation to any straight line which is parallel to (1) and on the same side of the origin, may be written thus

$$x \cos\alpha + y \sin\alpha - p' = 0 \quad\quad (2),$$

where p' is the perpendicular from the origin on this straight line. If this straight line pass through the point (x', y'), we must have

$$x' \cos\alpha + y' \sin\alpha - p' = 0;$$

therefore $$p' = x' \cos\alpha + y' \sin\alpha.$$

The length of the perpendicular from (x', y') on (1) will be $p' - p$ if the point and the origin are on different sides of the straight line, and $p - p'$ if they are on the same side; that is,

$$x' \cos\alpha + y' \sin\alpha - p$$

in the former case, and in the latter case

$$p - x' \cos\alpha - y' \sin\alpha.$$

If the straight line parallel to (1) be on the opposite side of the origin, its equation will be

$$x \cos(\pi + \alpha) + y \sin(\pi + \alpha) - p' = 0,$$

where p' is the length of the perpendicular from the origin on it. If this straight line pass through the point (x', y') we must have

$$x' \cos\alpha + y' \sin\alpha + p' = 0;$$

therefore $$p' = -x' \cos\alpha - y' \sin\alpha.$$

The length of the perpendicular from (x', y') on (1) will be the sum of p' and p, that is,

$$p - x' \cos\alpha - y' \sin\alpha.$$

We may now suppress the accents on x and y, and we have the same conclusion as before.

53. Thus the length of the perpendicular from the point (x, y) on the straight line

$$x \cos\alpha + y \sin\alpha - p = 0$$

is $$x \cos\alpha + y \sin\alpha - p, \quad \text{or } p - x \cos\alpha - y \sin\alpha,$$

according as the point (x, y) and the origin are on different sides of the straight line or on the same side of it.

The student will perceive that we speak here of the point (x, y) and the straight line $x \cos\alpha + y \sin\alpha - p = 0$, and that we use the same *symbols* x, y, in speaking of the point and of the straight line. But we do not mean that the point (x, y) is to be *on* the straight line, that is, we do not mean the x and y which are co-ordinates of the point (x, y) to have the *same values* as they have for any point in the straight line

$$x \cos\alpha + y \sin\alpha - p = 0.$$

We might use x', y' as co-ordinates of the point to prevent confusion, but it is found convenient to adopt the notation here used, as the advantages more than compensate for the increased attention which is required from the student in distinguishing the different meanings of the symbols.

54. We have in Art. 51 left the student to convince himself by drawing the figures in different ways, that the length of the perpendicular from the point (x, y) on the straight line (p, α) is *always* $\pm (x \cos\alpha + y \sin\alpha - p)$, the upper or lower sign being taken according as (x, y) and the origin are on *different* sides, or on the *same* side of the straight line (p, α). We may also arrive at the result imperfectly, thus. We may

first shew, as in Art. 47, that the length of the perpendicular must always be equal to one of the two expressions $\pm(x\cos\alpha+y\sin\alpha-p)$, and may then proceed to distinguish the cases. Now the expression $x\cos\alpha+y\sin\alpha-p$ is *negative* when the point (x, y) is the origin, because it becomes then $-p$; also this expression cannot change its sign so long as (x, y) is taken on the same side of the straight line (p, α) as the origin, *because it cannot change its sign without passing through the value zero*, and it cannot vanish until the point (x, y) is on the straight line. Hence the expression remains negative so long as (x, y) is on the same side of the straight line (p, α) as the origin. Similarly, if the expression is positive when the point (x, y) has *any one* position on the *other* side of the straight line (p, α), it will continue positive so long as (x, y) is on that side of the straight line; and it may be easily shewn that the expression *can* be made positive by suitable values of x and y; hence it *is* always positive while (x, y) is on the opposite side from the origin. We call this an imperfect method, because the sentence in italics on which the method depends, has probably not sufficiently attracted the student's attention up to this period of his studies to produce perfect conviction.

55. If the equation to a straight line be $x\cos\alpha+y\sin\alpha=0$, so that $p=0$, we shall still have $\pm(x\cos\alpha+y\sin\alpha)$ as the length of the perpendicular from the point (x, y) on it. We may discriminate as follows: let the equation be so written that the coefficient of y is *positive;* then for points on the same side of the straight line as the *positive* part of the axis of y, the perpendicular is $x\cos\alpha+y\sin\alpha$; for points on the other side it is $-(x\cos\alpha+y\sin\alpha)$. This is easily shewn by comparing a few figures, or as in Art. 54.

Oblique Axes.

56. The results in Arts. 32...40 hold whether the axes are rectangular or oblique; in Art. 33, however, m must have that meaning which is required when the axes are oblique.

To find the angle between two straight lines referred to oblique axes.

Let ω be the angle between the axes; $y=m_1x+c_1$ the equation to one straight line; $y = m_2 x + c_2$ the equation to the other. Let α_1, α_2 be the angles which these straight lines make with the axis of x; and β the angle between them.

By Art. 24

$$\tan \alpha_1 = \frac{m_1 \sin \omega}{1 + m_1 \cos \omega}; \quad \tan \alpha_2 = \frac{m_2 \sin \omega}{1 + m_2 \cos \omega}.$$

Hence $\tan \beta$ or $\tan (\alpha_2 - \alpha_1) = \dfrac{\dfrac{m_2 \sin \omega}{1 + m_2 \cos \omega} - \dfrac{m_1 \sin \omega}{1 + m_1 \cos \omega}}{1 + \dfrac{m_1 m_2 \sin^2 \omega}{(1 + m_1 \cos \omega)(1 + m_2 \cos \omega)}}$

$$= \frac{(m_2 - m_1) \sin \omega}{1 + (m_1 + m_2) \cos \omega + m_1 m_2}.$$

Hence the condition that the straight lines may be at right angles is

$$1 + (m_1 + m_2) \cos \omega + m_1 m_2 = 0.$$

57. *To find the length of the perpendicular drawn from a given point on a given straight line.*

We shall proceed as in the latter part of Art. 47; the student may also obtain the result by the method in the former part of that Article.

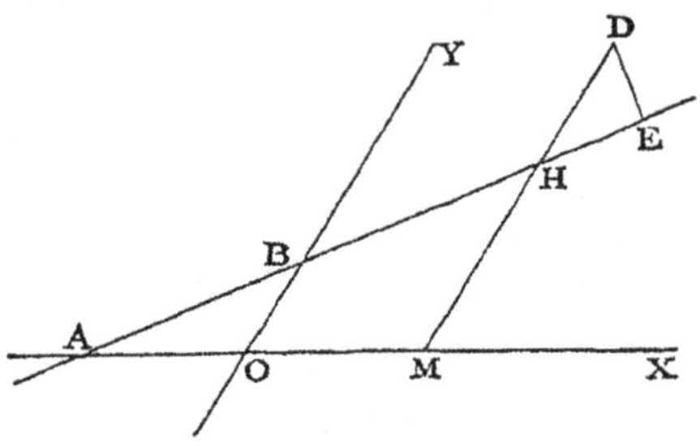

Let AB be the given straight line; D the given point; h, k its co-ordinates.

Let the equation to AB be $y = mx + c$.

Draw DHM parallel to OY, and DE perpendicular to AB; then

$$DE = DH \sin DHE.$$

Now $DH = DM - HM = k - (mh + c) = k - mh - c$.

Let $BAX = \alpha$, then DHE or $AHM = \omega - \alpha$,

$$\text{and } \frac{\sin \alpha}{\sin (\omega - \alpha)} = m \quad \text{(Art. 24)};$$

$$\text{therefore } \sin (\omega - \alpha) = \frac{\sin \alpha}{m} = \frac{\sin \omega}{\sqrt{(1 + 2m \cos \omega + m^2)}} \quad \text{(Art. 24)};$$

$$\text{therefore } DE = \frac{(k - mh - c) \sin \omega}{\sqrt{(1 + 2m \cos \omega + m^2)}}.$$

If a straight line be drawn from D to meet AB at an angle β, its length will be $DE \operatorname{cosec} \beta$, and will therefore be known since DE is known.

If the equation to a straight line be in the form given in Art. 26, namely, $x \cos \alpha + y \cos \beta - p = 0$, the length of the perpendicular on it from the point (x', y') will be

$$\pm (x' \cos \alpha + y' \cos \beta - p).$$

This may be deduced from the preceding expression, or it may be obtained in the manner of Art. 51.

Polar Co-ordinates.

58. *To find the polar equation to the straight line which passes through two given points.*

Let r_1, θ_1 be the co-ordinates of one point; and r_2, θ_2 those of the other; and suppose the equation to the straight line

$$r \cos (\theta - \alpha) = p,$$

that is, $$r \cos \theta \cos \alpha + r \sin \theta \sin \alpha = p \quad \text{.................(1)}.$$

Since this straight line passes through the two points, we have

$$r_1 \cos \theta_1 \cos \alpha + r_1 \sin \theta_1 \sin \alpha = p \quad \text{..................(2)},$$

$$r_2 \cos \theta_2 \cos \alpha + r_2 \sin \theta_2 \sin \alpha = p \quad \text{..................(3)}.$$

From (1) and (2)

$$(r \cos \theta - r_1 \cos \theta_1) \cos \alpha + (r \sin \theta - r_1 \sin \theta_1) \sin \alpha = 0 \ldots (4).$$

From (2) and (3)

$$(r_1 \cos \theta_1 - r_2 \cos \theta_2) \cos \alpha + (r_1 \sin \theta_1 - r_2 \sin \theta_2) \sin \alpha = 0 \ldots (5),$$

therefore $\frac{r\cos\theta - r_1\cos\theta_1}{r_1\cos\theta_1 - r_2\cos\theta_2} = \frac{r\sin\theta - r_1\sin\theta_1}{r_1\sin\theta_1 - r_2\sin\theta_2}$.

After reduction we obtain

$$rr_1\sin(\theta_1 - \theta) + r_1r_2\sin(\theta_2 - \theta_1) + r_2r\sin(\theta - \theta_2) = 0 \ldots (6).$$

This equation has a simple geometrical interpretation; for if we draw a figure and take O for the origin and A, B, P for the points (r_1, θ_1), (r_2, θ_2), (r, θ), respectively, we see that equation (6) is the expression of the fact that one of the triangles OAP, OBP, OAB, is equal in area to the sum of the other two.

59. We have seen that a straight line is the locus of an equation of the first degree; as we proceed it will appear that if an equation be of a degree higher than the first, the corresponding locus will be *generally* some curve; we may notice here some exceptional cases.

Suppose the equation

$$x^2 - 4ax + 4a^2 + y^2 = 0$$

be proposed; this equation may be written

$$(x - 2a)^2 + y^2 = 0.$$

Hence we see that the *only* solution is

$$y = 0, \quad x = 2a.$$

Thus the corresponding locus consists only of a single point on the axis of x at a distance $2a$ from the origin.

Again, suppose the equation to be

$$x^2 + y^2 + 1 = 0.$$

No real values of x and y will satisfy this equation; in this case then there is no corresponding locus, or as it is usually expressed, the locus is *impossible*. Thus, the locus corresponding to a given equation *may* reduce to a single point, or it may be impossible.

60. We have seen that the equation to a single straight line is always of the *first* degree; an equation of a higher degree than the first *may* however represent a locus consisting of two or more straight lines. For example, suppose

$$y^2 - x^2 = 0 \ldots\ldots\ldots\ldots\ldots\ldots (1);$$

therefore $y = x \ldots\ldots\ldots (2)$, or $y = -x \ldots\ldots\ldots (3)$.

If the co-ordinates of a point satisfy *either* (2) *or* (3), they will satisfy (1); that is, every point which is comprised in the locus (2) is comprised in (1), and every point which is comprised in (3) is also comprised in (1). Hence (1) represents *two* straight lines which pass through the origin, and make respectively angles of 45° and 135° with the axis of x.

Again take the general equation of the second degree between two variables

$$ax^2 + bxy + cy^2 + dx + ey + f = 0;$$

and let us determine when it represents two straight lines.

We have $cy^2 + (bx + e)\,y + ax^2 + dx + f = 0$. Hence considering this as a quadratic equation in y, and solving in the usual way we obtain

$$y = -\frac{bx+e}{2c} \pm \frac{\sqrt{\{(bx+e)^2 - 4c\,(ax^2 + dx + f)\}}}{2c}.$$

The expression under the radical sign is

$$(b^2 - 4ac)\,x^2 + 2\,(be - 2cd)\,x + e^2 - 4cf;$$

if this expression is an *exact square* with respect to x it is obvious that the proposed equation of the second degree breaks up into two equations of the first degree between x and y, and so represents two straight lines.

The condition which is necessary and sufficient to ensure that the expression under the radical sign is a perfect square with respect to x is, by *Algebra*, Chapter XXII,

$$(be - 2cd)^2 = (b^2 - 4ac)\,(e^2 - 4cf).$$

61. An equation which involves only *one* of the variables, represents a series of straight lines parallel to one of the axes. Thus, if there be an equation $f(x) = 0$, where $f(x)$ denotes any expression which involves x and known quantities, we obtain by solving it a series of values for x, as $x = a_1$, $x = a_2$, ... and each of these equations represents a straight line parallel to the axis of y. Similarly $f(y) = 0$ represents a series of straight lines parallel to the axis of x.

An equation of the form $f\left(\frac{y}{x}\right) = 0$ represents a series of straight lines passing through the origin; for by solving the equation we obtain a series of values for $\frac{y}{x}$, as $\frac{y}{x} = m_1$, $\frac{y}{x} = m_2, \ldots$ and each of these equations represents a straight line passing through the origin. Of course if an equation $f(x) = 0$, $f(y) = 0$, or $f\left(\frac{y}{x}\right) = 0$ have no real roots, the corresponding locus is impossible.

The equation $Ay^2 + Bxy + Cx^2 = 0$ may be put in the form

$$A\left(\frac{y}{x}\right)^2 + B\frac{y}{x} + C = 0.$$

Since this is a quadratic in $\frac{y}{x}$ we obtain *two* values for it, suppose $\frac{y}{x} = m_1$ and $\frac{y}{x} = m_2$; hence the equation generally represents *two* straight lines passing through the origin. If B^2 be less than $4AC$, then m_1 and m_2 are impossible, and the *only* solution of the given equation is $x = 0$, $y = 0$; that is, the locus is a single point, namely, the origin.

62. It is obvious that if the locus represented by an equation $f(x, y) = 0$ passes through the origin, the values $x = 0$, $y = 0$ must satisfy the equation. We can thus immediately determine by inspection whether a proposed locus passes through the origin or not.

63. In Art. 39 we determined the co-ordinates of the point of intersection of two given straight lines: the proposition may obviously be generalised. Let $f_1(x, y) = 0$, $f_2(x, y) = 0$, represent two curves, then the co-ordinates of the points where they meet will be determined by solving these simultaneous equations. It may be shewn that if one equation be of the mth degree and the other of the nth degree, the number of common points cannot exceed mn. (See *Theory of Equations*, Chapter XX.)

64. We will exemplify the Articles of this Chapter by applying them to demonstrate some properties of a triangle.

The straight lines drawn from the angles of a triangle to the middle points of the opposite sides meet at a point.

Let ABC be a triangle, D, E, F the middle points of the sides; take A for the origin, AB for the direction of the axis

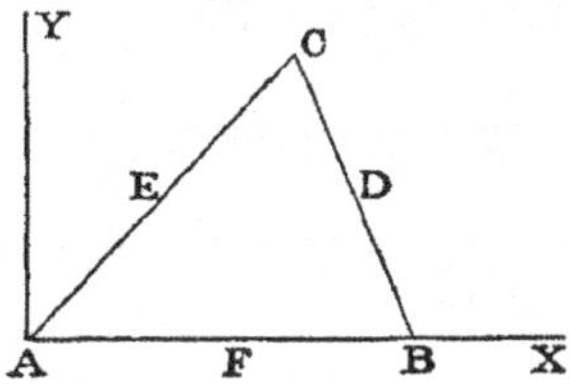

of x, and a straight line through A at right angles to AB for the axis of y. Let $AB = a$, and let x', y' be the co-ordinates of C. Since D is the middle point of CB, the abscissa of D is $\frac{1}{2}(x' + a)$ and its ordinate $\frac{y'}{2}$ (Art. 10); since E is the middle point of AC, the abscissa of E is $\frac{x'}{2}$ and its ordinate $\frac{y'}{2}$; since F is the middle point of AB, its abscissa is $\frac{a}{2}$ and its ordinate zero. Hence by Art. 35,

$$\text{the equation to } AD \text{ is } y = \frac{y'x}{x' + a} \quad\ldots\ldots\ldots\ldots\ldots(1);$$

$$\text{the equation to } BE \text{ is } y = \frac{y'(x - a)}{x' - 2a} \quad\ldots\ldots\ldots\ldots(2);$$

$$\text{the equation to } CF \text{ is } y = \frac{y'(2x - a)}{2x' - a} \quad\ldots\ldots\ldots(3).$$

To find the point of intersection of (2) and (3) we put

$$\frac{y'(x - a)}{x' - 2a} = \frac{y'(2x - a)}{2x' - a};$$

therefore $\quad (x - a)(2x' - a) = (2x - a)(x' - 2a)$;

therefore $\quad 3ax = a(x' + a)$; therefore $x = \frac{1}{3}(x' + a)$.

Substitute this value of x in (2) and we find $y = \frac{y'}{3}$.

We have thus determined the co-ordinates of the point of intersection of (2) and (3); moreover we see that these values satisfy (1); hence the straight line represented by (1) passes through the intersection of the straight lines represented by (2) and (3), which demonstrates the proposition.

The straight lines drawn from the angles of a triangle perpendicular to the opposite sides meet at a point.

The equation to BC is (Art. 35)

$$y = \frac{y'}{x' - a}(x - a);$$

hence the equation to the straight line through A perpendicular to BC is (Art. 44)

$$y = -\frac{x' - a}{y'} x \quad \text{.......................(4).}$$

The equation to AC is $y = \frac{y'}{x'} x$; hence the equation to the straight line through B perpendicular to AC is

$$y = -\frac{x'}{y'}(x - a) \quad \text{....................(5).}$$

The straight line through C perpendicular to AB will be parallel to the axis of y, and its equation will be (Art. 15)

$$x = x' \quad \text{.........................(6).}$$

Now at the point of intersection of (5) and (6) we have

$$x = x', \qquad y = -\frac{x'}{y'}(x' - a);$$

and as these values satisfy (4), the straight line represented by (4) passes through the intersection of the straight lines represented by (5) and (6).

The straight lines drawn through the middle points of the sides of a triangle respectively at right angles to those sides meet at a point.

The equation to the straight line through D at right angles to BC is

$$y - \frac{y'}{2} = -\frac{x' - a}{y'}\left(x - \frac{a + x'}{2}\right) \quad \text{........... (7).}$$

The equation to the straight line through E at right angles to CA is

$$y - \frac{y'}{2} = -\frac{x'}{y'}\left(x - \frac{x'}{2}\right) \quad \text{..................... (8).}$$

The equation to the straight line through F at right angles to AB is

$$x = \frac{a}{2} \quad\text{...............................(9).}$$

Now at the point of intersection of (8) and (9) we have

$$x = \frac{a}{2}, \qquad y = \frac{y'}{2} - \frac{x'}{y'}\left(\frac{a}{2} - \frac{x'}{2}\right);$$

these values satisfy (7); hence the straight lines represented by (7), (8), and (9), meet at a point.

Let us denote by P the point of intersection of the three straight lines in the first proposition, by Q the point of intersection of the three straight lines in the second proposition, and by R the point of intersection of the three straight lines in the third proposition; we will now shew that P, Q, and R lie on one straight line. The co-ordinates

of P are $x = \frac{1}{3}(x' + a)$, $\quad y = \frac{y'}{3}$;

of Q are $x = x'$, $\quad y = \frac{x'}{y'}(a - x')$;

of R are $x = \frac{a}{2}$, $\quad y = \frac{y'}{2} - \frac{x'(a - x')}{2y'}$.

Hence the equation to the straight line passing through P and Q is

$$y - \frac{y'}{3} = \frac{\frac{x'}{y'}(a - x') - \frac{y'}{3}}{x' - \frac{1}{3}(x' + a)}\left(x - \frac{x' + a}{3}\right) \quad\text{............(10).}$$

In this equation put $x = \frac{a}{2}$, then

$$y - \frac{y'}{3} = \frac{\frac{x'}{y'}(a - x') - \frac{y'}{3}}{\frac{1}{3}(2x' - a)}\left(\frac{a}{6} - \frac{x'}{3}\right) = -\frac{1}{2}\left\{\frac{x'}{y'}(a - x') - \frac{y'}{3}\right\};$$

therefore $y = -\frac{x'(a - x')}{2y'} + \frac{y'}{3} + \frac{y'}{6} = \frac{y'}{2} - \frac{x'(a - x')}{2y'}$.

Hence the point R is on the straight line represented by (10), for the co-ordinates of R satisfy (10).

EXAMPLES.

1. Find the equations to the straight lines which pass through the following pairs of points:

(1) $(0, 1)$, and $(1, -1)$. (2) $(2, 3)$, and $(2, 4)$.

(3) $(1, 1)$, and $(-2, -2)$. (4) $(0, -a)$, and $(0, -b)$.

2. Find the equations to the straight lines which pass through the point $(4, 4)$ and are inclined at an angle of 45° to the straight line $y = 2x$.

3. Find the equations to the straight lines which pass through the point $(0, 1)$ and are inclined at an angle of 30° to the straight line $y + x = 2$.

4. Find the equations to the straight lines which pass through the origin and are inclined at an angle of 45° to the straight line $x = 2$.

5. Find the equations to the straight lines which pass through the origin and are inclined at an angle of 60° to the straight line $x + y\sqrt{3} = 1$.

6. Find the angle between the straight lines $x + y = 1$, $y = x + 2$; also find the co-ordinates of the point of intersection.

7. Find the angle between the straight lines $x + y\sqrt{3} = 0$ and $x - y\sqrt{3} = 2$.

8. Find the angle between $x + 3y = 1$ and $x - 2y = 1$.

9. Find the equations to the straight lines passing through a given point in the axis of x, and making an angle of 45° with the axis of x.

10. Find the equation to the straight line which passes through the origin and is perpendicular to the straight line $x + y = 2$.

11. Find the perpendicular distance of the point $(1, -2)$ from the straight line $x + y - 3 = 0$.

12. Find the length of the perpendicular from the point (a, b) on the straight line $\frac{x}{a} + \frac{y}{b} = 1$.

13. Find the co-ordinates of the point of intersection of the straight lines $\frac{x}{a}+\frac{y}{b}=1$ and $\frac{x}{b}+\frac{y}{a}=1$.

14. Find the equation to the straight line which passes through the point (a, b), and through the intersection of the straight lines $\frac{x}{a}+\frac{y}{b}=1$ and $\frac{x}{b}+\frac{y}{a}=1$.

15. Shew what loci are represented by the equations:

(1) $x^2+y^2=0$, (2) $x^2-y^2=0$,

(3) $x^2+xy=0$, (4) $xy=0$,

(5) $x^2+y^2+a^2=0$, (6) $x(y-a)=0$.

16. Interpret the equations:

(1) $(x-a)(y-b)=0$,

(2) $(x-a)^2+(y-b)^2=0$,

(3) $(x-y+a)^2+(x+y-a)^2=0$.

17. Determine what straight lines are represented by the equation $y^2-4xy+3x^2=0$.

18. Shew that $3y^2-8xy-3x^2+30x-27=0$ represents two straight lines at right angles to one another.

19. Find the equations to the diagonals of the four-sided figure, the sides of which are represented by the equations

$$x=4, \quad y=5, \quad y=x, \quad y=2x.$$

20. $ABCDEF$ is a regular hexagon; take A for the origin, AB as axis of x, and a straight line through A at right angles to AB as axis of y: find the equations to all the straight lines joining the angular points of the hexagon.

21. Given the co-ordinates of the angular points of a triangle, find the equation to the straight line which joins the middle points of two sides.

22. Find the tangent of the angle between the straight lines

$$y-mx=0 \text{ and } my+x=0,$$

when referred to oblique axes.

23. Shew that whether the axes be rectangular or oblique the straight lines $y + x = 0$ and $y - x = 0$ are at right angles.

24. Given the lengths of two sides of a parallelogram and the angle between them, write down the equations to the two diagonals and find the angle between them; taking one of the corners as origin, and the two sides which meet at that corner as axes.

25. In the figure to Art. 75, take BA and BC as the axes of x and y; suppose $BA = a$, $BC = c$; and let h, k be the co-ordinates of D: then form the equations to AC, BD, AD, CD.

26. With the notation of the preceding Example, find the co-ordinates of the middle point of AC and those of the middle point of BD, and form the equation to the straight line passing through these two points.

27. With the same notation find the co-ordinates of the middle point of EF, and thus shew that this point lies on the straight line joining the middle points of AC and BD.

28. If $\frac{x}{a} + \frac{y}{b} = 1$ and $\frac{x}{a'} + \frac{y}{b'} = 1$ be the equations to two straight lines, which with the co-ordinate axes (rectangular or oblique) contain equal areas, and x', y' be the co-ordinates of the point of their intersection; shew that

$$\frac{y'}{x'} = \frac{b - b'}{a' - a}.$$

29. Determine what points on the axis of x are at a perpendicular distance a from the straight line $\frac{x}{a} + \frac{y}{b} = 1$.

30. Form the equation for determining the abscissa of a point, in the straight line of which the equation is $\frac{x}{a} + \frac{y}{b} = 1$, whose distance from a given point (α, β) shall be equal to a given straight line c. Shew that there are in general two such points, and in the particular case in which those points coincide

$$c^2 (a^2 + b^2) = (a\beta + b\alpha - ab)^2.$$

31. Find the tangent of the angle between the two straight lines represented by the equation $Ay^2 + Bxy + Cx^2 = 0$.

32. Find the points of intersection of the straight lines $x+2y-5=0$, $2x+y-7=0$, and $y-x-1=0$; and shew that the area of the triangle formed by them is $\frac{3}{2}$.

33. The area of the triangle formed by the straight lines

$$y = x \tan \alpha, \quad y = x \tan \beta, \quad y = x \tan \gamma + c,$$

is
$$\frac{c^2}{2} \frac{\sin(\alpha-\beta)\cos^2\gamma}{\sin(\alpha-\gamma)\sin(\beta-\gamma)}.$$

34. Given the equations to two parallel straight lines, find the distance between them.

35. Determine the angle between the straight lines

$$\frac{a}{r} = 4\cos\theta + 3\sin\theta, \quad \frac{b}{r} = 3\cos\theta - 4\sin\theta.$$

36. Interpret $F(\theta) = 0$; for example, $\sin 3\theta = 0$.

37. If the axes be inclined at an angle ω, the condition that the straight lines $Ax + By + C = 0$, $A'x + B'y + C' = 0$, may be equally inclined to the axis of x in opposite directions is

$$\frac{B}{A} + \frac{B'}{A'} = 2\cos\omega.$$

38. In the preceding Example, if besides being equally inclined to the axis of x the straight lines pass through the origin and are perpendicular to one another, the equation to the straight lines is $x^2 + 2xy\cos\omega + y^2\cos 2\omega = 0$.

39. Two parallel straight lines are drawn at an inclination θ to the axis of x through the two points whose co-ordinates are a, b, and a', b'; shew that the distance between these straight lines is $(b'-b)\cos\theta - (a'-a)\sin\theta$. Hence determine the rectangle whose sides pass through four given points, and whose area is given.

40. A square is moved so as always to have the two extremities of one of its diagonals on two fixed straight lines at right angles to each other in the plane of the square: shew that the extremities of the other diagonal will at the

same time move on two other fixed straight lines at right angles to each other.

41. AB and BC are two straight lines at right angles to each other, A is a fixed point, B moves along a given straight line, and AB to BC is a given ratio: determine the locus of C.

42. OX and OY are fixed straight lines meeting at any angle; a straight line of given length slides along OX, and another straight line of given length slides along OY. Find the locus of a point which is so taken that the sum of the areas formed by joining it to the ends of the moving straight lines is constant.

43. Shew that the straight lines FC, KB, AL, in the figure to Euclid I. 47, meet at a point.

44. If on the sides of a triangle as diagonals, parallelograms be described, having their sides parallel to two given straight lines, the other diagonals of the parallelograms will meet at a point.

45. If from a fixed point O a straight line be drawn $OABCD$... meeting at A, B, C, D,... any given fixed straight lines in one plane, and if

$$\frac{1}{OX} = \frac{1}{OA} + \frac{1}{OB} + \frac{1}{OC} + \dots$$

X being a point in OA, the locus of X is a straight line.

46. Shew that the area of the triangle contained by the axis of y and the straight lines $y = m_1x + c_1$, $y = m_2x + c_2$, is

$$\frac{(c_2 - c_1)^2}{2(m_2 - m_1)}.$$

47. Determine the area of the triangle contained by the straight lines $y = m_1x + c_1$, $y = m_2x + c_2$, $y = m_3x + c_3$.

48. The area of the triangle formed by the three straight lines $y = ax - \frac{kbc}{2}$, $y = bx - \frac{kac}{2}$, $y = cx - \frac{kab}{2}$, is

$$\frac{(a-b)(b-c)(c-a)k^2}{8}.$$

CHAPTER IV.

STRAIGHT LINE CONTINUED.

65. We have seen that each of the equations

$$Ax + By + C = 0, \quad A'x + B'y + C' = 0,$$

represents a straight line. We will now interpret the equation

$$Ax + By + C + \lambda (A'x + B'y + C') = 0 \text{.........(1)},$$

where λ is some constant quantity.

I. Equation (1) must represent *some* straight line, because it is of the first degree in the variables x, y. (Art. 16.)

II. The straight line represented by (1) passes through the intersection of the straight lines represented by

$$Ax + By + C = 0 \text{....................(2)},$$
$$A'x + B'y + C' = 0 \text{....................(3)}.$$

For the values of x and y which satisfy *simultaneously* (2) and (3) will obviously satisfy (1); that is, the point at which (2) and (3) intersect lies on (1).

III. By giving a suitable value to the constant λ the equation (1) may be made to represent *any* straight line which passes through the intersection of (2) and (3).

For let x_1, y_1 denote the co-ordinates of the point of intersection of (2) and (3); suppose *any* straight line drawn through this point, and let x_2, y_2 be the co-ordinates of another point in it. Now we have already shewn in II. that the straight line (1) passes through (x_1, y_1); we have therefore only to shew that by giving a suitable value to λ the straight line

(1) can be made to pass through (x_2, y_2), because two straight lines which have two common points must coincide. Substitute x_2, y_2 for x and y respectively in (1), and determine λ so as to satisfy the equation. Thus

$$\lambda = -\frac{Ax_2 + By_2 + C}{A'x_2 + B'y_2 + C'}.$$

Now use this value of λ in (1); then the equation

$$Ax + By + C - \frac{Ax_2 + By_2 + C}{A'x_2 + B'y_2 + C'}(A'x + B'y + C') = 0 \ldots (4)$$

represents a straight line passing through (x_1, y_1) and (x_2, y_2).

We have thus shewn that by giving a suitable value to λ, the equation (1) will represent *any* straight line passing through the intersection of (2) and (3).

66. The preceding Article is very important, and commonly presents difficulties to beginners. The student should not leave it until he is thoroughly familiar with the three propositions which are contained in it. The first proposition is obvious. To demonstrate the second proposition the student may, if he pleases, actually find the values of x and y which satisfy simultaneously $Ax + By + C = 0$, and $A'x + B'y + C' = 0$, and convince himself, by substituting these values, that they *do* satisfy $Ax + By + C + \lambda(A'x + B'y + C') = 0$. There is, however, no necessity for solving the first equations, because it is evident that values of x and y which make $Ax + By + C$ and $A'x + B'y + C'$ vanish simultaneously must also make $Ax + By + C + \lambda(A'x + B'y + C')$ vanish, because they make each of the two members of the expression vanish. The third proposition of the preceding Article is usually the most difficult: the student is apt to think it needs no demonstration. It may be obvious, however, that by giving different values to λ, different straight lines are represented, and that we can thus obtain *as many straight lines as we please*, but this does not shew that we can by a suitable value of λ in (1) represent *any* straight line passing through the intersection of (2) and (3).

For example, if the straight lines (2) and (3) be DSE and FSG respectively, *it might have happened* that all the straight lines represented by (1) fell within the angle FSD and none within FSE. It requires to be demonstrated then that by

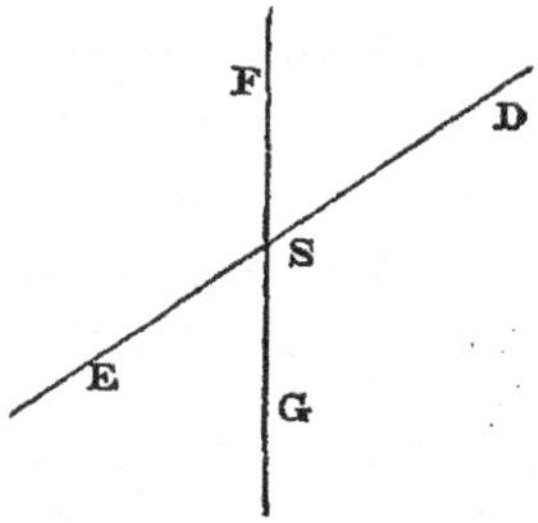

giving to λ a suitable value in (1) we can obtain the equation to *any* straight line through S.

67. It is often convenient to denote by a single symbol the expression which we equate to zero in our investigations in this subject; for example, in Art. 51 we have used the symbol α as an abbreviation for $x\cos\alpha + y\sin\alpha - p$. In like manner we may denote such expressions as $Ax + By + C$, $y - mx - c$, $\frac{x}{a} + \frac{y}{b} - 1$,... by single symbols, as u, v,... u',... Now it will be seen that the demonstration in Art. 65 applies to *any* form of the equation to a straight line as well as to the form $Ax + By + C = 0$ which we have used. Hence the result may be enunciated thus: if $u = 0$ and $v = 0$ be the equations to two straight lines, and λ a constant quantity, the equation $u + \lambda v = 0$ will represent a straight line passing through the intersection of the two straight lines; and by giving a suitable value to λ, the equation will represent *any* straight line passing through the intersection of the two straight lines.

68. If $u = 0$ and $v = 0$ be the equations to two straight lines, then as we have shewn, $u + \lambda v = 0$ will represent a straight line passing through their intersection; it is sometimes convenient to use the more symmetrical form $lu + mv = 0$, where l and m are both constants. It is obvious that what has been said respecting the first form applies to the second; in fact the second is deducible from the first by writing $\frac{m}{l}$ for λ. It must be remembered throughout this Chapter that l, m,

n,... λ,... are *constants*, though for shortness we may omit to state it specially in every Article.

69. Similarly if $u=0$, $v=0$, $w=0$, be the equations to three straight lines, and l, m, n be constants, the equation

$$lu + mv + nw = 0 \text{..................}(1)$$

will represent a straight line. Moreover, by giving suitable values to l, m, n we may in general make this equation represent *any* straight line whatever. For suppose we wish this equation to represent the straight line passing through (x_1, y_1) and (x_2, y_2). Let u_1, v_1, w_1 denote the values of u, v, w respectively when we put x_1 for x and y_1 for y; and let u_2, v_2, w_2 denote the respective values when x_2 and y_2 are put for x and y respectively. Determine the values of $\frac{m}{l}$ and $\frac{n}{l}$ from the equations $lu_1 + mv_1 + nw_1 = 0$ and $lu_2 + mv_2 + nw_2 = 0$; suppose we thus find $\frac{m}{l} = \frac{\mu}{\lambda}$, and $\frac{n}{l} = \frac{\nu}{\lambda}$; substitute these values in the equation $u + \frac{m}{l}v + \frac{n}{l}w = 0$, and we obtain

$$u + \frac{\mu}{\lambda}v + \frac{\nu}{\lambda}w = 0, \quad \text{or } \lambda u + \mu v + \nu w = 0,$$

which represents the straight line passing through the points (x_1, y_1) and (x_2, y_2).

We have said above that the equation (1) can *in general* be made to represent any straight line, because there are exceptions which we now proceed to notice.

When the straight lines represented by $u=0$, $v=0$, and $w=0$ meet at a point, the equation (1) represents a straight line which necessarily passes through that point. For since the three given straight lines meet at a point, u, v, and w vanish simultaneously at that point; therefore $lu + mv + nw$ also vanishes at that point, so that the straight line represented by equation (1) passes through that point.

When the three given straight lines are parallel the equations $u=0$, $v=0$, $w=0$ will be of the forms

$$Ax + By + C_1 = 0, \quad Ax + By + C_2 = 0, \quad Ax + By + C_3 = 0,$$

and thus equation (1) may be reduced to

$$Ax + By + \frac{lC_1 + mC_2 + nC_3}{l + m + n} = 0,$$

and this equation represents a straight line parallel to the given straight lines.

Thus if the three given straight lines meet at a point or are parallel, equation (1) will not represent *any* straight line; for the straight line represented by equation (1), in the former case passes through the point at which the given straight lines meet, and in the latter case is parallel to the given straight lines.

We may shew that there is no other exception. For the general investigation is always conclusive except when λ, μ, and ν all vanish, that is, when

$$v_1w_2 - v_2w_1 = 0, \quad w_1u_2 - w_2u_1 = 0, \quad u_1v_2 - u_2v_1 = 0 \ldots\ldots(2).$$

We shall now shew that when equations (2) are satisfied, the three given straight lines either all meet at a point or are parallel.

First suppose that the points (x_1, y_1) and (x_2, y_2) are not on any of the three given straight lines; so that none of the quantities u_1, v_1, w_1, u_2, v_2, w_2 vanish.

From the first of equations (2) we have $\frac{v_1}{v_2} = \frac{w_1}{w_2}$; hence by Art. 47 it follows that the ratio of the perpendiculars from (x_1, y_1) and (x_2, y_2) on the straight line $v = 0$, is the same as the ratio of the perpendiculars from the same points on the straight line $w = 0$. Hence it will follow geometrically either that the straight lines $v = 0$ and $w = 0$ are both parallel to the straight line joining (x_1, y_1) and (x_2, y_2), or else that these three straight lines meet at a point. Similar results follow from the second of equations (2), and from the third of equations (2). Hence in this case if equations (2) are satisfied, the three given straight lines either meet at a point or are parallel.

Next suppose that one of the two given points is situated on one of the three given straight lines; suppose for example that $w_1 = 0$. Then from the first of equations (2) it follows that either $v_1 = 0$ or $w_2 = 0$. Suppose we take $v_1 = 0$.

Then from the second and third of equations (2) we deduce either that $u_1 = 0$ or else that $w_2 = 0$ and $v_2 = 0$; in the former case the three given straight lines all pass through the point (x_1, y_1); in the latter case the straight lines $v = 0$ and $w = 0$ both pass through the two points (x_1, y_1) and (x_2, y_2), that is, two of the given straight lines coincide so that all three will reduce either to two intersecting straight lines or to two parallel straight lines. Suppose we take $w_2 = 0$ in conjunction with $w_1 = 0$. Then the straight line $w = 0$ passes through the given points (x_1, y_1) and (x_2, y_2). From the third of equations (2) we have $\frac{u_1}{u_2} = \frac{v_1}{v_2}$; and thus the straight lines $u = 0$ and $v = 0$ either meet on the straight line joining the points (x_1, y_1) and (x_2, y_2), or are parallel to this straight line; that is, the straight lines $u = 0$, $v = 0$, and $w = 0$ either meet at a point or are parallel.

70. Let $\alpha = 0$, $\beta = 0$ be the equations to two straight lines expressed in terms of the perpendiculars from the origin and their inclinations to the axis of x (see Art. 50), so that α is an abbreviation for $x \cos \alpha + y \sin \alpha - p_1$, and β is an abbreviation for $x \cos \beta + y \sin \beta - p_2$; we proceed to shew the meaning of the equations $\alpha - \beta = 0$ and $\alpha + \beta = 0$.

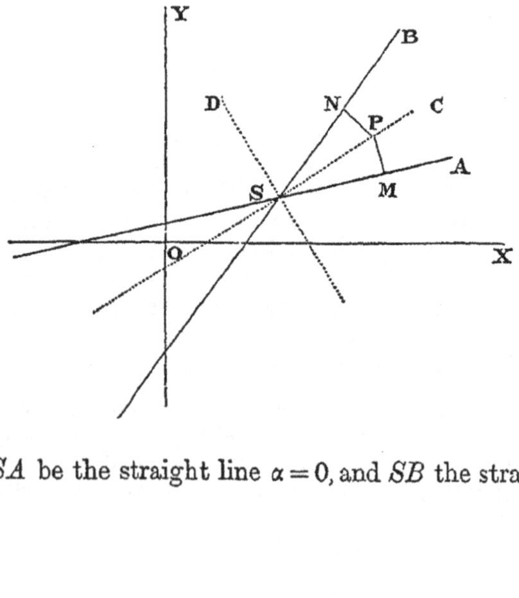

Let SA be the straight line $\alpha = 0$, and SB the straight line

$\beta = 0$; let SC bisect the angle ASB, and SD bisect the supplement of ASB; the angle DSC is therefore a right angle. Take any point P in SC and draw the perpendiculars PM, PN on SA, SB respectively. If x, y be the co-ordinates of P, the length of PM is α by Art. 54, and the length of PN is β. Since SC bisects the angle ASB, $PM = PN$; therefore for any point in SC we have $\beta = \alpha$; that is, the equation to SC is $\alpha = \beta$.

Similarly, the equation to SD is $\alpha = -\beta$.

Thus $\alpha - \beta = 0$ and $\alpha + \beta = 0$ represent the two straight lines which pass through the intersection of $\alpha = 0$ and $\beta = 0$ and bisect the angles formed by these straight lines.

The student must distinguish between the straight lines $\alpha - \beta = 0$ and $\alpha + \beta = 0$; the following rule may be used: the two straight lines $\alpha = 0$, $\beta = 0$, will divide the plane in which they lie into four compartments; ascertain in which of these compartments the origin of co-ordinates is situated; $\alpha - \beta = 0$ bisects that angle between $\alpha = 0$ and $\beta = 0$ in which the origin of co-ordinates lies. This is obvious from the investigation in the present Article and the remarks made in Arts. 53, 54.

The equation $\alpha + \lambda\beta = 0$ represents a straight line such that λ is numerically equal to the ratio of the perpendicular from any point of it on $\alpha = 0$ to the perpendicular from the same point on $\beta = 0$. If λ is positive the straight line $\alpha + \lambda\beta = 0$ lies in the same two of the four compartments just alluded to as the straight line $\alpha + \beta = 0$; if λ be negative the straight line $\alpha + \lambda\beta = 0$ lies in the same two compartments as the straight line $\alpha - \beta = 0$. From the figure we see that $PM = PS \sin PSM$, and $PN = PS \sin PSN$; hence λ or $\frac{PM}{PN} = \frac{\sin PSM}{\sin PSN}$; that is, λ expresses the ratio of the sine of the angle between $\alpha = 0$ and $\alpha + \lambda\beta = 0$ to the sine of the angle between $\beta = 0$ and $\alpha + \lambda\beta = 0$.

71. We shall continue to express the equation to a straight line by the abbreviation $\alpha = 0$ when the equation is of the form $x \cos \alpha + y \sin \alpha - p = 0$; when we do not wish to

restrict ourselves to this form, we shall use such notation as $u=0$, $v=0$, $u'=0$, ...

Let $u=0$, $v=0$ be the equations to two straight lines, the axes being *rectangular or oblique.;* then $u-\lambda v=0$ and $u+\lambda v=0$ represent two straight lines passing through the intersection of the first two. Suppose, as in Art. 70, that SA, SB are the first two straight lines and SC, SD the second two; *then will*

$$\frac{\sin CSA}{\sin CSB}=\frac{\sin DSA}{\sin DSB}.$$

For by Art. 57 it appears that if p be the perpendicular from a point (x, y) on the straight line $u=0$, then $p=\mu u$, where μ *is a constant quantity;* similarly if p' denote the perpendicular from the same point on $v=0$, then $p'=\mu' v$, where μ' is a constant quantity. Hence the equation $u-\lambda v=0$, or $\frac{p}{\mu}-\frac{\lambda p'}{\mu'}=0$ shews that $\frac{p}{p'}=\frac{\lambda\mu}{\mu'}$; thus we see that numerically without regard to algebraical sign

$$\frac{\sin CSA}{\sin CSB}=\frac{\lambda\mu}{\mu'}.$$

Similarly $$\frac{\sin DSA}{\sin DSB}=\frac{\lambda\mu}{\mu'};$$

therefore $$\frac{\sin CSA}{\sin CSB}=\frac{\sin DSA}{\sin DSB}.$$

72. We will apply the principles of the preceding Articles to some examples.

Let $\alpha=0$, $\beta=0$, $\gamma=0$ be the equations to three straight lines which meet and form a triangle, and suppose the origin of co-ordinates *within* the triangle; then the equations to the three straight lines bisecting the interior angles of the triangle are, by Art. 70,

$$\beta-\gamma=0\ldots(1);\quad \gamma-\alpha=0\ldots(2);\quad \alpha-\beta=0\ldots(3).$$

These three straight lines meet at a point; for it is obvious that the values of x and y which simultaneously satisfy (1) and (2) will also satisfy (3).

Again the equations to the three straight lines which pass through the angles of the triangle and bisect the angles supplemental to those of the triangle are

$$\beta+\gamma=0\ldots(4);\quad \gamma+\alpha=0\ldots(5);\quad \alpha+\beta=0\ldots(6).$$

It is obvious that (3), (4), and (5) meet at a point; similarly (5), (6), and (1) meet at a point; so likewise (4), (6), and (2) meet at a point.

In all our propositions and examples of this kind, we shall always suppose the origin of co-ordinates *within* the triangle, unless the contrary be stated.

73. If $\alpha=0$, $\beta=0$, $\gamma=0$ be the equations to three straight lines which form a triangle, then *any* straight line may be represented by an equation of the form $l\alpha+m\beta+n\gamma=0$; for the exceptional cases noticed in Art. 69 cannot occur here.

Let a, b, c denote the lengths of the sides of the triangle which form parts of the straight lines $\alpha=0$, $\beta=0$, $\gamma=0$ respectively. Take any point within the triangle and join it with the three angular points; thus we obtain three triangles the areas of which are respectively $-\frac{a\alpha}{2}$, $-\frac{b\beta}{2}$, and $-\frac{c\gamma}{2}$.

Hence $$a\alpha+b\beta+c\gamma=\text{a constant};$$

the constant being in fact twice the area of the triangle taken negatively.

This result holds obviously for any point *within* the triangle determined by $\alpha=0$, $\beta=0$, $\gamma=0$. It will be found on examining the different cases which arise that it is also true for any point *without* the triangle. Hence it is universally true.

Suppose we require the equation to a straight line parallel to the straight line $l\alpha+m\beta+n\gamma=0$.

This required equation may be written $l\alpha+m\beta+n\gamma+k=0$, where k is a constant. (Art. 38.)

Or, since $a\alpha+b\beta+c\gamma$ is a constant, the required equation may be written, $l\alpha+m\beta+n\gamma+k'(a\alpha+b\beta+c\gamma)=0$, where k' is a constant.

74. The straight lines represented by the equations $u=0$, $v=0$, $w=0$, will meet at a point, provided $lu+mv+nw$ is *identically* $=0$; l, m, n being constants. For if $lu+mv+nw=0$ *identically*, we have $w=-\frac{lu+mv}{n}$ always.

Hence the equation $w=0$ may be written $-\frac{lu+mv}{n}=0$, that is, the straight line $w=0$ is a straight line passing through the intersection of $u=0$ and $v=0$.

75. The following example will furnish a good exercise in the subject.

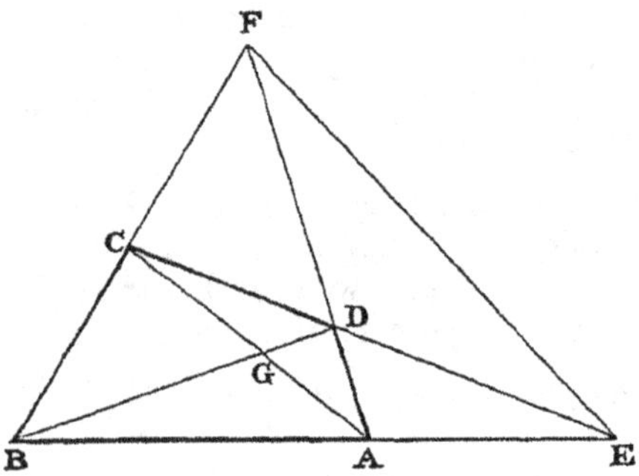

Let $ABCD$ be a quadrilateral; draw the diagonals AC, BD; produce BA and CD to meet at E, and AD and BC to meet at F; join EF, forming what is called the *third diagonal* of the quadrilateral. Suppose

$u=0$, the equation to AB, ... (1),

$v=0$, the equation to BC, ... (2),

$w=0$, the equation to CD, ... (3).

We propose to express the equations to the other straight lines of the figure in terms of u, v, w, and constant quantities. Assume for the equation to BD

$lu-mv=0$ .... (4),

and for the equation to CA

$mv-nw=0$ .... (5).

These assumptions are legitimate, because (4) represents *some* straight line passing through B, whatever be the values of the constants l and m; by properly assuming these constants, we may therefore make (4) represent BD. Also (5) represents *some* straight line through C, and by giving a suitable value to n, we may make it represent CA. We may if we please suppose one of the three constants l, m, n, equal to unity, but for the sake of symmetry we will not make this supposition. The equation to AD is

$$lu - mv + nw = 0 \quad\text{.................. (6)};$$

for (6) represents a straight line passing through the intersection of $lu - mv = 0$ and $w = 0$, that is, a straight line through D; also (6) represents a straight line passing through the intersection of $u = 0$ and $mv - nw = 0$, that is, a straight line through A. Hence (6) represents AD. The equation to EF is

$$lu + nw = 0 \quad\text{...................... (7)};$$

for (7) obviously represents some straight line through E, and since $lu + nw = lu - mv + nw + mv$, (7) represents some straight line through F. Hence (7) represents EF.

Let G be the intersection of AC and BD. The equation to EG is

$$lu - nw = 0 \quad\text{...................... (8)};$$

for (8) represents a straight line passing through the intersection of (1) and (3), and also through the intersection of (4) and (5). The equation to FG is

$$lu - 2mv + nw = 0 \quad\text{.................. (9)};$$

for (9) represents a straight line passing through the intersection of (4) and (5), and also through the intersection of (2) and (6).

Suppose BD produced to meet EF at H, and AC and EF produced to meet at K; then it may be shewn that the equation to AH is $2lu - mv + nw = 0$, that to CH is $mv + nw = 0$, that to KB is $lu + mv = 0$, that to KD is $lu - mv + 2nw = 0$.

We have introduced this example, not on account of any importance in the results, but as an exercise in forming the equations to straight lines. We proceed to another example.

76. *If there be two triangles such that the straight lines joining the corresponding angles meet at a point, then the intersections of the corresponding sides lie on one straight line.*

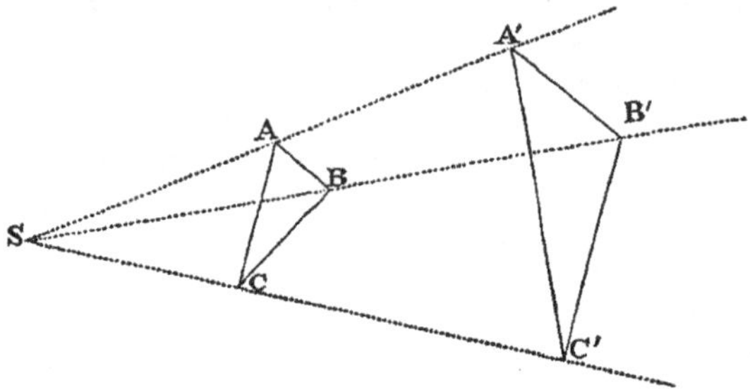

Let ABC be one triangle, $A'B'C'$ the other triangle; let S be the point at which the straight lines AA', BB', CC' meet. Let the equation to BC be $u=0$, to CA $v=0$, and to AB $w=0$. Assume for the equation to

$$B'C' \quad l'u+mv+nw=0 \qquad (1),$$

$$\text{and to } C'A' \quad lu+m'v+nw=0 \qquad (2).$$

It is shewn in Art. 69 that the equation to $B'C'$ may be written in the above form, and by the method of that Article it may be shewn that by giving suitable values to the constants l, m', we may make (2) represent $C'A'$. We will now shew that the equation to $A'B'$ may be written in the form

$$lu+mv+n'w=0 \qquad (3).$$

The constant n' may be obviously determined, so as to make the straight line represented by (3) pass through A'; let n' be so determined; it remains to shew that the straight line (3) will pass through B'. From (1) and (2) it follows that the equation

$$(l'-l)\,u+(m-m')\,v=0 \qquad (4)$$

represents *some* straight line through C'; but (4) obviously represents a straight line passing through the intersection of BC and CA. Hence (4) is the equation to CC'.

Again, the straight line represented by (3) by supposition passes through A'; hence from (2) and (3) we see that

$$(m'-m)\,v+(n-n')\,w=0 \qquad (5)$$

is the equation to AA'.

The equation $(l'-l)u+(n-n')w=0$.................(6)

represents a straight line passing through the intersection of BC and AB, that is, through B; and from (4) and (5) it follows that this straight line passes through the intersection of CC' and AA', that is, through S. Hence (6) is the equation to SB.

Now from (1) and (3) it follows that the straight lines represented by these equations meet on the straight line (6). Hence (3) is the equation to $A'B'$.

The required proposition now easily follows: for the straight line represented by

$$lu+mv+nw=0 \quad \text{.................(7)}$$

passes through the intersection of BC and $B'C'$, of CA and $C'A'$, and of AB and $A'B'$; that is, these three intersections lie on one straight line.

Conversely, if there be two triangles such that the intersections of the corresponding sides lie on one straight line, then the straight lines joining the corresponding angles meet at a point. To prove this we may begin with the equations to BC, CA, AB, $B'C'$, $C'A'$ as before, and assume (3) as the equation to *some* straight line through A'. Then (7) will represent the straight line passing through the intersection of BC and $B'C'$, and of CA and $C'A'$; now (3) is the equation to a straight line passing through the intersection of AB and (7); hence (3) must be the equation to $A'B'$. Then from the form of (1), (2), and (3), it follows immediately that CC' passes through the intersection of AA' and BB'.

It may be shewn also that the equation to the straight line which passes through the intersection of AB and $A'C'$, and of AC and $A'B'$, is

$$lu+m'v+n'w=0 \quad \text{.................(8).}$$

And the intersection of (8) with BC will lie on the straight line

$$l'u+m'v+n'w=0 \quad \text{.................(9).}$$

Similarly the straight line joining the intersection of BA and $B'C'$ with the intersection of BC and $B'A'$ meets CA on (9). And also the straight line joining the intersection of CA and $C'B'$ with the intersection of CB and $C'A'$ meets AB on (9).

The two triangles considered in this Article are said to be *homologous*; the point at which the straight lines joining the corresponding angles meet is called the *centre of homology*, and the straight line which contains the intersections of the corresponding sides is called the *axis of homology*.

77. The equation $u+\lambda v=0$ represents a straight line passing through the intersection of the straight lines $u=0$, $v=0$. Hence if there be a series of straight lines the equations of which are all of the *form* $u+\lambda v=0$, and differ merely in having different values of the constant λ, all these straight lines pass through a point, namely, the intersection of $u=0$ and $v=0$.

78. The student is recommended to make himself very familiar with the preceding Articles of the present Chapter, as they contain the essential principles of a subject which has received much attention during the last few years. When these principles are mastered no difficulty will be found in following the numerous investigations in which they have been applied.

The name *trilinear co-ordinates* is often applied to the subject which has been brought before the notice of the student in the present Chapter; and it is easy to explain the appropriateness of the term. Let there be any fixed triangle ABC, which may be called the *triangle of reference*; take any point P in the plane of the triangle, and let α, β, γ denote the perpendicular distances of P from BC, CA, AB respectively: then α, β, γ may be called the three co-ordinates of the point P. We shall consider α as positive when P is on the same side of BC as A is, and as negative when P is on the opposite side of BC; and a similar rule will be adopted with respect to the signs of β and γ.

The three co-ordinates of a point are connected by a relation; for $a\alpha+b\beta+c\gamma$ is equal to twice the area of the triangle ABC. See Art. 73.

It will be seen that the meanings here assigned to α, β, γ correspond with those already adopted in this Chapter, except that the *signs are reversed*. Thus to connect trilinear co-ordinates with the common co-ordinates we may suppose α to stand for $p - x\cos\alpha - y\sin\alpha$, and make similar suppositions with respect to β and γ.

Formulæ which involve trilinear co-ordinates may be investigated immediately from the definitions without any reference to the common co-ordinates; or they may be investigated with the aid of the common co-ordinates. The latter method is naturally suggested by the plan of an elementary work like the present; and accordingly we have in substance adopted this method in the present Chapter. We will now discuss briefly a few more applications of trilinear co-ordinates; the student should also exercise himself by the examples at the end of the Chapter.

I. *To find the angle between two given straight lines.*

Let $\lambda\alpha + \mu\beta + \nu\gamma = 0$ and $\lambda'\alpha + \mu'\beta + \nu'\gamma = 0$ be the equations to the straight lines.

If we express the first equation in rectangular co-ordinates it becomes

$$C - (\lambda\cos\alpha + \mu\cos\beta + \nu\cos\gamma)\,x - (\lambda\sin\alpha + \mu\sin\beta + \nu\sin\gamma)\,y = 0,$$

where C is a constant.

The second equation may be put into a similar form.

Let ϕ denote the angle between the two straight lines; then, by Art. 41, $\tan\phi = \dfrac{m - m'}{1 + mm'}$, where

$$m = -\frac{\lambda\cos\alpha + \mu\cos\beta + \nu\cos\gamma}{\lambda\sin\alpha + \mu\sin\beta + \nu\sin\gamma},$$

and
$$m' = -\frac{\lambda'\cos\alpha + \mu'\cos\beta + \nu'\cos\gamma}{\lambda'\sin\alpha + \mu'\sin\beta + \nu'\sin\gamma}.$$

Hence, substituting and reducing, we find $-\tan\phi$ is equal to a fraction of which the numerator is

$$(\mu\nu' - \mu'\nu)\sin(\gamma - \beta) + (\nu\lambda' - \nu'\lambda)\sin(\alpha - \gamma) + (\lambda\mu' - \lambda'\mu)\sin(\beta - \alpha),$$

and the denominator is

$$\lambda\lambda' + \mu\mu' + \nu\nu' + (\mu\nu' + \mu'\nu)\cos(\gamma - \beta)$$
$$+ (\nu\lambda' + \nu'\lambda)\cos(\alpha - \gamma) + (\lambda\mu' + \lambda'\mu)\cos(\beta - \alpha).$$

Now we can express the angles $\gamma - \beta$, $\alpha - \gamma$, $\beta - \alpha$ in terms of the angles of the triangle of reference. For suppose we take any point within the triangle of reference and draw perpendiculars on the sides; then the angle between the perpendiculars on AB and AC is the supplement of A: thus either $\gamma-\beta=180^\circ-A$ or $\beta-\gamma=180^\circ-A$. It depends on the position of the axis of x in the rectangular co-ordinates which of these cases holds.

We shall thus find that

$$\cos(\gamma-\beta)=-\cos A,\ \cos(\alpha-\gamma)=-\cos B,\ \cos(\beta-\alpha)=-\cos C,$$
$$\sin(\gamma-\beta)=\pm\sin A,\ \sin(\alpha-\gamma)=\pm\sin B,\ \sin(\beta-\alpha)=\pm\sin C;$$

and in the second line we must take the upper sign in *all three cases*, or the lower sign in *all three cases*.

Thus finally $\tan\phi$ is equal to the following expression with the double sign prefixed

$$\frac{(\mu\nu'-\mu'\nu)\sin A+(\nu\lambda'-\nu'\lambda)\sin B+(\lambda\mu'-\lambda'\mu)\sin C}{\lambda\lambda'+\mu\mu'+\nu\nu'-(\mu\nu'+\mu'\nu)\cos A-(\nu\lambda'+\nu'\lambda)\cos B-(\lambda\mu'+\lambda'\mu)\cos C}.$$

Again, by Art. **41**,

$$\sin\phi=\frac{m-m'}{\sqrt{(1+m^2)}\sqrt{(1+m'^2)}}.$$

Proceeding in the same way we find that $\sin\phi$ is equal to a fraction of which the numerator is

$$\pm\{(\mu\nu'-\mu'\nu)\sin A+(\nu\lambda'-\nu'\lambda)\sin B+(\lambda\mu'-\lambda'\mu)\sin C\},$$

and the denominator is the product of

$$\sqrt{(\lambda^2+\mu^2+\nu^2-2\mu\nu\cos A-2\nu\lambda\cos B-2\lambda\mu\cos C)}$$

and $\sqrt{(\lambda'^2+\mu'^2+\nu'^2-2\mu'\nu'\cos A-2\nu'\lambda'\cos B-2\lambda'\mu'\cos C)}$.

II. *To find the condition that two straight lines may be at right angles.*

The value of $\tan\phi$ must be infinite, and thus the denominator of the fraction obtained for $\tan\phi$ must be zero.

III. Let ABC be the triangle of reference; and suppose the straight line denoted by $l\alpha+m\beta+n\gamma=0$ to cut the sides of the triangle at D, E, F respectively.

At D we have $\alpha=0$, and therefore $m\beta+n\gamma=0$. Here β denotes the length of the perpendicular from D on AC, so that $\beta=CD\sin C$; and γ denotes the length of the perpendicular from D on AB, so that $\gamma=BD\sin B$.

Thus $$mCD \sin C = -nBD \sin B.$$

Similarly at E we have

$$\beta = 0, \quad \gamma = AE \sin A, \quad \alpha = CE \sin C;$$

therefore $$nAE \sin A = -lCE \sin C.$$

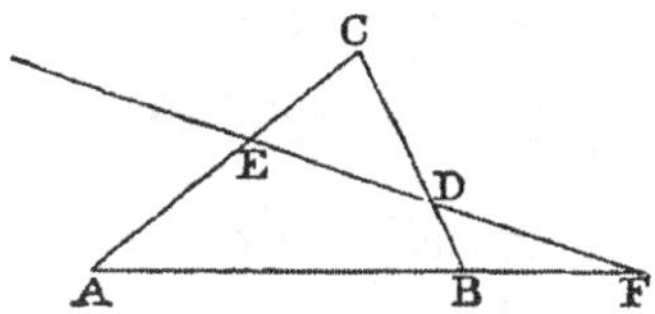

And at F we have

$$\gamma = 0, \quad \alpha = -BF \sin B, \quad \beta = AF \sin A;$$

therefore $$lBF \sin B = mAF \sin A.$$

Hence by multiplication we obtain

$$CD.AE.BF = BD.CE.AF.$$

See *Appendix to Euclid*, Arts. 56...58.

IV. Let ABC be the triangle of reference: we shall shew how the constants l, m, n in the equation to a straight line $l\alpha + m\beta + n\gamma = 0$ may be expressed in terms of the sides of the triangle and the perpendiculars from its angles on the straight line.

Let p, q, r denote the perpendiculars drawn from A, B, C respectively; any two of them will be considered to be of the *same* sign or of *contrary* signs according as they fall on the *same* side of the straight line or on *contrary* sides.

Proceeding as in III. we have

$$mCD \sin C = -nBD \sin B;$$

but $$\frac{CD}{BD} = -\frac{r}{q},$$

therefore $$mr \sin C = nq \sin B,$$

therefore $$mrc = nqb.$$

Similarly $$npa = lrc,$$

and $$lqb = mpa.$$

Hence
$$\frac{l}{pa} = \frac{m}{qb} = \frac{n}{rc};$$
and the equation to the straight line becomes
$$pa\alpha + qb\beta + rc\gamma = 0.$$

V. *To find the length of the perpendicular drawn from a given point on a given straight line.*

Let $(\alpha', \beta', \gamma')$ be the given point, and $\lambda\alpha + \mu\beta + \nu\gamma = 0$ the given straight line.

By Art. 49 the perpendicular distance is
$$\frac{\lambda\alpha' + \mu\beta' + \nu\gamma'}{\sqrt{(P^2 + Q^2)}},$$
where
$$-P = \lambda \cos \alpha + \mu \cos \beta + \nu \cos \gamma,$$
$$-Q = \lambda \sin \alpha + \mu \sin \beta + \nu \sin \gamma.$$

Thus $P^2 + Q^2 =$
$$\lambda^2 + \mu^2 + \nu^2 + 2\mu\nu \cos(\beta - \gamma) + 2\nu\lambda \cos(\gamma - \alpha) + 2\lambda\mu \cos(\alpha - \beta)$$
$$= \lambda^2 + \mu^2 + \nu^2 - 2\mu\nu \cos A - 2\nu\lambda \cos B - 2\lambda\mu \cos C.$$

VI. Suppose we take for the fixed point the vertex A of the triangle of reference, so that $\beta' = 0$ and $\gamma' = 0$; and use the values of l, m, n found in IV. Thus the length of the perpendicular from A on the straight line $pa\alpha + qb\beta + rc\gamma = 0$ is
$$\frac{pa\alpha'}{\sqrt{(p^2a^2 + q^2b^2 + r^2c^2 - 2qrbc \cos A - 2rpca \cos B - 2pqab \cos C)}};$$
and this perpendicular is equal to p. Moreover if Δ denote the area of the triangle of reference $\alpha' a = 2\Delta$. Hence finally
$$4\Delta^2 = p^2a^2 + q^2b^2 + r^2c^2 - 2qrbc \cos A - 2rpca \cos B - 2pqab \cos C.$$

This relation then must hold between the lengths of the perpendiculars drawn from A, B, C on any straight line. Substitute for $\cos A$, $\cos B$, and $\cos C$ their values in terms of the sides of the triangle; then the result may be put in the form
$$4\Delta^2 = a^2 (p - q)(p - r) + b^2 (q - r)(q - p) + c^2 (r - p)(r - q).$$

This may be easily verified. For we see that if it be true for one straight line it must be true for every parallel straight line, since it involves only the *differences* of the perpendiculars p, q, r. It will be sufficient then to shew that the result is true for every straight line which passes through an angular point of the triangle.

Take any straight line through A, suppose it to make an angle θ with AB, and an angle ϕ with AC; so that

$$\theta+\phi+A=180^\circ.$$

Then $$p=0,\ q=c\sin\theta,\ r=b\sin\phi.$$

We have then to shew that

$$q^2b^2+r^2c^2-2qrbc\cos A=4\Delta^2.$$

The left-hand member

$$=b^2c^2(\sin^2\theta+\sin^2\phi-2\sin\theta\sin\phi\cos A)$$
$$=b^2c^2\{\sin^2\theta+\sin^2\phi+2\sin\theta\sin\phi\cos(\theta+\phi)\}$$
$$=b^2c^2\{\sin^2\theta(1-\sin^2\phi)+\sin^2\phi(1-\sin^2\theta)$$
$$+2\sin\theta\sin\phi\cos\theta\cos\phi\}$$
$$=b^2c^2\{\sin^2\theta\cos^2\phi+\sin^2\phi\cos^2\theta+2\sin\theta\sin\phi\cos\theta\cos\phi\}$$
$$=b^2c^2\sin^2(\theta+\phi)=b^2c^2\sin^2 A.$$

This establishes the required relation.

VII. We have seen in Art. 69 that every straight line can be represented by an equation of the form

$$l\alpha+m\beta+n\gamma=0.$$

We shall now shew conversely that every equation of this form, *with a single exception*, will represent some straight line.

Develope the equation as in I.; then we see that it must represent a straight line except when

$$l\cos\alpha+m\cos\beta+n\cos\gamma=0,$$

and $$l\sin\alpha+m\sin\beta+n\sin\gamma=0.$$

Eliminate n; thus

$$l\sin(\gamma-\alpha)+m\sin(\gamma-\beta)=0;$$

therefore $$\frac{l}{\sin(\gamma-\beta)}=\frac{m}{\sin(\alpha-\gamma)};$$

and in the same way we find that each of these is equal to

$$\frac{n}{\sin(\beta-\alpha)}.$$

Thus by what is shewn in I. we have

$$\frac{l}{\sin A} = \frac{m}{\sin B} = \frac{n}{\sin C}.$$

It follows that $l\alpha + m\beta + n\gamma = 0$ *will always* denote a straight line *except* when l, m, and n are proportional to $\sin A$, $\sin B$, and $\sin C$, that is to a, b, and c.

And we have seen that $a\alpha + b\beta + c\gamma$ expresses double the area of the triangle of reference, so that it cannot be equal to zero.

VIII. In the equation $Ax + By + C = 0$, suppose that A and B diminish indefinitely while C remains constant. The straight line represented by the equation then moves away to an indefinite distance from the origin; for the intercepts on the axes are $-\frac{C}{A}$ and $-\frac{C}{B}$.

In like manner if l, m, n are in proportions to each other which differ infinitesimally from the proportions of a, b, c the straight line $l\alpha + m\beta + n\gamma = 0$ is situated at an indefinitely great distance from the triangle of reference. For abbreviation it is usual to speak of the equation $a\alpha + b\beta + c\gamma = 0$ as denoting *a straight line at an infinite distance, or a straight line at infinity;* very often the equation is said to represent *the straight line at infinity,* which is open to the objection that it seems to imply that there is some definite position towards which the straight line tends as it moves away from the triangle of reference.

IX. *To find the equation to the straight line which passes through two given points.*

Let $(\alpha_1, \beta_1, \gamma_1)$ and $(\alpha_2, \beta_2, \gamma_2)$ be the two points. Then, as in Art. 35, assume for the equation to the straight line

$$l\alpha + m\beta + n\gamma = 0.$$

Thus $$l\alpha_1 + m\beta_1 + n\gamma_1 = 0,$$

and $$l\alpha_2 + m\beta_2 + n\gamma_2 = 0.$$

Hence we deduce

$$\frac{l}{\beta_1\gamma_2 - \beta_2\gamma_1} = \frac{m}{\gamma_1\alpha_2 - \gamma_2\alpha_1} = \frac{n}{\alpha_1\beta_2 - \alpha_2\beta_1};$$

and the required equation is

$$\alpha(\beta_1\gamma_2-\beta_2\gamma_1)+\beta(\gamma_1\alpha_2-\gamma_2\alpha_1)+\gamma(\alpha_1\beta_2-\alpha_2\beta_1)=0.$$

Hence the condition which must hold in order that the point $(\alpha_3, \beta_3, \gamma_3)$ may be on the straight line which joins the points $(\alpha_1, \beta_1, \gamma_1)$ and $(\alpha_2, \beta_2, \gamma_2)$ is

$$\alpha_3(\beta_1\gamma_2-\beta_2\gamma_1)+\beta_3(\gamma_1\alpha_2-\gamma_2\alpha_1)+\gamma_3(\alpha_1\beta_2-\alpha_2\beta_1)=0.$$

X. Denote the condition just obtained by $T=0$ for abbreviation. The expression for the same condition in common rectangular co-ordinates is by Art. 36

$$x_1y_2-x_2y_1+x_2y_3-x_3y_2+x_3y_1-x_1y_3=0,$$

which we will denote by $C=0$.

We may infer that if we transform from trilinear co-ordinates to common rectangular co-ordinates the condition $T=0$ will become $C=0$; so that, whether the three points are in the same straight line or not, T can only differ from C by some constant factor which does not depend on the co-ordinates of the points. But, by Art. 11, when the three points are not in the same straight line C expresses double the area of the triangle which can be formed by joining them. Hence we conclude that the area of this triangle can also be expressed by kT, where k is some constant.

We may find the value of k by considering a particular case. Let the three points be the vertices of the triangle of reference; so that we may take $\beta_1=0$, $\gamma_1=0$, $\alpha_2=0$, $\gamma_2=0$, $\alpha_3=0$, $\beta_3=0$. Thus T reduces to $\alpha_1\beta_2\gamma_3$, which is equal to $\frac{8\Delta^3}{abc}$; therefore $k\,\frac{8\Delta^3}{abc}=\Delta$; therefore $k=\frac{abc}{8\Delta^2}$.

Hence the area of the triangle formed by joining the points $(\alpha_1, \beta_1, \gamma_1)$, $(\alpha_2, \beta_2, \gamma_2)$, and $(\alpha_3, \beta_3, \gamma_3)$ is

$$\frac{abc}{8\Delta^2}\left\{\alpha_3(\beta_1\gamma_2-\beta_2\gamma_1)+\beta_3(\gamma_1\alpha_2-\gamma_2\alpha_1)+\gamma_3(\alpha_1\beta_2-\alpha_2\beta_1)\right\}.$$

XI. The student should carefully notice in this subject that geometrical theorems may often be obtained by interpreting equations which naturally present themselves in our investigations. For example in Art. 72 we have shewn the meaning of the equations $\beta+\gamma=0$, $\gamma+\alpha=0$, $\alpha+\beta=0$:

we are naturally led to consider the meaning of the equation $\alpha+\beta+\gamma=0$. The straight line thus denoted passes through the intersection of $\beta+\gamma=0$ and $\alpha=0$, and through two analogous points. Hence we have this result: the straight lines which pass through the angles of a triangle and bisect the supplemental angles meet the respectively opposite sides in three points which lie on one straight line. Similarly we may interpret the following equations:

$$\beta+\gamma-\alpha=0,\ \gamma+\alpha-\beta=0,\ \alpha+\beta-\gamma=0.$$

XII. It is very easy to pass from trilinear co-ordinates to common oblique co-ordinates. Suppose we have any equation between α, β, and γ; we can express γ in terms of α and β, by means of the relation $a\alpha+b\beta+c\gamma=2\Delta$, and thus transform the given relation into one involving only α and β.

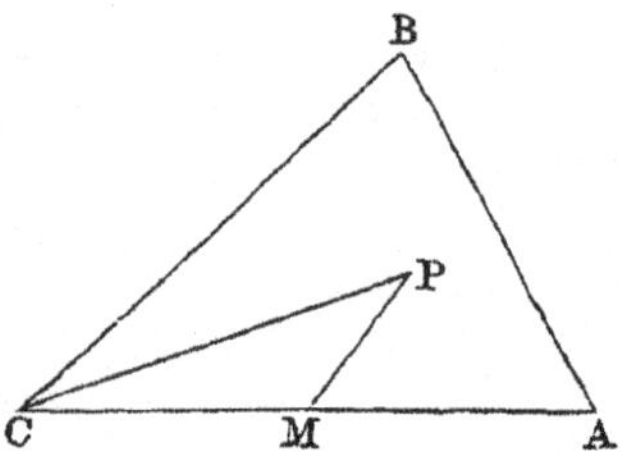

Let ABC be the triangle of reference. Suppose CA the axis of x, and CB the axis of y. Let P be any point; x, y its co-ordinates. Draw PM parallel to BC, meeting AC at M. Then if α and β refer to the point P, we have

$$\beta=PM\sin C=y\sin C;$$

and similarly $\alpha=x\sin C$. Thus if we substitute $x\sin C$ for α, and $y\sin C$ for β, we finally transform the equation into one involving the common oblique co-ordinates x and y.

EXAMPLES.

1. Find the equation to the straight line passing through the origin and the point of intersection of the straight lines

$$\frac{x}{a}+\frac{y}{b}=1, \quad \frac{x}{a'}+\frac{y}{b'}=1.$$

2. A, A' are two points on the axis of x, and B, B' two points on the axis of y, at given distances from the origin; AB and $A'B'$ intersect at P, and AB' and $A'B$ at Q; find the equation to the straight line PQ, and shew that the axes are divided harmonically by it.

3. If $\alpha = 0$, $\beta = 0$, $\gamma = 0$ be the equations to the sides of a triangle ABC opposite the angles A, B, C, prove that $\alpha \sin A - \beta \sin B = 0$ is the equation to the straight line bisecting AB from C.

4. Prove by means of such equations as that given in the preceding Example the first proposition in Art. 64.

5. Shew that $\alpha \cos A - \beta \cos B = 0$ is the equation to the perpendicular from C on AB.

6. Hence prove the second proposition in Art. 64.

7. If a, b, c be the lengths of the sides of a triangle opposite the angles A, B, C, respectively, prove that

$$\alpha \cos A - \beta \cos B - \frac{c}{2} (\sin B \cos A - \sin A \cos B) = 0$$

is the equation to the straight line which bisects AB and is at right angles to it. The equation may also be written

$$\left(\alpha - \frac{a \sin B \sin C}{2 \sin A}\right) \cos A - \left(\beta - \frac{b \sin C \sin A}{2 \sin B}\right) \cos B = 0.$$

8. Hence prove the third proposition in Art. 64.

9. Interpret the equation $a\alpha + b\beta = 0$.

10. Shew that $a\alpha + b\beta - c\gamma = 0$ is the equation to the straight line which joins the middle points of AC and BC.

11. Shew that $\alpha \cos A + \beta \cos B - \gamma \cos C = 0$ is the equation to the straight line which joins the feet of the perpendiculars from A on BC, and from B on AC.

12. If straight lines be drawn bisecting the angles of a triangle and the exterior angles formed by producing the sides, these straight lines will intersect at only four points besides the angles of the triangle.

13. If $u=0$, $v=0$, $w=0$ be the equations to three straight lines, find the equation to the straight line passing through the two points

$$\frac{u}{l}=\frac{v}{m}=\frac{w}{n}, \text{ and } \frac{u}{l'}=\frac{v}{m'}=\frac{w}{n'}.$$

14. Find the equation to the straight line passing through the intersections of the pairs of straight lines

$$2au+bv+cw=0, \qquad bv-cw=0;$$

and
$$2bu+av+cw=0, \qquad av-cw=0.$$

15. If $\alpha=0$, $\beta=0$, $\gamma=0$ be the equations to the sides of a triangle ABC, shew that the equation to the straight line which joins the centres of the inscribed circle and the circumscribed circle is

$$\alpha(\cos B-\cos C)+\beta(\cos C-\cos A)+\gamma(\cos A-\cos B)=0.$$

16. If the equations to the sides of a triangle ABC be $u=0$, $v=0$, $w=0$, and to the sides of a triangle $A'B'C'$, $u=a$, $v=b$, $w=c$, then AA', BB', and CC' meet at a point.

17. If the straight lines AA', BB', CC', in the last Example meet respectively the sides of the triangle ABC at D, E, F, shew that the intersections of DE and AB, of EF and BC, of FD and CA, will all lie on one straight line; and that a similar property will hold for the intersections of the same straight lines with the sides of the triangle $A'B'C'$.

18. In Art. 75, suppose the straight line joining F and G to meet AB at P and CD at Q; then find the equations to CP, DP, AQ, BQ, in terms of the notation of that Article.

19. From the middle points of the sides of a triangle straight lines are drawn at right angles (all internal or all external) and proportional to those sides: prove that the straight lines which join the angles with the extremities of the opposite perpendiculars pass through one point.

20. Let the three diagonals of a quadrilateral be produced to meet each other at three points, and let each of these points be joined with the two opposite corners of the quadrilateral: the six straight lines so drawn will meet each other three and three at four points.

21. In the figure constructed in the preceding Example the four straight lines which meet each other at any corner of the quadrilateral are so related that two of them are parallel to the sides, and two to the diagonals of some parallelogram.

22. Shew that the three points of intersection which are found in Examples 4, 6, 8, lie on the straight line

$$\alpha \sin A \cos A \sin (B - C) + \beta \sin B \cos B \sin (C - A) + \gamma \sin C \cos C \sin (A - B) = 0.$$

23. Let any point P be taken in the plane of a triangle ABC, and from the angular points A, B, C let straight lines be drawn through P cutting the opposite sides at D, E, F respectively: if the equations to BC, CA, AB be $u = 0$, $v = 0$, $w = 0$ respectively, shew that the equations to AP, BP, CP may be taken to be $mv - nw = 0$, $nw - lu = 0$, $lu - mv = 0$; and find the equations to EF, FD, DE.

24. With the notation of the preceding Example let EF and BC be produced to meet at A', let FD and CA be produced to meet at B', and DE and AB at C': then shew that A', B', C' lie on one straight line.

25. With the notation of the preceding Example shew that BB', CC', and AD meet at a point; also CC', AA', and BE; and AA', BB' and CF.

26. Three points A', B', C' in the sides BC, CA, AB of a triangle being joined form a second triangle of which any two sides make equal angles with the side of the former at which they meet. Shew that AA', BB', CC' are perpendiculars to BC, CA, AB.

27. ABC is any triangle, O the centre of the inscribed circle, O' the centre of the escribed circle which touches BC. The straight line OO' meets BC at D, and any straight line drawn through D meets AC at E and AB at F. The straight lines OF and $O'E$ meet at P, and the straight lines OE and $O'F$ at Q. Shew that A, P, and Q lie on one straight line perpendicular to OO'.

28. Find the equations to the two straight lines which bisect the angles formed by the straight lines

$$l\alpha + m\beta + n\gamma = 0, \text{ and } l'\alpha + m'\beta + n'\gamma = 0.$$

29. Shew that the co-ordinates of the point of intersection of $l'\alpha+m'\beta+n'\gamma=0$, and $l''\alpha+m''\beta+n''\gamma=0$, are given by

$$\frac{\alpha}{m'n''-m''n'}=\frac{\beta}{n'l''-n''l'}=\frac{\gamma}{l'm''-l''m'}$$

$$=\frac{2\Delta}{a(m'n''-m''n')+b(n'l''-n''l')+c(l'm''-l''m')}.$$

30. Find the length of the perpendicular drawn from the intersection of $l'\alpha+m'\beta+n'\gamma=0$, and $l''\alpha+m''\beta+n''\gamma=0$, on $l\alpha+m\beta+n\gamma=0$.

31. Shew that the area of the triangle formed by the straight lines

$l\alpha+m\beta+n\gamma=0$, $l'\alpha+m'\beta+n'\gamma=0$, and $l''\alpha+m''\beta+n''\gamma=0$,

is $$\frac{\Delta abc\,\{l(m'n''-m''n')+m(n'l''-n''l')+n(l'm''-l''m')\}^2}{DD'D''},$$

where $D=a(m'n''-m''n')+b(n'l''-n''l')+c(l'm''-l''m')$,

$D'=a(m''n-mn'')+b(n''l-nl'')+c(l''m-lm'')$,

$D''=a(mn'-m'n)+b(nl'-n'l)+c(lm'-l'm)$.

32. Find the condition which must hold in order that the equations $\frac{\alpha}{\lambda}=\frac{\beta}{\mu}$, $\frac{\beta}{\mu}=\frac{\gamma}{\nu}$, $\frac{\gamma}{\nu}=\frac{\alpha}{\lambda}$ may represent three parallel straight lines.

33. When the condition in the preceding Example is satisfied find the condition which must hold in order that the straight line $l\alpha+m\beta+n\gamma=0$ may be parallel to the three straight lines.

34. Find the condition which must hold in order that the equations $\frac{\alpha-\alpha'}{\lambda}=\frac{\beta-\beta'}{\mu}=\frac{\gamma-\gamma'}{\nu}$ may represent a straight line.

35. Determine what is represented by the equations in the preceding Example when the condition is not satisfied.

36. Through any point P within ABC, the triangle of reference, straight lines AP, BP, CP are drawn meeting the opposite sides at D, E, F respectively: if the equations to AP, BP, CP are $m\beta - n\gamma = 0$, $n\gamma - l\alpha = 0$, $l\alpha - m\beta = 0$, compare the areas of AEF and DEF with that of ABC.

37. Perpendiculars are drawn from the angles of a triangle on the opposite sides, and a second triangle is formed by joining the feet of these perpendiculars: shew that the two triangles are homologous, and that the equation to the axis of homology is

$$\alpha \cos A + \beta \cos B + \gamma \cos C = 0.$$

38. Investigate the condition which must hold in order that the following equation may represent two straight lines:

$$A\alpha^2 + B\beta^2 + C\gamma^2 + 2D\beta\gamma + 2E\gamma\alpha + 2F\alpha\beta = 0.$$

39. Investigate the following expressions for the square of the distance between the points $(\alpha_1, \beta_1, \gamma_1)$ and $(\alpha_2, \beta_2, \gamma_2)$:

$$\frac{(\alpha_1 - \alpha_2)^2 + (\beta_1 - \beta_2)^2 + 2(\alpha_1 - \alpha_2)(\beta_1 - \beta_2)\cos C}{\sin^2 C},$$

$$\frac{(\alpha_1 - \alpha_2)^2 \sin 2A + (\beta_1 - \beta_2)^2 \sin 2B + (\gamma_1 - \gamma_2)^2 \sin 2C}{2 \sin A \sin B \sin C},$$

$$-\frac{(\beta_1-\beta_2)(\gamma_1-\gamma_2)\sin A + (\gamma_1-\gamma_2)(\alpha_1-\alpha_2)\sin B + (\alpha_1-\alpha_2)(\beta_1-\beta_2)\sin C}{\sin A \sin B \sin C}.$$

40. If in Example 34 each of the fractions is equal to the distance between the points (α, β, γ) and $(\alpha', \beta', \gamma')$, shew that the following condition must also hold:

$$\lambda^2 \sin 2A + \mu^2 \sin 2B + \nu^2 \sin 2C = 2 \sin A \sin B \sin C.$$

CHAPTER V.

TRANSFORMATION OF CO-ORDINATES.

79. We have seen in the preceding Articles that the *general* equation to a straight line is of the form $y = mx + c$, but that the equation takes more simple forms in particular cases. If the origin is *on the straight line* the equation becomes $y = mx$; if the axis of x *coincides with the straight line*, the equation becomes $y = 0$. In a similar manner we shall see as we proceed that the equation to a curve often assumes a more or less simple form, according to the position of the origin and of the axes. It is consequently found convenient to introduce the propositions of the present Chapter, which enable us when we know the co-ordinates of a point with respect to any origin and axes, to express the co-ordinates of the same point with respect to any other given origin and axes. It will be seen that these propositions might have been placed at the end of the first Chapter, as they involve none of the results of the succeeding Chapters.

80. *To change the origin of co-ordinates without changing the direction of the axes, the axes being oblique or rectangular.*

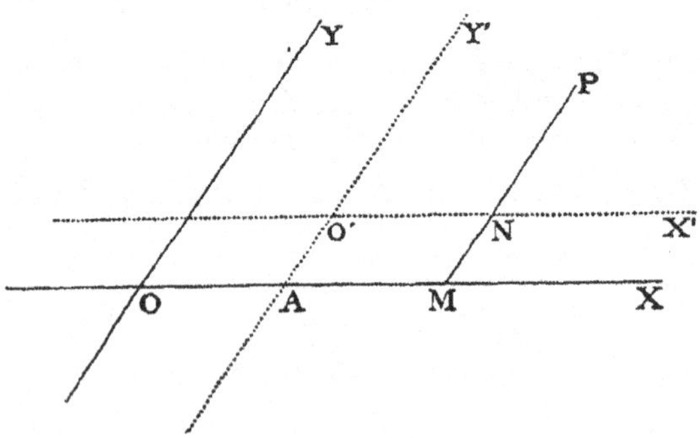

Let OX, OY be the original axes; $O'X'$, $O'Y'$ the new axes; so that $O'X'$ is parallel to OX, and $O'Y'$ parallel to

OY. Let h, k be the co-ordinates of O' with respect to O. Let P be any point; x, y its co-ordinates referred to the old axes; x', y' its co-ordinates referred to the new axes.

Let $Y'O'$ produced cut OX at A; draw PM parallel to OY meeting $O'X'$ at N; then

$$OA = h, \qquad AO' = k;$$

$$x = OM = AM + OA = O'N + OA = x' + h,$$

$$y = PM = PN + NM = PN + AO' = y' + k.$$

Hence the old co-ordinates of P are expressed in terms of its new co-ordinates.

81. *To change the direction of the axes without changing the origin, both systems being rectangular.*

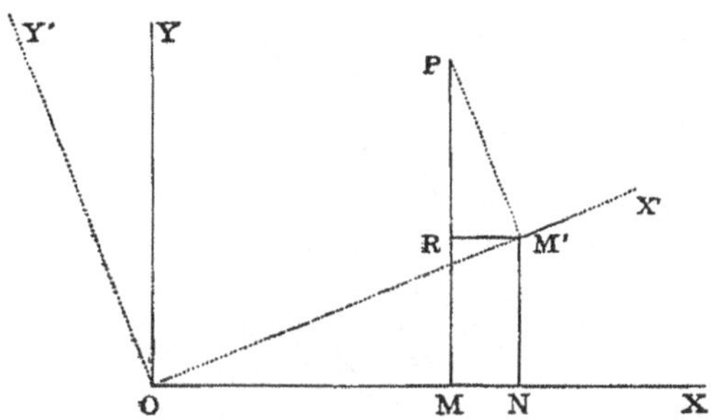

Let OX, OY be the old axes; OX', OY' the new axes, both systems being rectangular; let the angle $XOX' = \theta$. Let P be any point; x, y its co-ordinates referred to the old axes; x', y' its co-ordinates referred to the new axes. Draw PM parallel to OY, PM' parallel to OY', $M'N$ parallel to OY, and $M'R$ parallel to OX.

Then
$$\begin{aligned} x &= OM = ON - MN = ON - M'R \\ &= OM' \cos XOX' - PM' \sin M'PR \\ &= x' \cos\theta - y' \sin\theta; \\ y &= PM = RM + PR = M'N + PR \\ &= x' \sin\theta + y' \cos\theta. \end{aligned}$$

Hence the old co-ordinates of P are expressed in terms of its new co-ordinates.

82. In the preceding Article θ is measured from the positive part of the axis of x towards the positive part of the axis of y; therefore if in any example to which the formulæ are applied, OX' fall *on the other side* of OX, θ must be considered negative.

From the formulæ of the preceding Article, we see that

$$x^2 + y^2 = x'^2 + y'^2;$$

this of course should be the case, since the distance OP is the same whichever system of axes we use.

83. *To change the direction of the axes without changing the origin, both systems being oblique.*

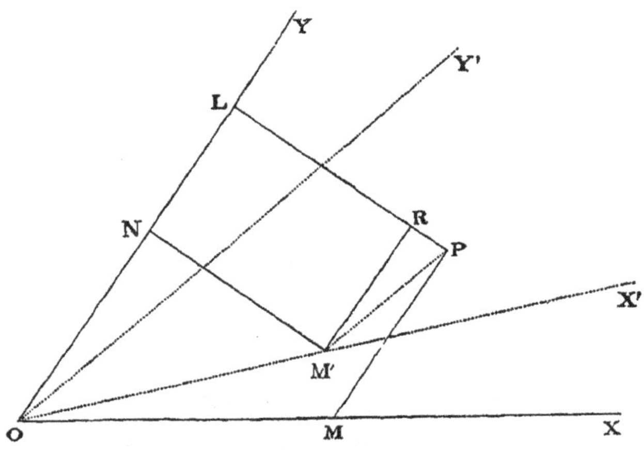

Let OX, OY be the old axes; OX', OY' the new axes. Let (XY) denote the angle between OX, OY; and let a similar notation be used to express the other angles which are formed by the straight lines meeting at O. Let P be any point; x, y its co-ordinates referred to the old axes; x', y' its co-ordinates referred to the new axes. Draw PM parallel to OY, and PM' parallel to OY'; from P and M' draw PL, $M'N$ perpendicular to OY; from M' draw $M'R$ perpendicular to PL. Then

$$x = OM, \qquad y = PM;$$
$$x' = OM', \qquad y' = PM'.$$

Now PL = perpendicular from M on $OY = x \sin(XY)$,

also
$$PL = RL + PR = M'N + PR$$
$$= OM' \sin X'OY + PM' \sin Y'OY$$
$$= x' \sin(X'Y) + y' \sin(Y'Y);$$

therefore $x \sin(XY) = x' \sin(X'Y) + y' \sin(Y'Y)$......(1).

Similarly by drawing from P and M' perpendiculars on OX we may shew that

$$y \sin(YX) = x' \sin(X'X) + y' \sin(Y'X) \text{......(2).}$$

Equations (1) and (2) express the old co-ordinates of P in terms of its new co-ordinates; (YX) and (XY) denote the *same angle*, but we use both forms for greater symmetry.

Let $XOX' = \alpha$, $XOY' = \beta$, $XOY = \omega$; then (1) and (2) become

$$x \sin \omega = x' \sin(\omega - \alpha) + y' \sin(\omega - \beta) \text{...........(3),}$$
$$y \sin \omega = x' \sin \alpha + y' \sin \beta \text{.................(4).}$$

84. Two particular cases of the general proposition in the preceding Article may be noticed.

If the original axes are rectangular $\omega = \frac{\pi}{2}$, and the equations (3) and (4) become

$$x = x' \cos \alpha + y' \cos \beta, \quad y = x' \sin \alpha + y' \sin \beta.$$

If the new axes be rectangular $\beta = \frac{\pi}{2} + \alpha$, and the equations (3) and (4) become

$$x \sin \omega = x' \sin(\omega - \alpha) - y' \cos(\omega - \alpha),$$
$$y \sin \omega = x' \sin \alpha + y' \cos \alpha.$$

85. Suppose we require to change both the origin and the direction of the axes; let x, y be the co-ordinates of a point referred to the old axes; x', y' the co-ordinates of the same point referred to the new axes. By Arts. 80 and 83 we have $x = x_1 + h$, $y = y_1 + k$, where h and k are the co-ordinates of the new origin referred to the old axes, and

$$x_1 = \frac{x' \sin(\omega - \alpha) + y' \sin(\omega - \beta)}{\sin \omega}, \quad y_1 = \frac{x' \sin \alpha + y' \sin \beta}{\sin \omega}.$$

The expressions for x_1 and y_1 will simplify when one or each of the systems is rectangular. (See Art. 84.)

86. The formulæ which connect the rectangular and polar co-ordinates of a point in the particular case in which the origin is the same in both systems, and the axis of x coincides with the initial line, have already been given. (See Art. 8.) The following is the general proposition.

To connect the polar and rectangular co-ordinates of a point.

Let OX, OY be the rectangular axes; let S be the *pole* and SA the *initial* line. Let h, k be the co-ordinates of S referred to O; draw SX' parallel to OX, and let the angle $ASX' = \alpha$.

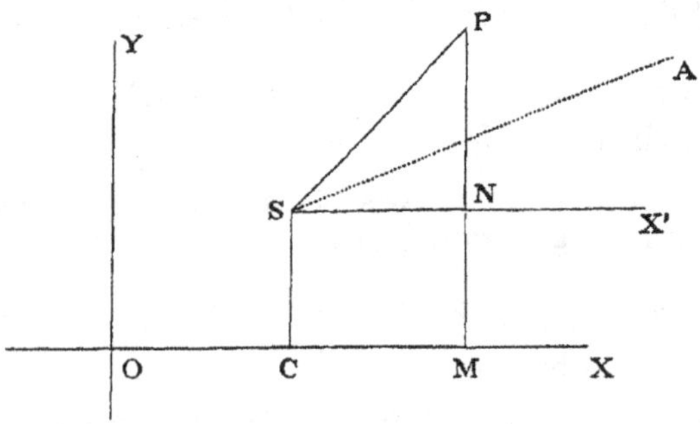

Let P be any point; x, y its co-ordinates referred to the rectangular axes; r, θ its polar co-ordinates. Draw PM, SC parallel to OY, the former cutting SX' at N, and join SP; then

$$x = OM, \qquad y = PM,$$

$$r = SP, \qquad \theta = \text{the angle } PSA.$$

And
$$x = OC + CM = OC + SN$$
$$= h + r\cos(\theta + \alpha) \quad \ldots\ldots\ldots\ldots(1),$$
$$y = MN + PN = SC + PN$$
$$= k + r\sin(\theta + \alpha) \quad \ldots\ldots\ldots\ldots(2).$$

If $\alpha = 0$ we have

$$x = h + r\cos\theta \quad \ldots\ldots\ldots\ldots(3),$$
$$y = k + r\sin\theta \quad \ldots\ldots\ldots\ldots(4).$$

87. By means of the formulæ of the present Chapter we shall sometimes be able to simplify the form of an equation; for example, the axes being rectangular, suppose we have

$$y^4 + x^4 + 6x^2y^2 = 2 \quad\ldots\ldots(1).$$

This equation represents some locus, and by ascribing different values to x and determining the corresponding values of y from the equation, we can find as many points of the locus as we please. The equation however will be simplified by turning the axes through an angle of 45°. In the formulæ of Art. 81 put $\frac{\pi}{4}$ for θ; thus

$$x = \frac{x' - y'}{\sqrt{2}}, \quad y = \frac{x' + y'}{\sqrt{2}} \quad\ldots\ldots(2).$$

Substitute these values in (1); thus

$$(x' + y')^4 + (x' - y')^4 + 6(x'^2 - y'^2)^2 = 8;$$

therefore $\quad 2(x'^4 + 6x'^2y'^2 + y'^4) + 6(x'^2 - y'^2)^2 = 8,$

or

$$x'^4 + y'^4 = 1 \quad\ldots\ldots(3).$$

Since (3) is a simpler form than (1), we shall find it easier to trace the locus by using (3) and the new axes, than by using (1) and the old axes. The student must observe that we make no change in the locus by thus changing the axes or the origin to which we refer it; that is, equation (1) represents precisely the same assemblage of points as (3); for instance, the point for which $x' = 1$ and $y' = 0$ is obviously situated on the locus (3); now *this point* will by (2) have for its co-ordinates referred to the old system $x = \frac{1}{\sqrt{2}}$, $y = \frac{1}{\sqrt{2}}$, and these values satisfy (1), that is, *this point* is on the locus (1).

We may remark that we cannot alter the *degree* of an equation by transforming the co-ordinates. For if in the expression $Ax^\alpha y^\beta$ we substitute the values of x and y in terms of x' and y' given in Arts. 80...84, we obtain

$$A(ax' + by' + h)^\alpha (cx' + ey' + k)^\beta,$$

where a, b, c, e, h, k are all constant quantities; by expanding this expression we shall obtain a series of terms of the form $A'x'^\gamma y'^\delta$, where $\gamma + \delta$ cannot be greater than $\alpha + \beta$. Hence the degree of an equation cannot be *raised* by transformation of co-ordinates. Neither can it be *depressed;* for if from a given equation we could by transformation obtain one of a *lower* degree, then by retracing our steps we should be able from the second equation to obtain one of a *higher* degree, which has been shewn to be impossible.

EXAMPLES.

1. Change the equation $r^2 = a^2 \cos 2\theta$ into one between x and y.

2. Shew that the equation $4xy - 3x^2 = a^2$ is changed into $x^2 - 4y^2 = a^2$, if the axes be turned through an angle whose tangent is 2.

3. Transform $\sqrt{x} + \sqrt{y} = \sqrt{c}$ so that the new axis of x may be inclined at 45° to the original axis.

4. The equation to a curve referred to rectangular axes is $y^2 + 4ay \cot \alpha - 4ax = 0$; find its equation referred to oblique axes inclined at an angle α, retaining the same axis of x.

5. Shew that the equation $x^2y^2 = a(x^3 + y^3)$ will admit of solution with respect to y' if the axes be moved through an angle of 45°.

6. If x, y be co-ordinates of a point referred to one system of oblique axes, and x', y' the co-ordinates of the same point referred to another system of oblique axes, and

$$x = mx' + ny', \qquad y = m'x' + n'y',$$

shew that

$$\frac{m^2 + m'^2 - 1}{n^2 + n'^2 - 1} = \frac{mm'}{nn'}.$$

CHAPTER VI.

THE CIRCLE.

88. We now proceed to the consideration of the loci represented by equations of the second degree; the simplest of these is the *circle*, with which we shall commence.

To find the equation to the circle referred to any rectangular axes.

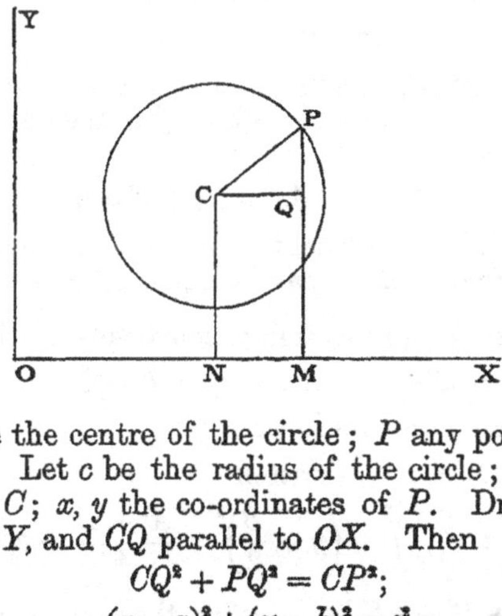

Let C be the centre of the circle; P any point on its circumference. Let c be the radius of the circle; a, b the co-ordinates of C; x, y the co-ordinates of P. Draw CN, PM parallel to OY, and CQ parallel to OX. Then

$$CQ^2 + PQ^2 = CP^2;$$

that is, $$(x-a)^2 + (y-b)^2 = c^2 \quad \ldots\ldots\ldots\ldots \quad (1),$$

or $$x^2 + y^2 - 2ax - 2by + a^2 + b^2 - c^2 = 0 \quad \ldots\ldots\ldots \quad (2).$$

This is the equation required.

The following varieties occur in the equation.

I. Suppose the origin of co-ordinates at the centre of the circle; then $a = 0$, and $b = 0$; thus (1) and (2) become

$$x^2 + y^2 - c^2 = 0 \quad \ldots\ldots\ldots\ldots\ldots \quad (3).$$

II. Suppose the origin *on* the circumference of the circle; then the values $x=0$, $y=0$, must satisfy (1) and (2); therefore

$$a^2+b^2-c^2=0,$$

which relation is also obvious from the figure, when O is *on* the circumference; hence (2) becomes

$$x^2+y^2-2ax-2by=0 \ldots\ldots\ldots\ldots (4).$$

III. Suppose the origin is *on* the circumference, and that the diameter which passes through the origin is taken for the axis of x; then $b=0$, and $a^2=c^2$; hence (2) becomes

$$x^2+y^2-2ax=0 \ldots\ldots\ldots\ldots (5).$$

Similarly if the origin be on the circumference and the axis of y coincide with the diameter through the origin, we have $a=0$, and $b^2=c^2$; hence (2) becomes

$$x^2+y^2-2by=0 \ldots\ldots\ldots\ldots (6).$$

Hence we conclude from (2) and the following equations, that the equation to a circle when the axes are rectangular is always of the form

$$x^2+y^2+Ax+By+E=0,$$

where A, B, E are constant quantities any one or more of which in particular cases may be equal to zero.

89. We shall next examine, conversely, if the equation

$$x^2+y^2+Ax+By+E=0 \ldots\ldots\ldots\ldots (1)$$

always has a circle for its locus.

Equation (1) may be written

$$\left(x+\frac{A}{2}\right)^2+\left(y+\frac{B}{2}\right)^2=\frac{A^2+B^2}{4}-E \ldots\ldots (2).$$

I. If A^2+B^2-4E be *negative*, the locus is impossible.

II. If $A^2+B^2-4E=0$, equation (2) represents a *point* the co-ordinates of which are $-\frac{A}{2}$, $-\frac{B}{2}$. This point may be considered as a circle which has an indefinitely small radius.

III. If A^2+B^2-4E be positive, we see by comparing equation (2) with equation (1) of the preceding Article that it represents a circle, such that the co-ordinates of its centre are $-\frac{A}{2}$, $-\frac{B}{2}$, and its radius $\frac{1}{2}(A^2+B^2-4E)^{\frac{1}{2}}$.

It will be a useful exercise to construct the circles represented by given equations of the form

$$x^2+y^2+Ax+By+E=0.$$

For example, suppose $x^2+y^2+4x-8y-5=0$,

or $$(x+2)^2+(y-4)^2=5+4+16=25.$$

Here the co-ordinates of the centre are -2, 4, and the radius is 5.

Tangent and Normal to a Circle.

90. Definition. Let two points be taken on a curve and a secant drawn through them; let the first point remain fixed and the second point move on the curve up to the first; the secant in its limiting position is called the tangent to the curve at the first point.

91. *To find the equation to the tangent at any point of a circle.*

Let the equation to the circle be

$$x^2+y^2=c^2 \quad \ldots\ldots\ldots\ldots\ldots\ldots\ldots\ldots (1).$$

Let x', y' be the co-ordinates of the point on the circle at which the tangent is drawn; and x'', y'' the co-ordinates of an adjacent point on the circle. The equation to the secant through (x', y') and (x'', y'') is, by Art. 35,

$$y-y'=\frac{y''-y'}{x''-x'}(x-x') \quad \ldots\ldots\ldots\ldots\ldots\ldots (2).$$

Now since (x', y') and (x'', y'') are both on the circumference of the circle,

$$x'^2+y'^2=c^2, \quad x''^2+y''^2=c^2;$$

therefore by subtraction, $x''^2-x'^2+y''^2-y'^2=0$,

or $$(x''-x')(x''+x')+(y''-y')(y''+y')=0:$$

therefore $$\frac{y''-y'}{x''-x'}=-\frac{x''+x'}{y''+y'}.$$

Hence (2) may be written

$$y - y' = -\frac{x'' + x'}{y'' + y'}(x - x') \dots\dots\dots\dots\dots (3).$$

Now in the limit when (x'', y'') coincides with (x', y'), we have $x'' = x'$, and $y'' = y'$; hence (3) becomes

$$y - y' = -\frac{2x'}{2y'}(x - x') = -\frac{x'}{y'}(x - x').$$

Thus the equation to the tangent at the point (x, y') is

$$y - y' = -\frac{x'}{y'}(x - x') \dots\dots\dots\dots\dots\dots (4).$$

This equation may be simplified; by multiplying by y' and transposing we have $xx' + yy' = x'^2 + y'^2$;

therefore $$xx' + yy' = c^2 \dots\dots\dots\dots\dots\dots\dots\dots (5).$$

92. The equation to the tangent can be conveniently expressed in terms of the tangent of the angle which the straight line makes with the axis of x. For the equation to the tangent at (x', y') is $yy' + xx' = c^2$, or $y = -\frac{x'}{y'}x + \frac{c^2}{y'}$.

Let $-\frac{x'}{y'} = m$; thus the equation becomes

$$y = mx + \frac{c^2}{y'}.$$

We have then to express $\frac{c^2}{y'}$ in terms of m.

Now $$x' = -my', \text{ and } x'^2 + y'^2 = c^2;$$

therefore $$y'^2(1 + m^2) = c^2,$$

and $$y' = \frac{c}{\sqrt{(1 + m^2)}}.$$

Hence the equation to the tangent may be written

$$y = mx + c\sqrt{(1 + m^2)}.$$

Conversely every straight line whose equation is of this form is a tangent to the circle.

93. The definition in Art. 90 may appear arbitrary to the student, and he may ask why we do not adopt that given by

Euclid (Def. 2, Book III.). To this we reply that the definition in Art. 90 will be convenient for *every* curve, which is not the case with Euclid's definition. The student however cannot at first be a judge of the necessity or propriety of any definition; he must confine himself to examining the consequences of the definition and the accuracy of the reasoning based upon it.

We may easily shew however that the straight line represented by the equation

$$xx' + yy' = c^2 \quad \ldots\ldots\ldots\ldots\ldots\ldots(1)$$

touches, according to Euclid's definition, the circle

$$x^2 + y^2 = c^2 \quad \ldots\ldots\ldots\ldots\ldots\ldots(2),$$

the point (x', y') being supposed to lie on the circle. To find the point or points of intersection of the straight line and circle we combine the equations (1) and (2); substitute in (2) the value of y from (1), then

$$x^2 + \left(\frac{c^2 - xx'}{y'}\right)^2 = c^2,$$

or $$x^2(x'^2 + y'^2) - 2c^2x'x + c^4 - c^2y'^2 = 0,$$

or $$c^2x^2 - 2c^2x'x + c^2x'^2 = 0;$$

therefore $$x^2 - 2xx' + x'^2 = 0;$$

therefore $$x = x';$$

therefore from (1), $$y = y'.$$

Hence (1) and (2) meet at *only one* point, the point (x', y'). Hence (1) *touches* the circle according to Euclid's definition.

94. Also every straight line which meets the circle at *one* point only is a tangent to the circle.

For suppose $x^2 + y^2 = c^2$ to be the equation to a circle and $y = mx + n$ the equation to a straight line; to find the points of intersection of the straight line and circle we combine the equations; thus we obtain, to determine the abscissæ of the points, $(mx + n)^2 + x^2 = c^2$ or $(m^2 + 1)x^2 + 2mnx + n^2 - c^2 = 0$. Now this quadratic equation will have *two* roots except when

$$(m^2 + 1)(n^2 - c^2) = m^2n^2,$$

that is, when $$n^2 = c^2(1 + m^2).$$

Hence if the straight line meets the circle it must meet it at *two* points unless this condition holds, and then, by Art. 92, the straight line is a tangent to the circle.

95. Instead of supposing one of the points on the circle fixed and the other to move along the circle as in the definition of Art. 90 we may suppose *both* to move along the circle until they meet at some fixed point of the circle, and the secant in its limiting position will be the tangent at that fixed point. For let (x', y') and (x'', y'') denote the two moving points on the circle, and (x_1, y_1) the fixed point. Then as in equation (3) of Art. 91, we shall have for the equation to the secant

$$y - y' = -\frac{x'' + x'}{y'' + y'}(x - x').$$

In the limit x' and x'' each $= x_1$, and y' and y'' each $= y_1$, and we obtain for the equation to the tangent at (x_1, y_1)

$$y - y_1 = -\frac{x_1}{y_1}(x - x_1),$$

which agrees with the former result.

96. If the equation to a circle be given in the form

$$(x - a)^2 + (y - b)^2 - c^2 = 0,$$

we may find the equation to the tangent at any point in the same manner as in Art. 91.

Let (x', y') be the point on the circle at which the tangent is drawn; (x'', y'') an adjacent point on the circle; then

$$(x' - a)^2 + (y' - b)^2 - c^2 = 0, \qquad (x'' - a)^2 + (y'' - b)^2 - c^2 = 0;$$

therefore $\quad (x'' - a)^2 - (x' - a)^2 + (y'' - b)^2 - (y' - b)^2 = 0,$

or $\quad (x'' - x')(x'' + x' - 2a) + (y'' - y')(y'' + y' - 2b) = 0 \ldots (1).$

Also the equation to the secant through (x', y') and (x'', y'') is

$$y - y' = \frac{y'' - y'}{x'' - x'}(x - x') \ldots\ldots\ldots\ldots (2).$$

By means of (1) this may be written

$$y - y' = -\frac{x'' + x' - 2a}{y'' + y' - 2b}(x - x') \ldots\ldots\ldots\ldots (3).$$

Now in the limit $x''=x'$ and $y''=y'$; hence we have for the equation to the tangent at (x', y')

$$y-y'=-\frac{x'-a}{y'-b}(x-x')\dots\dots\dots\dots\dots\dots(4).$$

This may be written

$$y-b-(y'-b)=-\frac{x'-a}{y'-b}\{x-a-(x'-a)\};$$

therefore $(x-a)(x'-a)+(y-b)(y'-b)$

$$=(x'-a)^2+(y'-b)^2=c^2\dots\dots(5).$$

97. Definition. The normal at any point of a curve is a straight line drawn through that point at right angles to the tangent to the curve at that point.

98. *To find the equation to the normal at any point of a circle.*

Let the equation to the circle be

$$x^2+y^2=c^2\dots\dots\dots\dots\dots\dots\dots\dots(1),$$

and let x', y' be the co-ordinates of a point on the circle, then the equation to the tangent at that point is $xx'+yy'=c^2$, or

$$y=-\frac{x'}{y'}x+\frac{c^2}{y'}.$$

Hence the equation to a straight line through (x', y') at right angles to the tangent at that point is

$$y-y'=\frac{y'}{x'}(x-x'), \quad \text{or } y=\frac{y'}{x'}x.$$

Since this equation is satisfied by the values $x=0$, $y=0$, the normal at any point passes through the origin of co-ordinates, that is, through the centre of the circle.

99. *From any external point two tangents can be drawn to a circle.*

Let the equation to a circle be

$$x^2+y^2=c^2\dots\dots\dots\dots\dots\dots\dots\dots(1),$$

and let h, k be the co-ordinates of an external point. Suppose x', y' the co-ordinates of a point on the circle such that the tangent at this point passes through (h, k). The equation to the tangent at (x', y') is

$$xx' + yy' = c^2 \text{ ....(2).}$$

Since this tangent passes through (h, k)

$$hx' + ky' = c^2 \text{......................(3).}$$

Also since (x', y') is on the circle

$$x'^2 + y'^2 = c^2 \text{ ....(4).}$$

Equations (3) and (4) determine the values of x' and y'. Substitute from (3) in (4), thus $x'^2 + \left(\frac{c^2 - hx'}{k}\right)^2 = c^2$;

therefore $\quad x'^2 (h^2 + k^2) - 2c^2hx' + c^2 (c^2 - k^2) = 0.$

The roots of this quadratic equation will be found to be both possible since (h, k) is an *external* point and therefore $h^2 + k^2$ greater than c^2. To each value of x' corresponds one value of y' by (3); hence *two* tangents can be drawn from any external point.

The straight line which passes through the points where these tangents meet the circle is called the *chord of contact.*

100. *Tangents are drawn to a circle from a given external point; to find the equation to the chord of contact.*

Let h, k be the co-ordinates of the external point; x_1, y_1 the co-ordinates of the point where *one* of the tangents from (h, k) meets the circle; x_2, y_2 the co-ordinates of the point where the other tangent from (h, k) meets the circle.

The equation to the tangent at (x_1, y_1) is

$$xx_1 + yy_1 = c^2 \text{ ....(1).}$$

Since this tangent passes through (h, k), we have

$$hx_1 + ky_1 = c^2 \text{ ....(2).}$$

Similarly, since the tangent at (x_2, y_2) passes through (h, k),

$$hx_2 + ky_2 = c^2 \text{ ....(3).}$$

Hence it follows that the equation to the *chord of contact* is

$$xh + yk = c^2 \text{ ....(4).}$$

For (4) is obviously the equation to *some straight line;*

also this straight line passes through (x_1, y_1), for (4) is satisfied by the values $x=x_1$, $y=y_1$, as we see from (2); similarly from (3) we conclude that this straight line passes through (x_2, y_2). Hence (4) is the required equation.

Thus we may use the following process to draw tangents to a circle from a given external point: draw the straight line which is represented by (4); join the points where it meets the circle with the given external point, and the straight lines thus obtained are the required tangents.

101. *Through any fixed point chords are drawn to a circle, and tangents to the circle drawn at the extremities of each chord: the locus of the intersection of the tangents is a straight line.*

Let h, k be the co-ordinates of the point through which the chords are drawn; let tangents to the circle be drawn at the extremities of one of these chords, and let (x_1, y_1) be the point at which they meet. The equation to the corresponding chord of contact is, by Art. 100, $xx_1 + yy_1 = c^2$. But this chord passes through (h, k); therefore $hx_1 + ky_1 = c^2$.

Hence the point (x_1, y_1) lies on the straight line

$$xh + yk = c^2;$$

that is, the locus of the intersection of the tangents is a straight line: this straight line is at right angles to that which joins the point (h, k) with the centre.

We will now demonstrate the converse of this proposition.

102. *If from any point in a straight line a pair of tangents be drawn to a circle, the chords of contact will all pass through a fixed point.*

Let $$Ax + By + C = 0 \quad\text{.....................(1)}$$

be the equation to the straight line; let (x', y') be a point in this straight line from which tangents are drawn to the circle; then the equation to the corresponding chord of contact is

$$xx' + yy' = c^2 \quad\text{.....................(2).}$$

Since (x', y') is on (1) we have $Ax' + By' + C = 0$; therefore (2) may be written $xx' - y\dfrac{Ax' + C}{B} = c^2$, or $$\left(x - \frac{Ay}{B}\right) x' - \frac{yC}{B} - c^2 = 0.$$

Now, whatever be the value of x', this straight line passes through the point whose co-ordinates are found by the simultaneous equations $x - \frac{Ay}{B} = 0$, $\frac{yC}{B} + c^2 = 0$; that is, the point for which $y = -\frac{Bc^2}{C}$, $x = -\frac{Ac^2}{C}$: this point is on the straight line drawn from the centre perpendicular to (1).

103. The student should observe the different interpretations that can be assigned to the equation $xh + yk - c^2 = 0$.

I. If (h, k) be any point whatever, the equation represents the locus of the intersection of tangents at the extremities of each chord through (h, k). (Art. 101.)

II. If (h, k) be an external point, the equation represents the *chord of contact*. (Art. 100.)

III. If (h, k) be *on* the circle, the equation represents the tangent at that point. (Art. 91.)

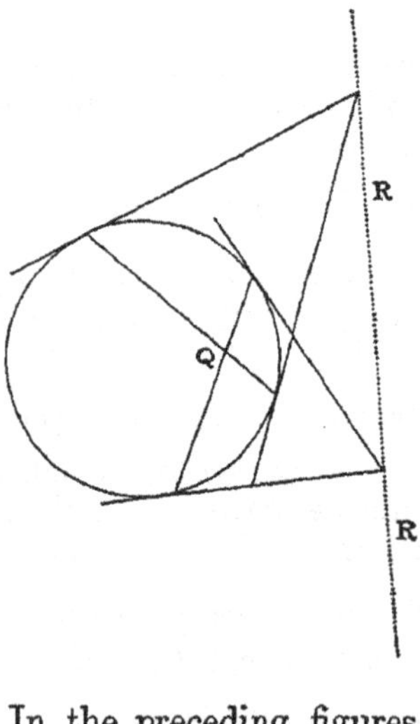

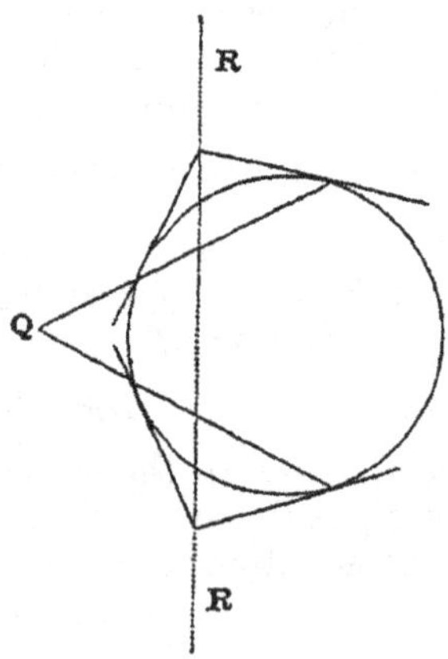

In the preceding figures Q denotes the point (h, k), and RR the straight line $xh + yk = c^2$.

In the first figure Q is *within* the circle, and the straight line RR receives only the interpretation I.

In the second figure Q is *without* the circle, hence the straight line RR receives both interpretations I. and II.; if therefore tangents be drawn from Q to the circle *they will meet it at the points where RR intersects it.*

If Q be *on* the circle, then RR becomes the tangent at Q.

Oblique Axes.

104. *To find the equation to the circle referred to any oblique axes.*

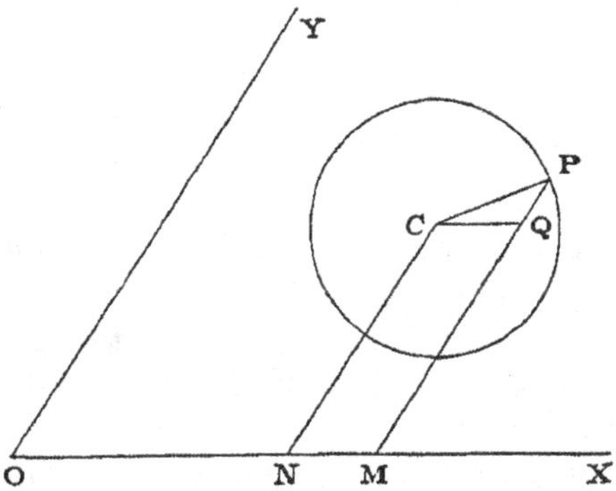

Let ω be the inclination of the axes; let C be the centre of the circle; P any point on its circumference. Let c be the radius of the circle; a, b the co-ordinates of C; x, y the co-ordinates of P. Draw CN, PM parallel to OY, and CQ parallel to OX. Then

$$CP^2 = CQ^2 + PQ^2 - 2CQ \,.\, PQ \cos CQP$$

$$= CQ^2 + PQ^2 + 2CQ \,.\, PQ \cos \omega;$$

that is, $(x-a)^2 + (y-b)^2 + 2(x-a)(y-b)\cos\omega = c^2$;

or, $x^2 + y^2 + 2xy\cos\omega - 2(a + b\cos\omega)x - 2(b + a\cos\omega)y$
$+ a^2 + b^2 + 2ab\cos\omega - c^2 = 0.$

Hence the equation to the circle referred to oblique axes is of the form

$$x^2 + y^2 + 2xy \cos \omega + Ax + By + E = 0,$$

where A, B, E are constant quantities.

Polar Equation.

105. *To find the polar equation to the circle.*

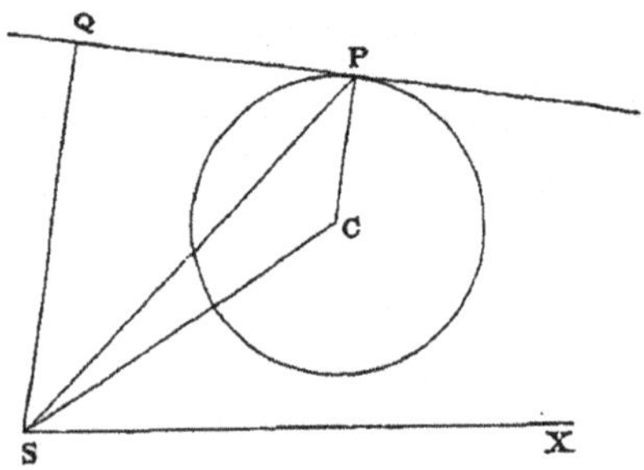

Let S be the pole, SX the initial line; C the centre of the circle, P any point on its circumference.

Let $SC = l$, $CSX = \alpha$, so that l, α are the polar co-ordinates of C; let c be the radius of the circle; and let r, θ be the polar co-ordinates of P.

Then $\quad CP^2 = PS^2 + CS^2 - 2PS \,.\, CS \,.\, \cos PSC$;

that is, $\quad c^2 = r^2 + l^2 - 2lr \cos(\theta - \alpha) \qquad \text{.................(1)},$

or $\quad r^2 - 2rl(\cos\alpha \cos\theta + \sin\alpha \sin\theta) + l^2 - c^2 = 0 \qquad \text{......(2)}.$

Hence the polar equation to the circle is of the form

$$r^2 + Ar\cos\theta + Br\sin\theta + E = 0 \qquad \text{...........(3)},$$

where A, B, E are constant quantities.

The polar equation may also be deduced from the equation referred to rectangular axes in Art. 88, by putting $r\cos\theta$ and $r\sin\theta$ for x and y respectively.

If the initial line be a diameter we have $\alpha = 0$, hence (1) becomes

$$r^2 - 2lr\cos\theta + l^2 - c^2 = 0 \qquad \text{................(4)}.$$

If, in addition, the origin be on the circumference $l^2 = c^2$,

therefore $$r = 2l \cos \theta \qquad \text{..........(5)}.$$

106. *To express the perpendicular from the origin on the tangent at any point in terms of the radius vector of that point.*

Let SQ be the perpendicular from the origin on the tangent at P, and suppose $SQ = p$; then

$$SC^2 = SP^2 + PC^2 - 2SP \,.\, PC \cos SPC$$
$$= SP^2 + PC^2 - 2SP \,.\, PC \sin SPQ;$$

that is, $$l^2 = r^2 + c^2 - 2cp.$$

In the figure S and C are on the same side of the tangent at P. If we take P so that the tangent at P falls between S and C, we shall find $l^2 = r^2 + c^2 + 2cp$.

107. These equations are sometimes useful in the solution of problems, or demonstration of properties of the circle. For example, take the equation (4) in Art. 105,

$$r^2 - 2rl \cos \theta + l^2 - c^2 = 0;$$

by the theory of quadratic equations we see that the product of the two values of r corresponding to any value of θ is $l^2 - c^2$, which is independent of θ. This agrees with Euclid III. 35, 36.

Also the sum of the two values of r is $2l \cos \theta$; hence if a straight line be drawn through the pole at an inclination θ to the initial line, the polar co-ordinates of the middle point of the chord which the circle cuts off from this straight line are $\frac{2l \cos \theta}{2}$, and θ; that is, $l \cos \theta$, and θ.

Hence the polar equation of the locus of the middle point of the chord is $r = l \cos \theta$; this by (5) in Art. 105, is a circle, of which the diameter is l.

EXAMPLES.

1. Determine the position and magnitude of the circles

(1) $x^2+y^2+4y-4x-1=0$,

(2) $x^2+y^2+6x-3y-1=0$.

2. Find the points of intersection of the straight lines

$$y+x=-1, \qquad y+x=-5, \qquad \text{and } 3y+4x=-25,$$

with the circle $x^2+y^2=25$.

3. A circle passes through the origin and intercepts lengths h and k respectively from the positive parts of the axes of x and y: determine the equation to the circle.

4. A circle passes through the points (h, k) and (h', k'): shew that its centre must lie on the straight line

$$(h-h')\left(x-\frac{h+h'}{2}\right)+(k-k')\left(y-\frac{k+k'}{2}\right)=0.$$

5. On the straight line joining (x', y') and (x'', y'') as a diameter a circle is described: find its equation.

6. A and B are two fixed points, and P a point such that $AP=mBP$, where m is a constant: shew that the locus of P is a circle, except when $m=1$.

7. The locus of the point from which two given unequal circles subtend equal angles is a circle.

8. Find the equation which determines the points of intersection of the straight line $\frac{x}{h}+\frac{y}{k}-1=0$, and the circle $x^2+y^2-2ax-2by=0$. Deduce the relation that must hold in order that the straight line may *touch* the circle.

9. Find the equation to the tangent at the origin to the circle $x^2+y^2-2y-3x=0$.

10. Shew that the length of the common chord of the circles whose equations are

$$(x-a)^2+(y-b)^2=c^2, \qquad (x-b)^2+(y-a)^2=c^2,$$

is $$\sqrt{\{4c^2-2(a-b)^2\}}.$$

11. A point moves so that the sum of the squares of its distances from the four sides of a square is constant: shew that the locus of the point is a circle.

12. A point moves so that the sum of the squares of its distances from the sides of an equilateral triangle is constant: shew that the locus of the point is a circle.

13. A point moves so that the sum of the squares of its distances from any given number of fixed points is constant: shew that the locus is a circle.

14. Shew what the equation to the circle becomes when the origin is a point on the perimeter, and the axes are inclined at an angle of 120°, and the parts of them intercepted by the circle are h and k respectively.

15. Find the inclination of the axes in order that the equation $x^2 + y^2 - xy - hx - hy = 0$ may represent a circle. Determine the position and the magnitude of the circle.

16. Find the inclination of the axes in order that the equation $x^2 + y^2 + xy - hx - hy = 0$ may represent a circle. Determine the position and the magnitude of the circle.

17. Determine the equation to the circle which has its centre at the origin, and its radius $= 3$, the axes being inclined at an angle of 45°.

18. Determine the equation to the circle which has each of the co-ordinates of its centre $= -\frac{1}{3}$ and its radius $= \frac{2}{\sqrt{3}}$, the axes being inclined at an angle of 60°.

19. The axes being inclined at an angle ω, find the radius of the circle $x^2 + y^2 + 2xy \cos \omega - hx - ky = 0$.

20. Shew that the equation to a circle of radius c referred to two tangents inclined at an angle ω as axes is

$$x^2 + y^2 + 2xy \cos \omega - 2(x + y) c \cot \frac{\omega}{2} + c^2 \cot^2 \frac{\omega}{2} = 0.$$

21. Shew that the equation in the preceding Example may also be written $x + y - 2\sqrt{(xy)} \sin \frac{\omega}{2} = c \cot \frac{\omega}{2}$.

22. Find the value of c in order that the circles

$$(x-a)^2+(y-b)^2=c^2, \quad \text{and} \quad (x-b)^2+(y-a)^2=c^2,$$

may touch each other.

23. ABC is an equilateral triangle; take A as origin, and AB as axis of x: find the rectangular equation to the circle which passes through A, B, C. Deduce the polar equation to this circle.

24. If the centre of a circle be the pole, shew that the polar equation to the chord of the circle which subtends an angle 2β at the centre is $r=c\cos\beta\sec(\theta-\alpha)$, where α is the angle between the initial line and the straight line from the centre which bisects the chord. Deduce the polar equation to a straight line touching the circle at a given point.

25. Find the polar equation to the circle, the origin being on the circumference and the initial line a tangent. Shew that with this origin and initial line, the polar equation to the tangent at the point θ' is $r\sin(2\theta'-\theta)=2c\sin^2\theta'$.

26. Shew that if the origin be on the circumference and the diameter through that point make an angle α with the initial line, the equation to the circle is $r=2c\cos(\theta-\alpha)$.

27. Determine the locus of the equation

$$r=A\cos(\theta-\alpha)+B\cos(\theta-\beta)+C\cos(\theta-\gamma)+\ldots\ldots$$

28. AB is a given straight line; through A two indefinite straight lines are drawn equally inclined to AB, and any circle passing through A and B meets those straight lines at L, M: shew that the *sum* of AL and AM is constant when L and M are on *opposite* sides of AB, and that the *difference* of AL and AM is constant when L and M are on the *same* side of AB.

29. ABC is an equilateral triangle: find the locus of P when $PA=PB+PC$.

30. There are n given straight lines making with another fixed straight line angles $\alpha, \beta, \gamma, \ldots\ldots$; a point P is taken such that the sum of the squares on the perpendiculars from

it on these n straight lines is constant: find the conditions that the locus of P may be a circle.

31. A point moves so that the sum of the squares of its distances from the sides of a regular polygon is constant: shew that the locus of the point is a circle.

32. A straight line moves so that the sum of the perpendiculars AP, BQ, from the fixed points A and B is constant: find the locus of the middle point of PQ.

33. O is a fixed point and AB a fixed straight line; a straight line is drawn from O meeting AB at P; in OP a point Q is taken so that $OP \cdot OQ = k^2$: find the locus of Q.

34. A straight line is drawn from a fixed point O, meeting a fixed circle at P; in OP a point Q is taken so that $OP \cdot OQ = k^2$: find the locus of Q.

35. Shew that $(hy - kx)^2 = c^2 \{(x-h)^2 + (y-k)^2\}$ represents the two tangents to the circle, $x^2 + y^2 = c^2$, which pass through the point (h, k).

36. Determine what is represented by the equation

$$r^2 - ra \cos 2\theta \sec \theta - 2a^2 = 0.$$

37. The polar equation to a circle being $r = 2c \cos \theta$, shew that the equation $2c \cos \beta \cos \alpha = r \cos (\beta + \alpha - \theta)$ represents a chord such that the radii drawn to its extremities from the pole, make angles α, β with the initial line.

38. Tangents to a circle at the points P and Q intersect at T; if the straight lines joining these points with the extremity of a diameter cut a second diameter perpendicular to the former at the points p, q, t, respectively, shew that $pt = qt$.

39. Find the equation to the circle which passes through three points whose co-ordinates are given.

40. Shew that the co-ordinates of the centre and the radius of the circle in the preceding Example are always finite except when the three given points are on a straight line.

CHAPTER VII.

RADICAL AXIS. POLE AND POLAR.

Radical Axis.

108. WE have shewn in Art. 88 that the equation to a circle is $(x-a)^2+(y-b)^2-c^2=0$. We shall write this for abbreviation $S=0$. If the point (x, y) be not on the circumference of the circle, S is not $=0$; we may in that case give a simple geometrical meaning to S.

I. Let (x, y) be *without* the circle; draw a tangent from (x, y) to the circle; join the point of contact with the centre of the circle (a, b); also join (x, y) with (a, b). Let C represent the point (a, b), Q the point (x, y), and T the point of contact of the tangent. Thus we have a right-angled triangle formed, and since $(x-a)^2+(y-b)^2=QC^2$, it follows that $S=QT^2$; that is, S expresses the square of the tangent from (x, y) to the circle. By Euclid III. 36, the square of the tangent is equal to the rectangle of the segments made by the circle on any straight line drawn from (x, y), and thus S will also express the value of this rectangle.

II. Let (x, y) be *within* the circle; then S is negative. Let C and Q have the same meaning as before, and produce CQ to meet the circle at T and T'; then

$$-S=CT^2-CQ^2=(CT-CQ)(CT+CQ)=TQ\,.\,T'Q.$$

Hence by Euclid III. 35, if *any* straight line PQP' be drawn meeting the circle at P and P', the value of the rectangle $PQ\,.\,P'Q$ is $-S$.

109. Let $\quad S$ denote $(x-a)^2+(y-b)^2-c^2$,

and $\quad S'$ denote $(x-a')^2+(y-b')^2-c'^2$;

so that $\quad S=0$...........(1), and $S'=0$...........(2).

are the equations to two circles: we proceed to interpret the equation

$$S - S' = 0 \quad\text{.......................}(3).$$

$S - S'$ contains only the first powers of x and y; therefore $S - S' = 0$ is the equation to *some straight line.* Also if values of x and y can be found to satisfy simultaneously (1) and (2), these values will satisfy (3). Hence when the circles represented by (1) and (2) intersect, (3) is the equation to the straight line which joins their *points of intersection.*

Also suppose that from any point in (3), external to both circles, we draw tangents to (1) and (2); then, by Art. 108, these tangents are equal in length. Hence whether (1) and (2) intersect or not, the straight line (3) has the following property: *if from any point of it straight lines be drawn to touch both circles, the lengths of these straight lines are equal.*

110. An equation of the form

$$A(x^2 + y^2) + Bx + Dy + E = 0$$

will represent a circle; for after division by A we obtain the ordinary form of the equation to a circle. We shall say that the equation to a circle is in its *simplest form* when the coefficient of x^2 and y^2 is unity.

DEFINITION. If $S = 0$, $S' = 0$, be the equations to two circles in their *simplest forms*, the straight line $S - S' = 0$ is called the *radical axis* of the circles.

The axes of co-ordinates may here be rectangular or oblique.

Or we may give a geometrical definition thus. A straight line can always be found such that if from any point of it tangents be drawn to two given circles, these tangents are equal; this straight line is called the radical axis of the circles.

111. *The three radical axes belonging to three given circles meet at a point.*

Let the equations to the three circles be

$$S_1 = 0 \;\ldots\ldots\; (1), \qquad S_2 = 0 \;\ldots\ldots\; (2), \qquad S_3 = 0 \;\ldots\ldots\; (3).$$

The equations to the radical axes are

$$S_1 - S_2 = 0, \text{ belonging to (1) and (2)},$$
$$S_2 - S_3 = 0, \text{ .... (2) and (3)},$$
$$S_3 - S_1 = 0, \text{ .... (3) and (1)}.$$

These three straight lines meet at a point; since it is obvious that the values of x and y which simultaneously satisfy two of the equations, will also satisfy the third.

112. A large number of inferences may be drawn from the preceding Articles by examining the special cases which fall under the general propositions. (See Plücker *Analytisch-Geometrische Entwickelungen*, Vol. I. pp. 49...69.) We notice a few of these respecting the radical axis of two circles.

113. *The radical axis is perpendicular to the straight line joining the centres of the two circles.*

Let the equations to the circles be

$$(x-a)^2 + (y-b)^2 - c^2 = 0, \quad (x-a')^2 + (y-b')^2 - c'^2 = 0;$$

then the equation to the radical axis is

$$(x-a)^2 - (x-a')^2 + (y-b)^2 - (y-b')^2 - c^2 + c'^2 = 0;$$

that is,

$$x(a'-a) + y(b'-b) + \tfrac{1}{2}(a^2 - a'^2 + b^2 - b'^2 - c^2 + c'^2) = 0 \ldots (1).$$

And the equation to the straight line joining the centres of the circles is (Art. 35)

$$y - b = \frac{b'-b}{a'-a}(x-a) \ldots\ldots\ldots\ldots\ldots\ldots (2);$$

(1) and (2) are at right angles by Art. 42.

114. When two circles touch, their radical axis is the common tangent at the point of contact. For the radical axis passes through the common point and is perpendicular to the straight line joining the centres of the circles.

115. Suppose the radius of one of the circles to become indefinitely small, that is, the circle to become a point; the radical axis then has the following property: if from any

point of the radical axis we draw a straight line to the given point, and a tangent to the given circle, the straight line and the tangent will be equal in length.

116. The radical axis of a point and a circle falls *without* the circle, whether the point be *without* or *within* the circle. For if the radical axis met the circle, the co-ordinates of the points of intersection would satisfy the *equation to the point* as well as the equation to the circle. But the equation to the point can be satisfied by no co-ordinates except the co-ordinates of that point; therefore the radical axis cannot meet the circle. If the point be *on* the circle, the radical axis is the tangent to the circle at this point.

117. Suppose *both* circles to become points. Then the straight lines drawn from any point in the radical axis to the two fixed points are equal in length. Hence the radical axis belonging to two given points is the straight line which bisects at right angles the distance between the two given points.

118. Suppose in Art. 111 that each circle becomes a point; the theorem proved is then the following: the straight lines drawn from the middle points of the sides of a triangle at right angles to the sides meet at a point.

119. It is a well-known geometrical problem *to draw a straight line which shall touch two given circles:* see *Appendix to Euclid,* Art. 4. If the circles do not intersect, *four* common tangents can be drawn; two of them will be equally inclined to the straight line joining the centres, and will intersect on that straight line *between* the circles; the other two will also be equally inclined to the straight line joining the centres, and will intersect on that straight line *beyond* the smaller circle. These two points of intersection are called *centres of similitude.*

We will briefly explain some of the properties of *centres of similitude.*

I. Let a centre of similitude of two circles be taken as the pole, and the straight line passing through the centres of the circles as the initial line. By Art. 105 the equations to the two circles will be of the forms

$$r^2 - 2rl\cos\theta + l^2 - c^2 = 0,\ r^2 - 2rl'\cos\theta + l'^2 - c'^2 = 0 \ldots (1).$$

From the first equation

$$r = l\cos\theta \pm \sqrt{(c^2 - l^2\sin^2\theta)} \text{ .... } (2).$$

When the two values of r are equal the radius vector becomes a tangent: this takes place when $l^2\sin^2\theta = c^2$. Since the circles have common tangents passing through the pole $\frac{c^2}{l^2} = \frac{c'^2}{l'^2}$, and therefore $\frac{c'}{l'} = \pm\frac{c}{l}$. If the lower sign is taken the centre of similitude is *between* the centres of the two circles; if the upper sign is taken the centre of similitude is *on the production* of the straight line which joins the centres: we may call the former the *inner* centre of similitude, and the latter the *outer* centre of similitude.

Since $\frac{c'^2}{l'^2} = \frac{c^2}{l^2}$ the second of equations (1) may be written

$$r = \frac{l'}{l}\{l\cos\theta \pm \sqrt{(c^2 - l^2\sin^2\theta)}\} \text{ .... } (3).$$

From (2) and (3) we have the following result: Let A be the centre of one circle, and B the centre of another, and let T be a centre of similitude; let any straight line through T cut the former circle at K and L, and the latter at M and N, so that TK is less than TL, and TM less than TN: then

$$\frac{TK}{TM} = \frac{TL}{TN} = \frac{TA}{TB}.$$

II. When two circles intersect only one pair of common tangents can be drawn; and when one circle is entirely within the other no common tangent can be drawn. Nevertheless two points always exist such as the point T just considered; so that we may take the following as the most general definition of the centre of similitude of two circles: A centre of similitude is a point on the straight line joining the centres or on this straight line produced such that its distances from the centres are proportional to the radii of the corresponding circles. The essential property of a centre of similitude may be considered to be that expressed by the final result in I.

III. Let T be a centre of similitude of two circles; draw from T two straight lines, one cutting the circles at K, L, M, N; and the other at k, l, m, n.

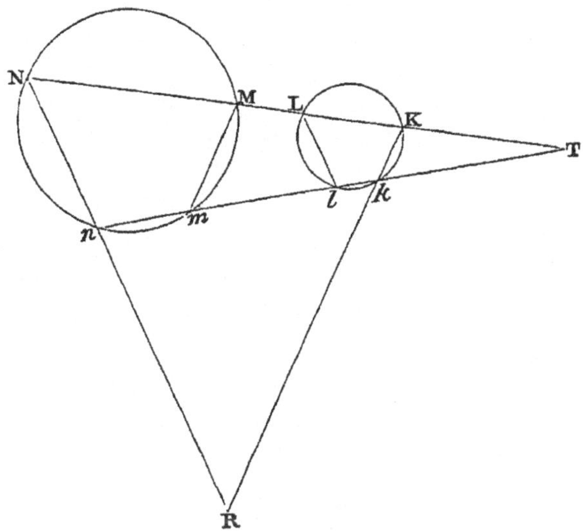

Now we have just shewn that

$$\frac{TK}{TM} = \frac{Tk}{Tm};$$

therefore the triangles TKk and TMm are similar, and Mm is parallel to Kk.

Hence the angle Kkl = the angle Mmn; and therefore the angles MNn and Kkl are supplemental, by Euclid III. 22, so that a circle would pass round $NKkn$: let Nn and Kk be produced to meet at R, then $RK.Rk = RN.Rn$, by Euclid III. 36. Cor. Hence the tangents from R to the two circles are equal, by Euclid III. 36; and therefore R is on the radical axis of the two circles.

Similarly Nn is parallel to Ll; and Mm and Ll if produced meet on the radical axis.

IV. Suppose there are three circles; since each pair has two centres of similitude there will be six centres of similitude on the whole: we shall shew that four straight lines can be drawn each containing three centres of similitude.

Let A, B, C be the centres of three circles; let p, q, r be their radii.

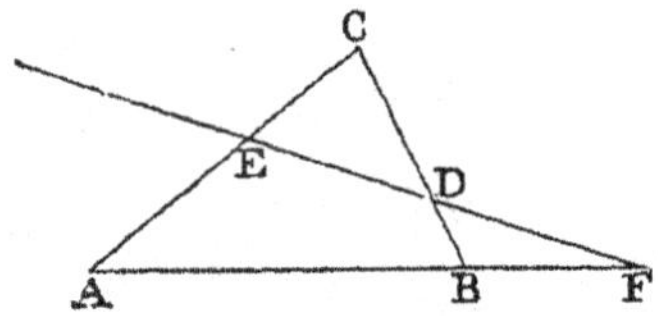

Let F be a centre of similitude of the circles which have their centres at A and B; draw a straight line through F meeting CA and CB at E and D respectively.

By page 71 we have

$$AE \,.\, CD \,.\, BF = CE \,.\, BD \,.\, AF.$$

But
$$\frac{AF}{BF} = \frac{p}{q};$$

thus
$$\frac{CD}{BD} = \frac{p}{q}\frac{CE}{AE}.$$

Now suppose that E is a centre of similitude of the circles which have their centres at A and C; then

$$\frac{CE}{AE} = \frac{r}{p}; \quad \text{therefore} \quad \frac{CD}{BD} = \frac{r}{q}.$$

Hence D is a centre of similitude of the circles which have their centres at B and C. In this way we obtain results which can be enunciated definitely thus: the outer centre of similitude of two circles, and the two inner centres of similitude of these two circles and any third circle lie on a straight line; also the three outer centres of similitude lie on a straight line.

Pole and Polar.

120. Definition. If the equation to a given circle be $x^2 + y^2 = c^2$, and h, k be the co-ordinates of any point, then the straight line $xh + yk = c^2$ is called the *polar* of the point (h, k) with respect to the given circle, and the point (h, k) is

called the *pole* of the straight line $xh + yk = c^2$ with respect to the given circle.

We may also express our definition thus: the *polar* of a given point with respect to a given circle is the straight line whose equation involves the co-ordinates of the given point in the same manner as the equation to the tangent at any point of the circle involves the co-ordinates of the point of contact; and the given point is the *pole* of the straight line.

This definition might be misunderstood. For the equation to the tangent to a circle at a given point might be expressed in different forms by using the relation which holds between the co-ordinates of the given point by virtue of the equation to the circle. We might for example express the equation to the tangent in terms of *either* of the co-ordinates of the given point alone. But in the above definition we mean that the equation to the tangent is to be in the form which it naturally assumes, involving the co-ordinates of the given point *rationally*.

Or we may define the *polar* of a point by means of the properties which it possesses (Art. 103). The *polar* of a given point with respect to a given circle is the straight line which is the locus of the intersection of tangents drawn at the extremities of every chord through the given point; and the given point is called the *pole* of this straight line.

If the given point be without the circle, its polar coincides with the *chord of contact* of tangents drawn from that point.

121. *If one straight line pass through the pole of another straight line, the second straight line will pass through the pole of the first straight line.*

Let (x', y') be the pole of the *first* straight line, and therefore the equation to the *first* straight line

$$xx' + yy' = c^2 \text{......................(1)}.$$

Let (x'', y'') be the pole of the *second* straight line, and therefore the equation to the *second* straight line

$$xx'' + yy'' = c^2 \text{......................(2)}.$$

Since (1) passes through (x'', y'') we have $x''x' + y''y' = c^2$; and since this equation holds, (2) passes through (x', y').

122. *The intersection of two straight lines is the pole of the straight line which joins the poles of those straight lines.*

Denote the two straight lines by A and B, and the straight line joining their poles by C; since C passes through the pole of A, therefore, by Art. 121, A passes through the pole of C; similarly B passes through the pole of C; therefore the intersection of A and B is the pole of C.

MISCELLANEOUS EXAMPLES.

1. Find the tangent of the angle between the two straight lines whose intercepts on the axes are respectively a, b, and a', b'.

2. If the two straight lines represented by the equation $x^2(\tan^2\phi + \cos^2\phi) - 2xy\tan\phi + y^2\sin^2\phi = 0$, make angles α, β with the axis of x, shew that $\tan\alpha \sim \tan\beta = 2$.

3. One side of a square a corner of which is at the origin makes an angle α with the axis of x: find the equations to the four sides and the two diagonals.

4. Find the equations to the diagonals of the parallelogram formed by the straight lines

$$\frac{x}{a} + \frac{y}{b} = 1,\ \frac{x}{a} + \frac{y}{b} = 2,\ \frac{x}{b} + \frac{y}{a} = 1,\ \frac{x}{b} + \frac{y}{a} = 2;$$

and shew that the diagonals are at right angles.

5. The distance of a point (x_1, y_1) from each of two straight lines which pass through the origin of co-ordinates is δ: shew that the two straight lines are represented by the equation

$$(x_1y - xy_1)^2 = (x^2 + y^2)\,\delta^2.$$

6. Find the condition that one of the straight lines represented by $Ay^2 + Bxy + Cx^2 = 0$ may coincide with one of those represented by $ay^2 + bxy + cx^2 = 0$.

7. If $\alpha = 0$, $\beta = 0$, $\gamma = 0$ be the equations to the three sides of a triangle, and a, b, c be the perpendicular distances

between these sides and those of another triangle parallel to them respectively, the straight line joining the centres of the inscribed circles will be represented by any of the equations

$$\frac{\alpha-\beta}{a-b}=\frac{\beta-\gamma}{b-c}=\frac{\gamma-\alpha}{c-a}.$$

8. Shew that the equation to the straight line passing through the middle point of the side BC of a triangle ABC and parallel to the external bisector of the angle A is

$$\beta+\gamma-\frac{a}{2}(\sin B+\sin C)=0.$$

9. The equation to the straight line drawn parallel to BC through the centre of the escribed circle which touches BC is

$$(\alpha+\beta)\sin B+(\alpha+\gamma)\sin C=0.$$

10. Find the equations to the straight lines which pass through the intersection of the straight lines

$$l\alpha+m\beta+n\gamma=0, \qquad l'\alpha+m'\beta+n'\gamma=0,$$

and divide the angles between them into parts having their sines in a given ratio.

11. Find the equations to the two straight lines which bisect the angles between the straight lines represented by $Ay^2+Bxy+Cx^2=0$.

12. Find the condition in order that the straight lines $Ay^2+Bxy+Cx^2=0$ and $ay^2+bxy+cx^2=0$ may have their angles bisected by the same pair of straight lines.

13. If $u=0$, $v=0$, be the equations to two circles, shew that by giving a suitable value to the constant λ, the equation $u+\lambda v=0$ will represent any circle passing through the points of intersection of the given circles.

14. A fixed circle is cut by a series of circles, all of which pass through two given points: shew that the straight lines which join the points of intersection of the fixed circle with each circle of the series all meet at a point.

CHAPTER VIII.

THE PARABOLA.

123. THERE are three curves which we now proceed to define; we shall then deduce their equations from the definitions, and investigate some of their properties from their equations.

DEFINITION. A *conic section* is the locus of a point which moves so that its distance from a fixed point bears a constant ratio to its distance from a fixed straight line. If this ratio be *unity*, the curve is called a *parabola*, if *less* than unity, an *ellipse*, if *greater* than unity, an *hyperbola*.

The fixed point is called the focus, and the fixed straight line the directrix.

124. It will be shewn hereafter that if a cone be cut by a plane, the curve of intersection will be one of the following; a parabola, an ellipse, an hyperbola, a circle, two straight lines, one straight line, or a point. Hence the term *conic section* is applied to the parabola, ellipse, and hyperbola, and may be extended to include the circle, two straight lines, one straight line and point. We shall also shew that every curve of the second degree must be a conic section in this larger sense of the term.

At present we confine ourselves to tracing the consequences of the definitions in Art. 123.

125. *To find the equation to the Parabola.*

A parabola is the locus of a point which moves so that its distance from a fixed point is *equal* to its distance from a fixed straight line.

Let S be the fixed point, YY' the fixed straight line. Draw SO perpendicular to YY'; take O as the origin, OS

as the direction of the axis of x, OY as that of the axis of y. Suppose $OS = 2a$.

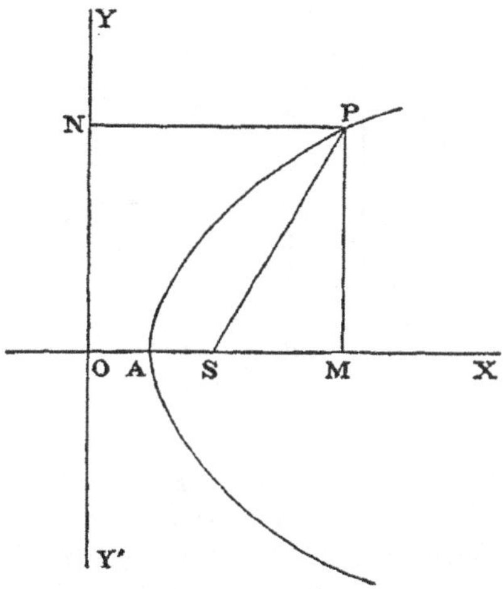

Let P be any point on the locus; join SP; draw PM parallel to OY and PN parallel to OX; let $OM = x$, $PM = y$.

By definition $SP = PN$; therefore $SP^2 = PN^2$; therefore $PM^2 + SM^2 = PN^2$, that is, $y^2 + (x - 2a)^2 = x^2$;

$$\text{therefore } y^2 = 4a(x - a) \qquad (1).$$

This is the equation to the parabola with the assumed origin and axes. The curve cuts the axis of x at a point A which bisects OS; for when $y = 0$ in (1), we have $x = a$. The equation will be simplified if we put the origin at A; let $x' = AM$, then $x' = x - a$, and (1) becomes $y^2 = 4ax'$.

We may suppress the accent, if we remember that the origin is now at A; thus we have for the equation to the parabola

$$y^2 = 4ax \qquad (2).$$

126. *To trace the parabola from its equation* $y^2 = 4ax$.

From this equation we see that for every positive value of x there are *two* values of y, equal in magnitude, but of opposite sign. Hence for every point P on one side of the axis of x, there is a point P' on the other side, such that

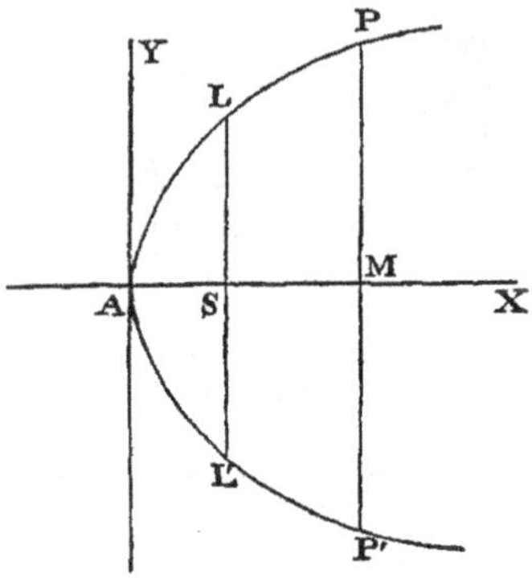

$P'M = PM$. Thus the curve is symmetrical with respect to the axis of x. Negative values of x do not give possible values of y; hence no part of the curve lies to the left of the origin. As x may have any positive value, the curve extends without limit on the right of the origin.

A is called the *vertex* of the curve, and AX the *axis* of the curve.

127. We have drawn the curve *concave* towards the axis of x; the following proposition will justify the figure.

The ordinate of any point of the curve which lies between the vertex and a fixed point of the curve is greater than the corresponding ordinate of the straight line joining the vertex and the fixed point.

Let P be the fixed point; x', y' its co-ordinates; then the equation to AP is $y = \frac{y'}{x'} x = \sqrt{\left(\frac{4a}{x'}\right)} \cdot x$, since $y'^2 = 4ax'$.

Let x denote any abscissa *less than* x', then since the ordinate of the curve is $\sqrt{(4ax)}$, and that of the straight line is $\sqrt{\left(\frac{4a}{x'}\right)} \cdot x$ or $\sqrt{\left(\frac{x}{x'}\right)} \times \sqrt{(4ax)}$, it is obvious that the ordinate of the curve is greater than that of the straight line.

All points may be said to be *outside* the curve for which

$y^2 - 4ax$ is positive; that is, points for which x is negative, or for which x is positive and less than $\frac{y^2}{4a}$. And all points may be said to be *inside* the curve for which $y^2 - 4ax$ is negative. Since the square of the distance of any point from the focus is $y^2 + (x-a)^2$, that is $(x+a)^2 + y^2 - 4ax$, it follows that the distance of any point, not on the curve, from the focus is greater or less than its distance from the directrix according as the point is outside or inside the curve.

128. DEFINITION. The double ordinate through the focus of a conic section is called the Latus Rectum.

Thus in the figure in Art. 126, LSL' is the Latus Rectum.

Let $x = a$, then from the equation $y^2 = 4ax$, $y = \pm 2a$. Hence $LS = L'S = 2a$; and $LL' = 4a$.

129. *To express the focal distance of any point of the parabola in terms of the abscissa of the point.*

The distance of any point on the curve from the focus is equal to the distance of the same point from the directrix. Hence (see figure to Art. 125), $SP = AM + AS$, $= x + a$.

Tangent and normal to a Parabola.

130. *To find the equation to the tangent at any point of a parabola.* (See Definition, Art. 90.)

Let x', y' be the co-ordinates of the point, x'', y'' the co-ordinates of an adjacent point on the curve.

The equation to the secant through these points is

$$y - y' = \frac{y'' - y'}{x'' - x'}(x - x') \quad\quad(1);$$

since (x', y') and (x'', y'') are on the parabola

$$y'^2 = 4ax', \quad y''^2 = 4ax'';$$

therefore $y''^2 - y'^2 = 4a(x'' - x')$;

therefore $\frac{y'' - y'}{x'' - x'} = \frac{4a}{y'' + y'}$;

hence (1) may be written $y - y' = \frac{4a}{y'' + y'}(x - x')$.

Now in the limit $y''=y'$; hence the equation to the tangent at the point (x', y') is

$$y-y'=\frac{2a}{y'}(x-x')\dots\dots\dots\dots\dots(2).$$

This equation may be simplified; multiply by y', thus

$$yy'=2a(x-x')+y'^2=2ax-2ax'+4ax'$$
$$=2a(x+x')\dots\dots\dots\dots\dots\dots\dots(3).$$

131. The equation to the tangent can be conveniently expressed in terms of the tangent of the angle which the straight line makes with the axis of the parabola.

For the equation to the tangent at (x', y') is

$$yy'=2a(x+x'),$$

or

$$y=\frac{2a}{y'}x+\frac{2ax'}{y'}=\frac{2a}{y'}x+\frac{4ax'}{2y'}$$
$$=\frac{2a}{y'}x+\frac{y'}{2}\dots\dots\dots\dots\dots\dots\dots(1).$$

Let $\frac{2a}{y'}=m$; therefore $\frac{y'}{2}=\frac{a}{m}$; thus (1) may be written

$$y=mx+\frac{a}{m}\dots\dots\dots\dots\dots\dots\dots(2);$$

this is the required equation. Conversely, every straight line whose equation is of this form is a tangent to the parabola.

132. It may be shewn as in Art. 93, that a tangent to the parabola meets it at only one point. Also, if a straight line meets a parabola at only one point, it will in general be the tangent at that point.

For suppose the equation to a parabola to be

$$y^2=4ax\dots\dots\dots\dots\dots\dots\dots(1),$$

and the equation to a straight line to be

$$y=mx+c\dots\dots\dots\dots\dots\dots\dots(2).$$

To determine the abscissæ of the points of intersection, we have the equation $(mx+c)^2=4ax$,

or

$$m^2x^2+(2mc-4a)x+c^2=0\dots\dots\dots\dots(3);$$

this quadratic equation will have two roots, except when $(mc-2a)^2 = m^2c^2$, that is, when $c = \frac{a}{m}$.

Hence if the straight line (2) meets the parabola, it will meet it at two points, unless $c = \frac{a}{m}$, and then the straight line is a tangent to the parabola by Art. 131.

If, however, the equation (2) be of the form $y = c$, so that the straight line is parallel to the axis of x, then instead of (3) we have the equation $c^2 = 4ax$, which has but one root; *hence a straight line parallel to the axis of the parabola* meets it at only one point, but is not a tangent.

133. The axis of y is a tangent to the curve at the vertex.

For the equation to the tangent at (x', y') is

$$yy' = 2a(x + x');$$

and when $x' = 0$ and $y' = 0$, this becomes $x = 0$.

134. *To find the equation to the normal at any point of a parabola.* (See Definition, Art. 97.)

Let x', y' be the co-ordinates of the point; the equation to the tangent at that point is

$$y = \frac{2a}{y'}(x + x') \quad\quad (1).$$

The equation to a straight line through (x', y') at right angles to (1) is

$$y - y' = -\frac{y'}{2a}(x - x') \quad\quad (2).$$

This is the equation to the normal at (x', y').

135. The equation to the normal may also be expressed in terms of the tangent of the angle which the straight line makes with the axis of the curve.

For the equation to the normal is $y = -\frac{y'}{2a}x + y' + \frac{y'x'}{2a}$,

or

$$y = -\frac{y'}{2a}x + y' + \frac{y'^3}{8a^2} \quad\quad (1).$$

Let $\quad -\frac{y'}{2a} = m;$ therefore $y' = -2am;$

thus (1) may be written

$$y = mx - 2am - am^3 \quad\text{.................... (2).}$$

136. We shall now deduce some properties of the parabola from the preceding Articles.

Let x', y' be the co-ordinates of P; let PT be the tangent at P and PG the normal at P.

The equation to the tangent at P is $yy' = 2a\,(x + x')$.

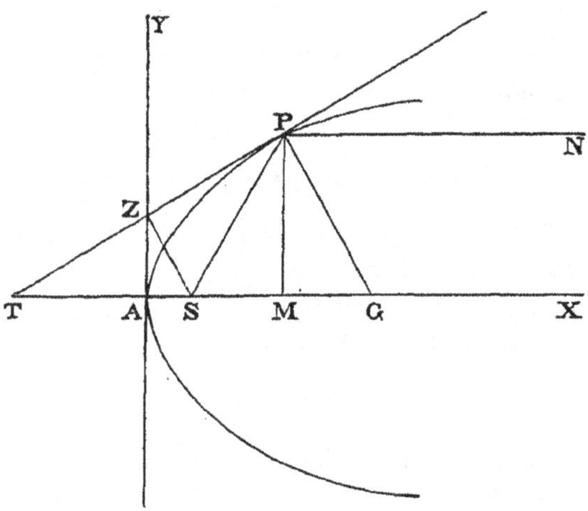

Let $y = 0$, then $x = -x'$; hence $AT = AM$.

Also $ST = AT + AS, \ = AM + AS, \ = SP$ (Art. 129).

Hence the triangle STP is isosceles, and the angle STP is equal to the angle SPT. Thus if PN be parallel to the axis of the curve, PN and PS are equally inclined to the tangent at P, so that the tangent bisects the angle between PS and NP produced.

Since the angle PTS is half the angle PSX, it follows that the angle between two tangents to a parabola is half the angle between the focal distances of the points of contact.

137. The equation to the normal at P is

$$y - y' = -\frac{y'}{2a}(x - x').$$

At the point G, where the normal cuts the axis, $y = 0$; hence from the above equation $x - x' = 2a$; thus

$MG = 2a =$ half the latus rectum. Also $SG = SP = ST$.

138. *To find the locus of the intersection of the tangent at any point with the perpendicular on it from the focus.*

Let x', y' be the co-ordinates of any point P on the curve; the equation to the tangent at P is

$$y = \frac{2a}{y'}(x + x') \qquad (1).$$

The equation to the straight line through the focus perpendicular to (1) is

$$y = -\frac{y'}{2a}(x - a) \qquad (2).$$

We have now to eliminate x' and y' by means of (1), (2), and

$$y'^2 = 4ax' \qquad (3).$$

From (3) we find x' in terms of y', and thus (1) may be written

$$y = \frac{2a}{y'}x + \frac{y'}{2} \qquad (4).$$

Thus the problem is reduced to the elimination of y' from (2) and (4); from (2)

$$y' = -\frac{2ay}{x - a} \qquad (5);$$

substitute in (4); then $y = -\frac{(x-a)x}{y} - \frac{ay}{x-a}$;

therefore $\quad y^2(x-a) + (x-a)^2 x + ay^2 = 0,$

or $\quad \{y^2 + (x-a)^2\} x = 0 \qquad (6).$

If the factor $y^2 + (x-a)^2$ be equated to zero, we have

$$y = 0, \qquad x = a \qquad (7).$$

The point thus determined is the focus; this however is *not* the locus of the intersection of (1) and (2), for the values in (7), although they satisfy (2), do not satisfy (1). We conclude therefore that the required locus is given by the equation $x=0$, which we obtain by considering the other factor in (6).

This result can be easily verified; for if we put $x=0$ in (1) we obtain $y=\frac{2ax'}{y'}=\frac{y'}{2}$; and if we put $x=0$ in (2), we also obtain $y=\frac{y'}{2}$; thus (1) and (2) intersect on the straight line $x=0$. Or we may equate the right-hand members of (1) and (2), and by reduction obtain $(4a^2+y'^2)\,x=ay'^2-4a^2x'$, which is zero: therefore $x=0$. Thus, if in the figure in Art. 136, Z be the intersection of the tangent at P with the axis of y, SZ is perpendicular to the tangent.

139. The process of the preceding Article is of frequent use and of great importance. We have in (1) and (2) the equations to two straight lines; if we obtain the values of x and y from these simultaneous equations, we thus determine the point of intersection of the straight lines; the values of x and y will depend upon those of x' and y', thus giving different points of intersection corresponding to the different straight lines represented by (1) and (2). If from (1), (2), and (3) we eliminate x' and y' we obtain an equation which holds for the co-ordinates of *every* point of intersection of (1) and (2). This is, by our definition of a locus, the equation corresponding to the locus of the intersection of (1) and (2).

Sometimes the elimination produces, as in the preceding Article, an equation which does not represent the required locus. The student has probably noticed in solving algebraical questions that he often arrives at other results besides that which he is especially seeking. We can frequently interpret these additional results; thus in the preceding Article, since, whatever x' and y' may be, the values $x=a$, $y=0$, satisfy one of the equations which we use in effecting the elimination, we might anticipate that our result would involve a corresponding factor.

140. If the straight line from the focus, instead of being perpendicular to the tangent, meet it at any constant angle, the locus of their intersection will still be a straight line. We will indicate the steps of the investigation. Suppose β the angle between the tangent and the straight line from the focus; equation (1) remains as in Art. 138; instead of (2) we have, by Art. 45,

$$y = \frac{\dfrac{2a}{y'} + \tan\beta}{1 - \dfrac{2a}{y'}\tan\beta}(x-a) = \frac{2a + y'\tan\beta}{y' - 2a\tan\beta}(x-a).$$

Instead of (5) in Art. 138, we shall find

$$y' = \frac{2a(x-a) + 2ay\tan\beta}{y - (x-a)\tan\beta}.$$

The result of the elimination is

$$y\{y-(x-a)\tan\beta\}\{x-a+y\tan\beta\} \\ -x\{y-(x-a)\tan\beta\}^2 - a(x-a+y\tan\beta)^2 = 0.$$

Now, guided by the result of Art. 138, we may anticipate that $y^2 + (x-a)^2$ will prove a factor of the left-hand member of the equation; and we shall find by reduction that the equation may be written $\{y^2 + (x-a)^2\}(y\tan\beta - x\tan^2\beta - a) = 0$.

Hence the required locus is determined by

$$y = x\tan\beta + a\cot\beta.$$

141. *To find the length of the perpendicular from the focus on the tangent at any point of the parabola.*

The equation to the tangent at the point (x', y') is

$$y = \frac{2a}{y'}(x+x').$$

The perpendicular on this from the point $(a, 0)$, by Art. 47,

$$= \frac{2a(a+x')}{\sqrt{(y'^2+4a^2)}} = \frac{2a(a+x')}{\sqrt{\{4a(a+x')\}}} = \sqrt{\{a(a+x')\}}.$$

Call the focal distance of the point of contact r, and the perpendicular p; then, by Art. 129, $r = a + x'$;

$$\text{therefore } p = \sqrt{(ar)}.$$

Also PG = twice $SZ = 2\sqrt{(ar)}$: see Art. 136.

142. *From any external point two tangents can be drawn to a parabola.*

Let the equation to the parabola be $y^2 = 4ax$; and let h, k be the co-ordinates of an external point. Suppose x', y' the co-ordinates of a point on the parabola such that the tangent at this point passes through (h, k). The equation to the tangent at (x', y') is $yy' = 2a(x + x')$.

Since this tangent passes through (h, k)

$$ky' = 2a(h + x') \quad \text{.....................(1)}.$$

Also since (x', y') is on the parabola

$$y'^2 = 4ax' \quad \text{.......................(2)}.$$

Equations (1) and (2) determine the values of x' and y'.

Substitute from (2) in (1), thus $ky' = 2ah + \frac{y'^2}{2}$, therefore $y'^2 - 2ky' + 4ah = 0$. The roots of this quadratic will be found to be both possible, since (h, k) is an *external* point and therefore k^2 greater than $4ah$. To each value of y' corresponds one value of x' by (1); hence *two* tangents can be drawn from any external point.

The straight line which passes through the points where these tangents meet the parabola is called the *chord of contact.*

143. *Tangents are drawn to a parabola from a given external point: to find the equation to the chord of contact.*

Let h, k be the co-ordinates of the external point; x_1, y_1 the co-ordinates of the point where one of the tangents from (h, k) meets the parabola; x_2, y_2 the co-ordinates of the point where the other tangent from (h, k) meets the parabola.

The equation to the tangent at (x_1, y_1) is

$$yy_1 = 2a(x + x_1) \quad \text{....................(1)}.$$

Since this tangent passes through (h, k) we have

$$ky_1 = 2a(h + x_1) \quad \text{....................(2)}.$$

Similarly, since the tangent at (x_2, y_2) passes through (h, k)

$$ky_2 = 2a(h + x_2) \text{..................(3).}$$

Hence it follows that the equation to the chord of contact is

$$ky = 2a(x + h) \text{..................(4).}$$

For (4) is obviously the equation to *some* straight line; also this straight line passes through (x_1, y_1), for (4) is satisfied by the values $x = x_1$, $y = y_1$, as we see from (2); similarly from (3) we conclude that this straight line passes through (x_2, y_2). Hence (4) is the required equation.

Thus we may use the following process to draw tangents to a parabola from a given external point. Draw the straight line which is represented by (4), join the points where it meets the parabola with the given external point, and the straight lines thus obtained are the required tangents.

144. *Through any fixed point chords are drawn to a parabola, and tangents to the parabola drawn at the extremities of each chord: the locus of the intersection of the tangents is a straight line.*

Let h, k be the co-ordinates of the point through which the chords are drawn; let tangents to the parabola be drawn at the extremities of one of these chords, and let (x_1, y_1) be the point at which they meet. The equation to the corresponding chord of contact by Art. 143 is $yy_1 = 2a(x + x_1)$. But this chord passes through (h, k); therefore $ky_1 = 2a(h + x_1)$. Hence the point (x_1, y_1) lies on the straight line $ky = 2a(x + h)$; that is, the locus of the intersection of the tangents is a straight line.

We will now prove the converse of this proposition.

145. *If from any point in a straight line a pair of tangents be drawn to a parabola, the chords of contact will all pass through a fixed point.*

Let

$$Ax + By + C = 0 \text{..................(1)}$$

be the equation to the straight line; let (x', y') be a point in this straight line from which tangents are drawn to the parabola; then the equation to the corresponding chord of contact is

$$yy' = 2a(x + x') \text{..................(2).}$$

Since (x', y') is on (1) we have $Ax' + By' + C = 0$; therefore (2) may be written $y(Ax' + C) + 2aB(x + x') = 0$,

or $$(Ay + 2aB)x' + Cy + 2aBx = 0.$$

Now whatever be the value of x', this straight line passes through the point whose co-ordinates are found by the simultaneous equations $Ay + 2aB = 0$, $Cy + 2aBx = 0$; that is the point for which $y = -\frac{2aB}{A}$, $x = \frac{C}{A}$.

The student should observe the different interpretations that can be assigned to the equation $ky = 2a(x + h)$. The statements in Art. 103 with respect to the circle may all be applied to the parabola.

146. Some interesting geometrical investigations relating to tangents to a parabola from an external point may be noticed.

To draw the two tangents to a parabola from any external point.

Let O denote the external point and S the focus. On OS as diameter describe a circle, and let it cut the tangent at the vertex at Z and z. Join OZ and Oz: these straight lines, produced if necessary, are the tangents from O by Art. 138 and Euclid III. 31.

Or we may proceed thus. Join OS. With centre O and

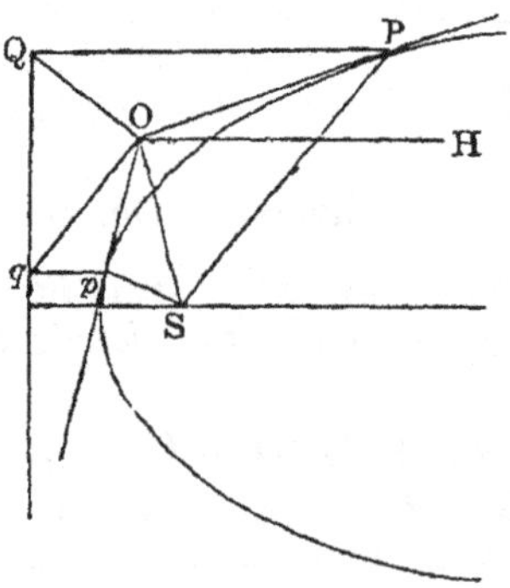

radius OS describe a circle, and let it cut the directrix at Q and q. Through these points draw parallels to the axis meeting the parabola at P and p. Then OP and Op are the required tangents.

For join OQ and SP. Then in the triangles OPS and OPQ we have $OS = OQ$ by construction, $PS = PQ$ by the nature of the parabola, and OP common. Therefore the angle $OPS =$ the angle OPQ; and OP is the tangent at P by Art. 136.

Similarly Op is the tangent at p.

The two tangents to a parabola from an external point subtend equal angles at the focus.

Since the triangles OPS and OPQ are equal in all respects, the angle $OSP =$ the angle OQP; and similarly the angle $OSp =$ the angle Oqp: and the angles OQP and Oqp are equal, for they are the complements of the equal angles OQq and OqQ.

The angle between a tangent and a straight line parallel to the axis is equal to the angle between the other tangent and the straight line from the external point to the focus.

Draw OH parallel to the axis.

The angle $QOH =$ the angle qOH; that is

twice the angle $POS -$ the angle SOH

$=$ twice the angle $pOS +$ the angle SOH;

therefore the angle $POS =$ the angle pOH, and therefore also the angle $POH =$ the angle pOS.

The student should observe the extension thus given to the result in Art. 136: at any point of the curve the straight line which bisects the angle between the focal distance of the point and the parallel to the axis is *at right angles* to the tangent, and at any external point the straight line which bisects the angle between the focal distance and the parallel to the axis is *equally inclined* to the two tangents.

The circle which passes through the intersections of three tangents to a parabola will pass through the focus.

Let P, Q, R be the points of contact, and pqr the triangle formed by the tangents.

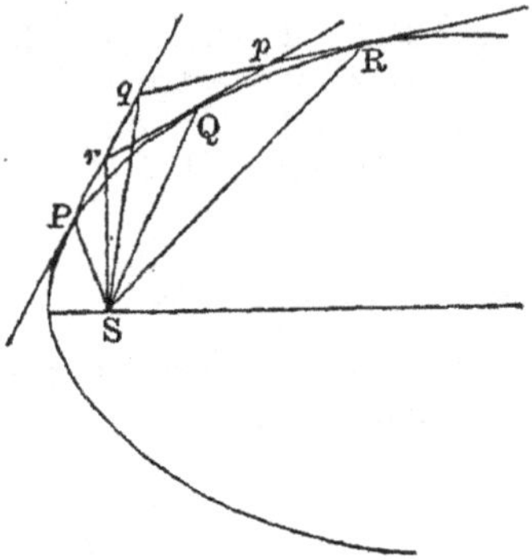

Since Pr and Qr subtend equal angles at S the angle PSr is half the angle PSQ.

Similarly the angle PSq is half the angle PSR. Hence the angle qSr is half the angle QSR; that is by Art. 136 the angle qSr is equal to the angle qpr: therefore S is on the circumference of the circle which passes round pqr.

Diameters.

147. *To find the length of a straight line drawn from any point in a given direction to meet a parabola.*

Let x', y' be the co-ordinates of the point from which the straight line is drawn; x, y the co-ordinates of the point to which the straight line is drawn; θ the inclination of the straight line to the axis of x; r the length of the straight line; then (Art. 27)

$$x = x' + r\cos\theta, \quad y = y' + r\sin\theta.$$

If (x, y) be on the parabola, these values may be substituted in the equation $y^2 = 4ax$; thus $(y' + r\sin\theta)^2 = 4a(x' + r\cos\theta)$;

or $$r^2\sin^2\theta + 2r(y'\sin\theta - 2a\cos\theta) + y'^2 - 4ax' = 0.$$

From this quadratic two values of r can be found, which are the lengths of the straight lines that can be drawn from (x', y') in the given direction to the parabola.

When the point (x', y') is *within* the parabola, the roots of the above quadratic will be of *different* signs; in this case the two straight lines that can be drawn from (x', y') to meet the curve are drawn in *different* directions. When the point (x', y') is *without* the parabola, the roots are of the *same* sign, and the straight lines are drawn in the *same* direction.

148. DEFINITION. *A diameter of a curve is the locus of the middle points of a series of parallel chords.*

149. *To find the diameter of a given system of parallel chords in a parabola.*

Let θ be the inclination of the chords to the axis of the parabola; let x', y' be the co-ordinates of the middle point of any one of the chords; the equation which determines the lengths of the straight lines drawn from (x', y') to the curve is (Art. 147)

$$r^2 \sin^2\theta + 2r(y' \sin\theta - 2a\cos\theta) + y'^2 - 4ax' = 0 \quad \text{.....(1)}.$$

Since (x', y') is the *middle* point of the chord, the values of r furnished by this quadratic must be *equal in magnitude* and *opposite in sign;* hence the coefficient of r must vanish;

thus $$y' \sin\theta - 2a\cos\theta = 0;$$

therefore $$y' = 2a\cot\theta \quad \text{.......................(2)};$$

thus the required diameter is a straight line parallel to the axis of the parabola.

Hence every diameter is parallel to the axis of the parabola.

Also every straight line parallel to the axis of the parabola is a diameter, that is, bisects some system of parallel chords; for by giving to θ a suitable value, the equation (2) may be made to represent *any* straight line parallel to the axis.

150. Let a tangent be drawn to the parabola at the point where the straight line $y' = 2a\cot\theta$ meets the curve; the equation to the tangent is $y = \frac{2a}{y'}(x + x')$; that is, $y = \tan\theta\,(x + x')$; hence, *the tangent at the extremity of any diameter of the parabola is parallel to the chords which that diameter bisects.*

151. *To find the equation to the parabola, the axes being any diameter and the tangent at the point where it meets the curve.*

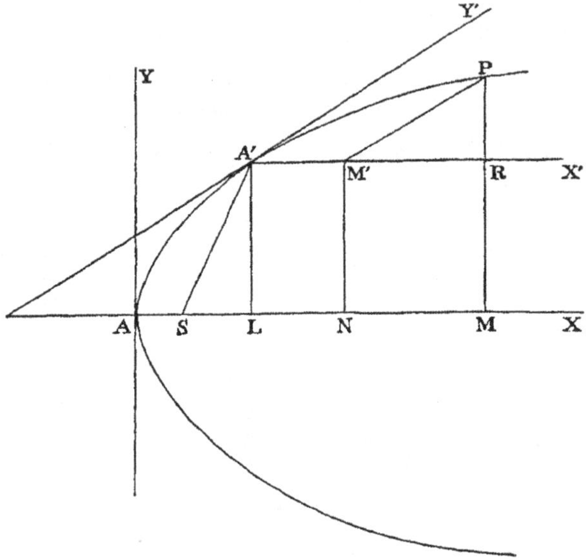

Let h, k be the co-ordinates of a point A' on the parabola; take this point for a new origin; draw through it a straight line $A'X'$ parallel to the axis of the curve for the new axis of x, and a tangent $A'Y'$ to the curve for the new axis of y. Let $Y'A'X' = \theta$; then (Art. 150) $\frac{2a}{k} = \tan\theta$.

Let x, y be the co-ordinates of a point P on the curve referred to the original axes; x', y' the co-ordinates of the *same* point referred to the new axes; draw PM parallel to AY and PM' parallel to $A'Y'$; also draw $A'L$, $M'N$ parallel to AY; let R denote the intersection of PM and $A'X'$; then

$$x = AM = AL + LN + NM = AL + A'M' + M'R$$
$$= h + x' + y' \cos\theta,$$
$$y = PM = RM + PR = A'L + PR$$
$$= k + y' \sin\theta.$$

Substitute these values in the equation $y^2 = 4ax$; thus

$$(k + y' \sin\theta)^2 = 4a(h + x' + y' \cos\theta),$$

or $$y'^2 \sin^2\theta + 2y'(k \sin\theta - 2a\cos\theta) + k^2 - 4ah = 4ax'.$$

But, $k = 2a \cot\theta$, and $k^2 = 4ah$; thus we have

$$y'^2 \sin^2\theta = 4ax',$$

or $$y'^2 = \frac{4a}{\sin^2\theta} x',$$

which is the required equation.

We may shew that $\frac{a}{\sin^2\theta} = SA'$; for $SA' = a + h$ (Art. 129); and $h = \frac{k^2}{4a} = a\cot^2\theta$; therefore $a + h = \frac{a}{\sin^2\theta}$.

Hence the equation may be written $y'^2 = 4a'x'$, where $a' = SA'$; or suppressing the accents on the variables

$$y^2 = 4a'x.$$

152. The equation to the tangent to the parabola will be of *the same form* whether the axes be rectangular, or the oblique system formed by a diameter and the tangent at its extremity; for the investigation of Art. 130 will apply without any change to the equation $y^2 = 4a'x$ which represents a parabola referred to such an oblique system.

153. *Tangents at the extremities of any chord of a parabola meet on the diameter which bisects that chord.*

Refer the parabola to the diameter bisecting the chord, and the corresponding tangent, as axes; let the equation to the parabola be $y^2 = 4a'x$; let x', y' be the co-ordinates of one extremity of the chord; then the equation to the tangent at this point is

$$yy' = 2a'(x + x') \qquad (1).$$

The co-ordinates of the other extremity of the chord are x', $-y'$; and the equation to the tangent there is

$$-yy' = 2a'(x + x') \qquad (2).$$

The straight lines represented by (1) and (2) meet at the point for which $y = 0$, $x = -x'$; this demonstrates the theorem.

Polar Equation.

154. *To find the Polar Equation to the parabola, the focus being the pole.*

Let $SP = r$, $ASP = \theta$, (see figure to Art. 125); then $SP = PN$, by definition; that is, $SP = OS + SM$;

or $$r = 2a + r\cos(\pi - \theta);$$

therefore $$r(1 + \cos\theta) = 2a,$$

and $$r = \frac{2a}{1 + \cos\theta}.$$

If we denote the angle XSP by θ, then we have as before $SP = OS + SM$; thus $r = 2a + r\cos\theta$, and $r = \dfrac{2a}{1 - \cos\theta}$.

155. The polar equation to the parabola when the *vertex* is the pole may be conveniently deduced from the equation $y^2 = 4ax$ by putting $r\cos\theta$ and $r\sin\theta$ for x and y respectively; we thus obtain $r = \dfrac{4a\cos\theta}{\sin^2\theta}$.

We add a few miscellaneous propositions on the parabola.

Definition. A chord passing through the focus of a conic section is called a focal chord.

156. *If tangents be drawn at the extremities of any focal chord of a parabola,* (1) *the tangents will intersect on the directrix,* (2) *the tangents will meet at right angles,* (3) *the straight line drawn from the point of intersection of the tangents to the focus will be perpendicular to the focal chord.*

(1) If the tangents to a parabola meet at the point (h, k) the equation to the chord of contact is, $ky = 2a(x + h)$ by Art. 143. Suppose the chord passes through the focus; then the values $x = a$, $y = 0$, must satisfy this equation;

$$\text{therefore } 0 = 2a(a + h);$$

$$\text{therefore } h = -a;$$

that is, the point of intersection of the tangents is on the directrix.

(2) The equation to the tangent to a parabola may be written (Art. 131) $y = mx + \frac{a}{m}$. Suppose (h, k) a point on the tangent; therefore $hm^2 - km + a = 0$. This quadratic will determine the inclinations to the axis of the parabola of the two straight lines that may be drawn through the point (h, k) to touch the parabola. Suppose m_1, m_2 the tangents of these inclinations, then by the theory of quadratic equations $m_1 m_2 = \frac{a}{h}$.

If $h = -a$, $m_1 m_2 = -1$; that is, the two tangents are at right angles.

(3) The equation to the straight line passing through the focus and (h, k) is $y = \frac{k}{h-a}(x-a)$. If $h = -a$, this becomes $y = -\frac{k}{2a}(x-a)$; the straight line is therefore perpendicular to the focal chord of which the equation is $yk = 2a(x-a)$.

157. *If through any point within or without a parabola, two straight lines be drawn parallel to two given straight lines to meet the curve, the rectangles of the segments will be to one another in an invariable ratio.*

Let (x', y') be the given point, and suppose α and β respectively the inclinations of the given straight lines to the axis of the parabola. By Art. 147, if a straight line be drawn through (x', y') to meet the curve and be inclined at an angle α to the axis, the lengths of its segments are given by the equation $r^2 \sin^2\alpha + 2r(y' \sin\alpha - 2a\cos\alpha) + y'^2 - 4ax' = 0$.

Therefore by the theory of quadratic equations the rectangle of the segments $= \frac{y'^2 - 4ax'}{\sin^2\alpha}$.

Similarly the rectangle of the segments of the straight line drawn through (x', y') at an angle $\beta = \frac{y'^2 - 4ax'}{\sin^2\beta}$.

Hence the ratio of the rectangles $= \frac{\sin^2\beta}{\sin^2\alpha}$, and this ratio is constant whatever x' and y' may be.

Let O be the point through which the straight lines OPp, OQq, are drawn inclined to the axis of the parabola at angles α, β, respectively; then we have shewn that

$$\frac{OP \cdot Op}{OQ \cdot Oq} = \frac{\sin^2\beta}{\sin^2\alpha}.$$

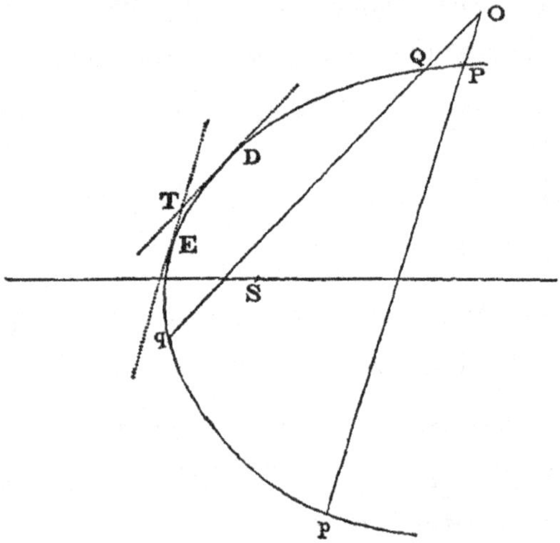

Let tangents to the parabola be drawn parallel to Pp, Qq, meeting the parabola at E and D respectively; let S be the focus; then by Art. 151,

$$\frac{SE}{SD} = \frac{\sin^2\beta}{\sin^2\alpha}; \text{ therefore } \frac{OP \cdot Op}{OQ \cdot Oq} = \frac{SE}{SD}.$$

Suppose O to coincide with T; then $OP \cdot Op$ becomes TE^2 and $OQ \cdot Oq$ becomes TD^2;

$$\text{therefore } \frac{TE^2}{TD^2} = \frac{SE}{SD}.$$

EXAMPLES.

1. Find the equation to the straight line joining A and L. (See figure to Art. 126.)

2. Find the equation to the circle which passes through A, L, L'. (See figure to Art. 126.)

3. A point moves so that its shortest distance from a given circle is equal to its distance from a given fixed diameter of that circle: find the locus of the point.

4. Trace the curves $y^2 = 4ax$, and $x^2 + 4ay = 0$; and determine their points of intersection.

5. Determine the equation to the tangent at L. (See figure to Art. 126.)

6. Find the angle between the straight lines in Examples 1 and 5.

7. Determine the equation to the normal at L.

8. Find the point where the normal at L meets the curve again, and the length of the intercepted chord.

9. Find the point in a parabola where the tangent is inclined at an angle of 30° to the axis of x.

10. The length of the perpendicular from the intersection of the directrix and axis on the tangent at (x', y') is

$$\frac{a(x'-a)}{\sqrt{\{a(x'+a)\}}}.$$

11. Find the points of contact of tangents the perpendiculars on which from the intersection of the directrix and axis are equal to one-fourth of the latus rectum.

12. A circle has its centre at the vertex A of a parabola whose focus is S, and the diameter of the circle is $3AS$: shew that the common chord bisects AS.

13. Trace the curve $y = x - x^2$, and determine whether the straight line $x + y = 1$ is a tangent to it.

14. The tangent at any point of a parabola will meet the directrix and latus rectum produced at two points equally distant from the focus.

15. PM is an ordinate of a point P on a parabola; a straight line is drawn parallel to the axis bisecting PM and cutting the curve at Q; MQ cuts the tangent at the vertex A at T: shew that $AT = \frac{2}{3}PM$.

16. If from any point P of a circle PC be drawn to the centre C, and a chord PQ be drawn parallel to the diameter ACB and bisected at R, shew that the locus of the intersection of CP and AR is a parabola.

17. Find the ordinates of the points where the straight line $y = mx + c$ meets the parabola; hence determine the ordinate of the middle point of the chord which the parabola intercepts on this straight line.

18. A is the origin, B is a point on the axis of y, BQ is a straight line parallel to the axis of x; in AQ, produced if necessary, P is taken such that its ordinate is equal to BQ: shew that the locus of P is a parabola.

19. From any point Q in the straight line BQ which is perpendicular to the axis CAB of a parabola whose vertex is A, PQ is drawn parallel to the axis to meet the curve at P: shew that if CA be taken equal to AB, the straight lines AQ and CP will intersect on the parabola.

20. At the point (x', y') a normal is drawn: find the co-ordinates of the point where the normal meets the curve again, and the length of the intercepted chord.

21. If the normal at any point P meet the curve again at Q, and $SP = r$, and p be the perpendicular from S on the tangent at P, then $PQ = \dfrac{4pr}{r - a}$.

22. P is any point on a parabola, A the vertex; through A is drawn a straight line perpendicular to the tangent at P,

and through P is drawn a straight line parallel to the axis; the straight lines thus drawn meet at a point Q: shew that the locus of Q is a straight line. Find also the equation to the locus of Q' the intersection of the perpendicular from A and the ordinate at P.

23. PQ is a chord of a parabola, PT the tangent at P. A straight line parallel to the axis of the parabola cuts the tangent at T, the arc PQ at E, and the chord PQ at F. Shew that $TE : EF :: PF : FQ$.

24. In a parabola whose equation is $y^2 = 4ax$, pairs of tangents are drawn at points whose abscissæ are in the ratio of $1 : \mu$; shew that the equation to the locus of their intersection will be $y^2 = (\mu^{\frac{1}{4}} + \mu^{-\frac{1}{4}})^2 \, ax$ when the points are on the *same* side of the axis, and $y^2 = -(\mu^{\frac{1}{4}} - \mu^{-\frac{1}{4}})^2 \, ax$ when they are on *different* sides.

25. Two straight lines are drawn from the vertex of a parabola at right angles to each other; the points where these straight lines meet the curve are joined, thus forming a right-angled triangle: find the least area of this triangle.

26. Let r and r' be the lengths of two radii vectores drawn at right angles to each other from the vertex of a parabola: then $(rr')^{\frac{4}{3}} = 16a^2 (r^{\frac{2}{3}} + r'^{\frac{2}{3}})$.

27. Find the polar equation to the parabola referred to the intersection of the directrix and axis as origin and the axis as initial line.

28. If a straight line be drawn from the intersection of the directrix and axis cutting the parabola, the rectangle of the intercepts made by the curve is equal to the rectangle of the parts into which the parallel focal chord is divided by the focus.

29. Find the polar equation to the parabola when the intersection of the directrix and the axis is the origin and the initial line the directrix.

30. A system of parallel chords is drawn in a parabola: find the locus of the point which divides each chord into segments whose product is constant.

31. In a triangle ABC if $\tan A \tan \frac{B}{2} = 2$, and AB be fixed, the locus of C will be a parabola whose vertex is A and focus B.

32. Find the equation to the parabola referred to tangents at the extremities of the latus rectum as axes.

33. Find the equation to the parabola referred to the normal and tangent at L as axes.

34. P is a point on a parabola; x', y' are its co-ordinates: find the equation to the circle described on SP as diameter.

35. Shew that the circle described on SP as diameter *touches* the tangent at the vertex.

36. If the straight line $y = m(x - a)$ meets the parabola at (x', y') and (x'', y''), shew that

$$x' + x'' = 2a + \frac{4a}{m^2}; \quad x'x'' = a^2; \quad y' + y'' = \frac{4a}{m}; \quad y'y'' = -4a^2.$$

37. A circle is described on a focal chord of a parabola as diameter; if m be the tangent of the inclination of this chord to the axis of x, the equation to the circle is

$$x^2 - 2ax\left(1 + \frac{2}{m^2}\right) + y^2 - \frac{4ay}{m} - 3a^2 = 0.$$

38. Any circle described on a focal chord as diameter *touches* the directrix.

39. If the focus of the parabola be the origin, shew that the equation to the tangent at (x', y') is $yy' = 2a(x + x' + 2a)$.

40. If the focus of a parabola be the origin, shew that the equation to a tangent to the parabola is $y = m(x + a) + \frac{a}{m}$.

41. Two parabolas have a common focus and axis, and a tangent to one intersects a tangent to the other at right angles: find the locus of the point of intersection.

42. If a chord of the parabola $y^2=4ax$ be a tangent of the parabola $y^2=8a(x-c)$, shew that the straight line $x=c$ bisects that chord.

43. From any point there cannot be drawn more than three normals to a parabola.

44. In a parabola whose equation is $y^2=4ax$, the ordinates of three points such that the normals pass through the same points are y_1, y_2, y_3; shew that $y_1+y_2+y_3=0$. Shew also that a circle described through these three points passes through the vertex of the parabola.

45. If two of the normals which can be drawn to a parabola through a point are at right angles, the locus of that point is a parabola.

46. If two equal parabolas have the same focus and their axes perpendicular to each other, they enclose a space whose length is $8a$, and breadth is $2a\sqrt{2}$, where $4a$ is the latus rectum of the parabolas.

47. Find the length of the perpendicular from an external point (h, k) on the corresponding chord of contact.

48. From an external point (h, k) two tangents are drawn to a parabola: shew that the length of the chord of contact is $\frac{(k^2+4a^2)^{\frac{1}{2}}(k^2-4ah)^{\frac{1}{2}}}{a}$.

49. From an external point (h, k) two tangents are drawn to a parabola: shew that the area of the triangle formed by the tangents and chord is $\frac{(k^2-4ah)^{\frac{3}{2}}}{2a}$.

50. Tangents to a parabola TP, Tp are drawn at the extremities of a focal chord; PG, pg are normals at the same points. Shew that $\frac{1}{PG^2}+\frac{1}{pg^2}$ is invariable; and that the normals subtend equal angles at T.

51. Two equal parabolas have the same axis, but their vertices do not coincide. If through any point O on the inner

curve two chords of the outer curve POp, QOq, be drawn at right angles to one another, then $\frac{1}{PO \cdot Op} + \frac{1}{QO \cdot Oq}$ is invariable.

52. A circle described upon a chord of a parabola as diameter just touches the axis: shew that if θ be the inclination of the chord to the axis, $4a$ the latus rectum of the parabola, and c the radius of the circle, $\tan\theta = \frac{2a}{c}$.

53. If θ, θ' be the inclinations to the axis of the parabola of the two tangents through (h, k), shew that

$$\tan\theta + \tan\theta' = \frac{k}{h}; \quad \tan\theta \tan\theta' = \frac{a}{h}.$$

54. If two tangents be drawn to a parabola so that the sum of the angles which they make with the axis is constant, the locus of their intersection will be a straight line passing through the focus.

55. Shew that the two tangents through (h, k) are represented by the equation

$$h(y-k)^2 - k(y-k)(x-h) + a(x-h)^2 = 0;$$

or $$(k^2 - 4ah)(y^2 - 4ax) = \{ky - 2a(x+h)\}^2.$$

56. Shew that the straight lines drawn from the vertex to the points of contact of the tangents from (h, k) are represented by the equation $hy^2 = 2x(ky - 2ax)$.

57. Determine the co-ordinates of the point of intersection of two tangents to a parabola

$$y = m_1 x + \frac{a}{m_1} \text{ and } y = m_2 x + \frac{a}{m_2}.$$

Also form the equation to the straight line drawn from this point of intersection perpendicular to a third tangent; and determine the ordinate of the point where this straight line meets the directrix.

58. A triangle is formed by three tangents to a parabola: shew that the perpendiculars from each angle on the opposite side intersect on the directrix.

CHAPTER IX.

THE ELLIPSE.

158. *To find the equation to the ellipse.*

The ellipse is the locus of a point which moves so that its distance from a fixed point bears a constant ratio to its distance from a fixed straight line, the ratio being less than unity.

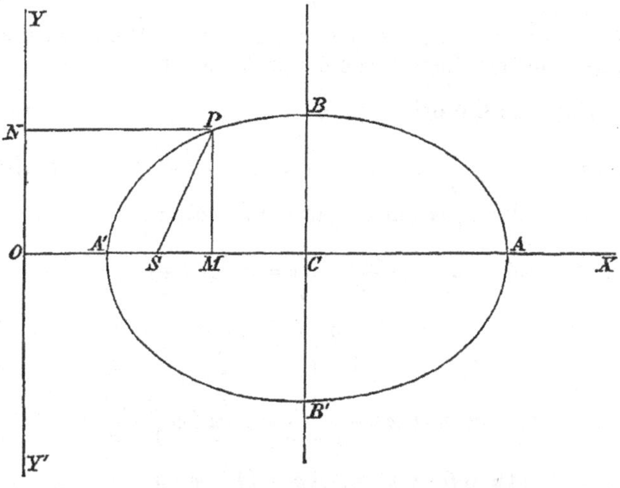

Let S be the fixed point, YY' the fixed straight line. Draw SO perpendicular to YY'; take O as the origin, OS as the direction of the axis of x, OY as that of the axis of y.

Let P be a point on the locus; join SP; draw PM parallel to OY and PN parallel to OX. Let $OS = p$, and let e be the ratio of SP to PN. Let x, y be the co-ordinates of P.

By definition, $SP = e \,.\, PN$; therefore $SP^2 = e^2 PN^2$;

$$\text{therefore } PM^2 + SM^2 = e^2 PN^2,$$

that is, $$y^2 + (x-p)^2 = e^2x^2.$$

This is the equation to the ellipse with the assumed origin and axes.

159. To find where the ellipse meets the axis of x, we put $y=0$ in the equation to the ellipse; thus $(x-p)^2 = e^2x^2$; therefore $x-p=\pm ex$; therefore $x = \frac{p}{1\mp e}$. Let $OA' = \frac{p}{1+e}$ and $OA = \frac{p}{1-e}$; then A and A' are points on the ellipse.

A and A' are called the *vertices* of the ellipse, and C, the point midway between A and A', is called the *centre* of the ellipse.

160. We shall obtain a simpler form of the equation to the ellipse by transferring the origin to A' or C.

I. Suppose the origin at A'.

Since $OA' = \frac{p}{1+e}$, we put $x = x' + \frac{p}{1+e}$ and substitute this value in the equation $y^2 + (x-p)^2 = e^2x^2$;

thus $$y^2 + \left(x' + \frac{p}{1+e} - p\right)^2 = e^2\left(x' + \frac{p}{1+e}\right)^2;$$

or $$y^2 + \left(x' - \frac{ep}{1+e}\right)^2 = e^2\left(x' + \frac{p}{1+e}\right)^2;$$

$$\text{therefore } y^2 + x'^2 - \frac{2x'ep}{1+e} = e^2\left(x'^2 + \frac{2px'}{1+e}\right);$$

$$\text{therefore } y^2 = 2pex' - (1-e^2)\,x'^2$$

$$= (1-e^2)\left(\frac{2pex'}{1-e^2} - x'^2\right).$$

The distance $A'A = \frac{p}{1-e} - \frac{p}{1+e} = \frac{2ep}{1-e^2}$; we shall denote this by $2a$; hence the equation becomes

$$y^2 = (1-e^2)(2ax' - x'^2).$$

We may suppress the accent if we remember that the origin is at the vertex A', and thus write the equation

$$y^2 = (1-e^2)(2ax - x^2) \text{..................(1)}.$$

II. Suppose the origin at C.

Since $A'C = a$, we put $x = x' + a$ and substitute this value in (1); thus $y^2 = (1-e^2)\{2a(x'+a) - (x'+a)^2\}$

$$= (1-e^2)(a^2 - x'^2).$$

We may suppress the accent if we remember that the origin is now at the centre C, and thus write the equation

$$y^2 = (1-e^2)(a^2 - x^2) \text{..................(2)}.$$

In (2) suppose $x = 0$, then $y^2 = (1-e^2)a^2$; if then we denote the ordinate CB by b we have $b^2 = (1-e^2)a^2$; thus (1) may be written

$$y^2 = \frac{b^2}{a^2}(2ax - x^2) \text{..................(3)},$$

and (2) may be written

$$y^2 = \frac{b^2}{a^2}(a^2 - x^2) \text{..................(4)},$$

or, more symmetrically,

$$\frac{x^2}{a^2} + \frac{y^2}{b^2} = 1, \text{ or } a^2y^2 + b^2x^2 = a^2b^2 \text{..........(5)}.$$

161. Since $A'S = e\,OA'$ and $OA' = \dfrac{p}{1+e}$, we have

$$A'S = \frac{ep}{1+e} = \frac{(1-e)\,ep}{1-e^2} = a(1-e),$$

$$OA' = \frac{p}{1+e} = \frac{a(1-e)}{e},$$

$$SC = A'C - A'S = a - a(1-e) = ae,$$

$$OC = A'C + OA' = a + \frac{a(1-e)}{e} = \frac{a}{e},$$

$$OS = p = \frac{a(1-e^2)}{e}.$$

162. We may now ascertain the form of the ellipse. Take the equation referred to the centre as origin

$$y^2 = \frac{b^2}{a^2}(a^2 - x^2) \text{.....................(1).}$$

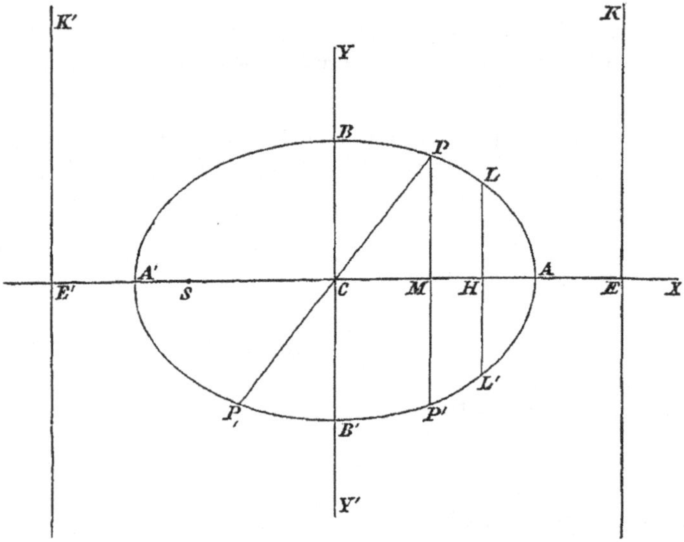

For every value of x less than a there are two values of y, equal in magnitude but of opposite sign. Hence if P be a point in the curve on one side of the axis of x there is a point P' on the other side of the axis such that $P'M = PM$. Thus the curve is symmetrical with respect to the axis of x. Values of x greater than a do not give possible values of y; hence, CA being equal to a, the curve does not extend to the right of A.

If we ascribe to x any negative value comprised between 0 and $-a$, we obtain for y the same pair of values as when we ascribe to x the corresponding positive value between 0 and a. Hence the portion of the curve to the left of YY' is similar to the portion to the right of YY'.

As the equation (1) may be put in the form

$$x^2 = \frac{a^2}{b^2}(b^2 - y^2) \text{..........................(2),}$$

we see that the axis of y also divides the curve symmetrically and that the curve does not extend beyond the points B and B', where CB and CB' each $= b$.

The straight line $E'K'$ is the directrix; S is the corresponding focus.

Since the curve is symmetrical with respect to the straight line YCY', it follows that if we take $CH = CS$ and $CE = CE'$, and draw EK at right angles to CE, the point H and the straight line EK will form respectively a second focus and directrix by means of which the curve might have been generated.

163. The point C is called the *centre* of the ellipse because *every chord of the ellipse which passes through C is bisected at C.* For suppose (h, k) to be a point on the curve, so that the equation $\frac{x^2}{a^2} + \frac{y^2}{b^2} = 1$ is satisfied by the values $x = h$, $y = k$; then $(-h, -k)$ is also a point on the curve, because since $x = h$, $y = k$, satisfy the above equation, it is obvious that $x = -h$, $y = -k$, will also satisfy it. Hence to every point P on the curve there corresponds another point $P_{,}$ in the opposite quadrant, such that $PCP_{,}$ is a straight line and $P_{,}C = PC$. Hence every chord passing through C is bisected at C.

164. We have drawn the curve concave towards the axis of x; the following proposition will justify the figure.

The ordinate of any point of the curve which lies between a vertex and a fixed point of the curve is greater than the corresponding ordinate of the straight line joining that vertex and the fixed point.

Let A' be the vertex, and take it for the origin; let P be the fixed point; x', y' its co-ordinates. Then the equation to the ellipse is (Art. 160) $y^2 = \frac{b^2}{a^2}(2ax - x^2)$.

The equation to $A'P$ is $y = \frac{y'}{x'}x$, or $y = \frac{b}{a}\sqrt{\left(\frac{2a}{x'} - 1\right)}\,x$, since (x', y') is on the ellipse.

Let x denote any abscissa less than x', then since the

ordinate of the curve is $\frac{b}{a}\sqrt{(2ax-x^2)}$ or $\frac{b}{a}\sqrt{\left(\frac{2a}{x}-1\right)}\,x$, and that of the straight line is $\frac{b}{a}\sqrt{\left(\frac{2a}{x'}-1\right)}\,x$, it is obvious that the ordinate of the curve is greater than that of the straight line.

165. AA' and BB' are called *axes* of the ellipse. The axis AA' which contains the two foci is called the *major* axis and sometimes the *transverse* axis; BB' is called the *minor* axis and sometimes the *conjugate* axis.

The ratio which the distance of any point in the ellipse from the focus bears to the distance of the same point from the corresponding directrix is called the *excentricity* of the ellipse. We have denoted it by the symbol e.

To find the *latus rectum* (see Art. 128) we put $x = CH$, that is $= ae$, in equation (1) of Art. 162; thus

$$y^2 = \frac{b^2a^2(1-e^2)}{a^2} = \frac{b^4}{a^2};$$

therefore $LH = \frac{b^2}{a}$, and the latus rectum $= \frac{2b^2}{a}$.

Since $b^2 = a^2 - a^2e^2$; therefore $b^2 + a^2e^2 = a^2$; that is,

$$CB^2 + CH^2 = a^2;$$

$$\text{therefore } BH = a;$$

similarly $$BS = a.$$

166. *To express the focal distances of any point of the ellipse in terms of the abscissa of the point.*

Let S be one focus, $E'K'$ the corresponding directrix; H the other focus, EK the corresponding directrix. Let P be a point on the ellipse; x, y its co-ordinates, the centre being the origin. Join SP, HP, and draw $N'PN$ parallel to the major axis, and PM perpendicular to it.

Then $SP = ePN' = e(E'C + CM) = e\left(\frac{a}{e} + x\right) = a + ex$.

Also $HP = ePN = e(CE - CM) = e\left(\frac{a}{e} - x\right) = a - ex$.

Hence $SP + HP = 2a$; that is, the *sum* of the focal distances of any point on the ellipse is equal to the major axis.

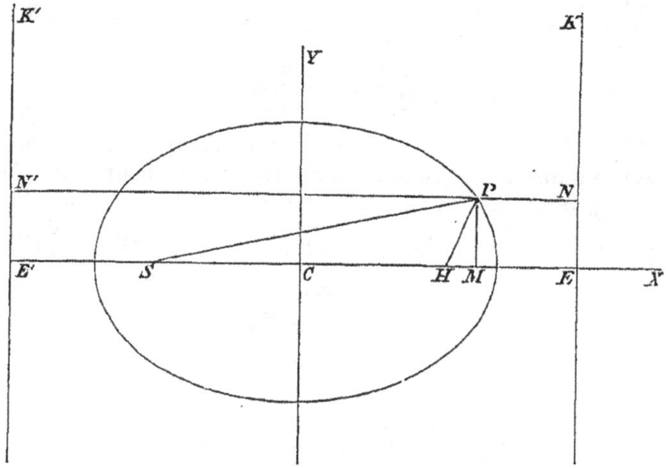

It is obvious from Euclid, I. 21, that the sum of the focal distances of any point *outside* the ellipse is *greater* than the major axis, and the sum of the focal distances of any point *inside* the ellipse is *less* than the major axis.

It is easily seen that $\frac{x^2}{a^2} + \frac{y^2}{b^2} - 1$ is *positive* for any point *outside* the ellipse, and *negative* for any point *inside* the ellipse.

Let the co-ordinates of any point Q be x and y; then

$$HQ^2 = y^2 + (x - ae)^2 = (ex - a)^2 + y^2 + (1 - e^2)(x^2 - a^2)$$

$$= e^2\left(x - \frac{a}{e}\right)^2 + y^2 + \frac{b^2}{a^2}(x^2 - a^2).$$

Thus HQ^2 is greater or less than $e^2\left(x - \frac{a}{e}\right)^2$ according as Q is outside or inside the ellipse; therefore the focal distance of any point not on the curve bears to the distance of the point from the corresponding directrix a ratio which is greater or less than e according as the point is outside or inside the ellipse.

167. The equation $y^2 = \frac{b^2}{a^2}(a^2 - x^2)$ may be written $y^2 = \frac{b^2}{a^2}(a - x)(a + x)$. Hence (see the figure to Art. 162)

$$\frac{PM^2}{A'M \cdot MA} = \frac{BC^2}{AC^2}.$$

168. Let a circle be described on the major axis of the ellipse as a diameter; its equation referred to the centre as origin will be $y^2 = a^2 - x^2$. Hence if any ordinate MP of the ellipse be produced to meet the circle at P' we have $PM^2 = \frac{b^2}{a^2} P'M^2$;

$$\text{therefore } \frac{PM}{P'M} = \frac{b}{a}.$$

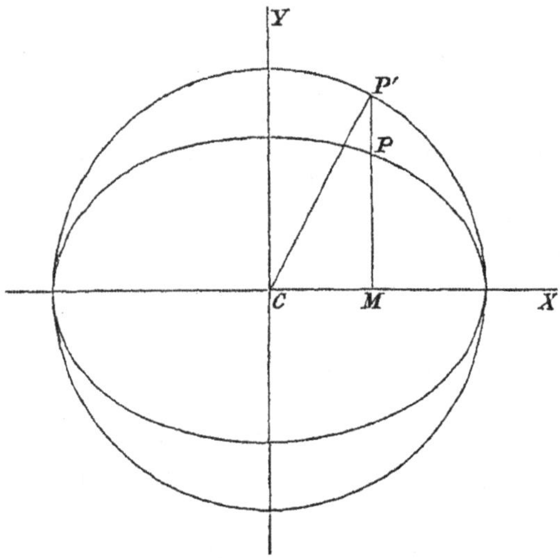

Join P' with C the centre of the ellipse; let $P'CM = \phi$, and let x, y be the co-ordinates of P; then

$$x = CP' \cos\phi = a\cos\phi, \quad y = \frac{b}{a} P'M = \frac{b}{a} a \sin\phi = b\sin\phi.$$

These values of x and y are sometimes useful in the solution of problems.

The angle $P'CM$ is called the *excentric angle* of the point P.

The circle described on the major axis of an ellipse as diameter is sometimes called *the auxiliary circle.*

169. From Art. 160 we see that the equation to the ellipse when the vertex is the origin is $y^2 = 2pex - (1 - e^2) x^2$.

If we suppose $e = 1$, this becomes $y^2 = 2px$, which is the equation to a parabola whose latus rectum is $2p$.

Also in the ellipse $a = \frac{ep}{1 - e^2}$, $b = a\sqrt{(1 - e^2)} = \frac{ep}{\sqrt{(1 - e^2)}}$, AH or $a(1 - e) = \frac{ep}{1 + e}$.

If we now make $e = 1$, we have a and b infinite, and $a(1 - e) = \frac{p}{2}$. Thus if we suppose the distance between the vertex and the nearer focus of an ellipse to remain constant while the excentricity approaches continually nearer to unity, the major and minor axes of the ellipse increase indefinitely and the ellipse about the vertex approximates to the form of a parabola.

Thus if any property is established for an ellipse we may seek for a corresponding property in the parabola by referring the ellipse to the vertex as origin and examining what the result becomes when e is made to approach continually to unity, while the distance between the vertex and the nearer focus remains constant.

Tangent and Normal to an Ellipse.

170. *To find the equation to the tangent at any point of an ellipse.* (See Definition, Art. 90.)

Let x', y' be the co-ordinates of the point, x'', y'' the co-ordinates of an adjacent point on the curve.

The equation to the secant through these points is

$$y - y' = \frac{y'' - y'}{x'' - x'}(x - x') \quad\ldots\ldots\ldots\ldots\ldots\ldots (1);$$

since (x', y') and (x'', y'') are points on the ellipse,

$$a^2y'^2 + b^2x'^2 = a^2b^2, \quad a^2y''^2 + b^2x''^2 = a^2b^2;$$

therefore $$a^2(y''^2 - y'^2) + b^2(x''^2 - x'^2) = 0;$$

therefore $$\frac{y''-y'}{x''-x'} = -\frac{b^2}{a^2}\cdot\frac{x''+x'}{y''+y'}.$$

Hence (1) may be written

$$y - y' = -\frac{b^2}{a^2}\cdot\frac{x''+x'}{y''+y'}(x-x').$$

Now in the limit $x''=x'$, and $y''=y'$; hence the equation to the tangent at the point (x', y') is

$$y - y' = -\frac{b^2x'}{a^2y'}(x-x')\dots\dots\dots\dots\dots\dots(2).$$

This equation may be simplified; multiply by a^2y', thus

$$a^2yy' + b^2xx' = a^2y'^2 + b^2x'^2 = a^2b^2.$$

171. The equation to the tangent can be conveniently expressed in terms of the tangent of the angle which the straight line makes with the major axis of the ellipse. For the equation to the tangent at (x', y') is

$$a^2yy' + b^2xx' = a^2b^2, \text{ or } y = -\frac{b^2x'}{a^2y'}x + \frac{b^2}{y'}.$$

Let $-\frac{b^2x'}{a^2y'} = m$; thus the equation becomes $y = mx + \frac{b^2}{y'}$; we have then to express $\frac{b^2}{y'}$ in terms of m.

Now $\quad b^2x' = -a^2y'm$, and $a^2y'^2 + b^2x'^2 = a^2b^2$;

therefore $$a^2y'^2 + \frac{a^4m^2y'^2}{b^2} = a^2b^2;$$

therefore $$y'^2(a^2m^2 + b^2) = b^4,$$

therefore $$\frac{b^2}{y'} = \sqrt{(a^2m^2 + b^2)}.$$

Hence the equation to the tangent may be written

$$y = mx + \sqrt{(a^2m^2 + b^2)}.$$

Conversely every straight line whose equation is of this form is a tangent to the ellipse.

It may be shewn as in Arts. 93, 94, that the tangent at any point of an ellipse meets it at only *one* point, and that a straight line which meets an ellipse at only one point is the tangent at that point.

172. The tangents at the extremities of either axis are parallel to the other axis.

For the co-ordinates of A are a, 0. (See the figure to Art. 162.) Hence, putting $x'=a$, and $y'=0$, the equation $a^2yy'+b^2xx'=a^2b^2$ becomes $x=a$, which is the equation to a straight line through A parallel to CY. Similarly the tangent at A' is parallel to CY, and the tangents at B and B' are parallel to CX.

173. *To find the equation to the normal at any point of an ellipse.* (See Definition, Art. 97.)

Let x', y' be the co-ordinates of the point; the equation to the tangent at that point is

$$y=-\frac{b^2x'}{a^2y'}x+\frac{b^2}{y'} \text{ ....(1).}$$

The equation to the straight line through (x', y') at right angles to (1) is

$$y-y'=\frac{a^2y'}{b^2x'}(x-x') \text{ ....(2).}$$

This is the equation to the normal at (x', y').

174. The equation to the normal may also be expressed in terms of the tangent of the angle which the straight line makes with the major axis of the ellipse. The equation to the normal at (x', y') is $y=\frac{a^2y'}{b^2x'}x-\left(\frac{a^2}{b^2}-1\right)y'$. Let $\frac{a^2y'}{b^2x'}=m$; thus the equation becomes

$$y=mx-\frac{a^2-b^2}{b^2}y' \text{ ....(1);}$$

we have then to express $\frac{a^2-b^2}{b^2}y'$ in terms of m.

Now, $b^2x'=\frac{a^2y'}{m}$, and $a^2y'^2+b^2x'^2=a^2b^2$;

$$\text{therefore } a^2y'^2+\frac{a^4y'^2}{b^2m^2}=a^2b^2;$$

therefore $y'^2(b^2m^2 + a^2) = b^4m^2$. Hence (1) becomes

$$y = mx - \frac{(a^2 - b^2)m}{\sqrt{(b^2m^2 + a^2)}} \quad \text{..................(2).}$$

175. We shall now deduce some properties of the ellipse from the preceding Articles.

Let x', y' be the co-ordinates of P; let PT be the tangent at P, and PG the normal at P; PM, PN perpendiculars on the axes.

The equation to the tangent at P is $a^2yy' + b^2xx' = a^2b^2$.

Let $y = 0$, then $x = \frac{a^2}{x'}$, hence $CT = \frac{CA^2}{CM}$; therefore

$$CM \, . \, CT = CA^2.$$

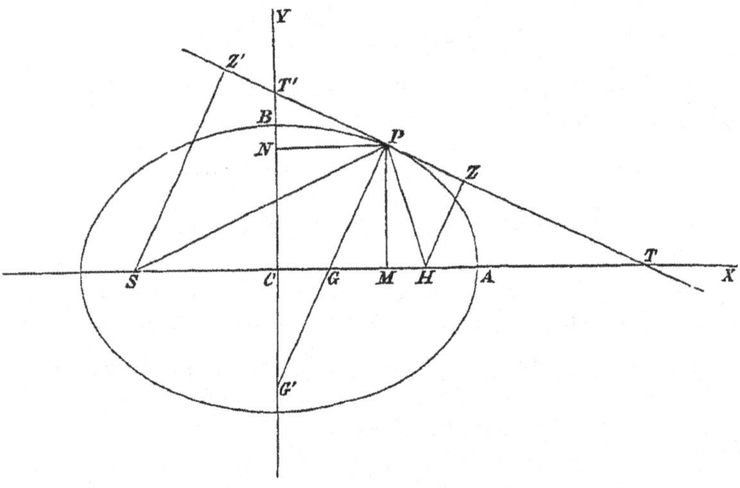

Similarly, if the tangent at P meet CY at T',

$$CN \, . \, CT' = CB^2.$$

176. The equation to the normal at P is

$$y - y' = \frac{a^2y'}{b^2x'}(x - x').$$

At the point G where the normal cuts the major axis,

$y = 0$, hence from the above equation $x - x' = -\dfrac{b^2x'}{a^2}$; therefore $x = x'\left(1 - \dfrac{b^2}{a^2}\right) = e^2x'$. Thus $CG = e^2 CM$.

At the point G' where the normal cuts the minor axis, $x = 0$, hence from the above equation $y = y' - \dfrac{a^2y'}{b^2} = -\dfrac{a^2e^2}{b^2}y'$. Thus $CG' = \dfrac{a^2e^2}{b^2}PM$.

Suppose the focal distance PH produced to meet the ellipse again at p. Let Q denote the middle point of Pp, and through Q draw a straight line parallel to the major axis meeting the normal PG at K. Then, by similar triangles,

$$\frac{QK}{QP} = \frac{HG}{HP} = \frac{ae - e^2x}{a - ex} = e;$$

$$\text{thus } QK = e\,.\,QP = \frac{e}{2}\,.\,Pp.$$

If K had denoted the point of intersection of the straight line through Q and the normal at p, we should have obtained the same value of QK; hence we have the following result: *the straight line parallel to the major axis which passes through the intersection of normals at the ends of a focal chord bisects that chord.*

177. The lengths of PG and PG' may be conveniently expressed in terms of the focal distances of P.

$$PG^2 = PM^2 + GM^2 = y'^2 + (x' - e^2x')^2$$

$$= y'^2 + x'^2(1 - e^2)^2 = y'^2 + \frac{b^4x'^2}{a^4} = \frac{b^2}{a^2}(a^2 - x'^2) + \frac{b^4}{a^4}x'^2$$

$$= \frac{b^2}{a^2}\left\{a^2 - \left(1 - \frac{b^2}{a^2}\right)x'^2\right\} = \frac{b^2}{a^2}(a^2 - e^2x'^2).$$

Let $SP = r'$, $HP = r$; then $r' = a + ex'$, $r = a - ex'$;

thus $$PG^2 = \frac{b^2rr'}{a^2}.$$

Similarly, it may be shewn that $PG'^2 = \dfrac{a^2rr'}{b^2}$.

178. *The normal at any point bisects the angle between the focal distances of that point.*

Let x', y' be the co-ordinates of P; the co-ordinates of S are $-ae$, 0; hence the equation to SP is (Art. 35)

$$y = \frac{y'}{x' + ae}(x + ae).$$

The equation to the normal at P is $y - y' = \frac{a^2 y'}{b^2 x'}(x - x')$.

Hence the tangent of the angle GPS

$$= \frac{\frac{a^2 y'}{b^2 x'} - \frac{y'}{x' + ae}}{1 + \frac{a^2 y'^2}{b^2 x'(x' + ae)}} = \frac{(a^2 - b^2)\, x' y' + a^3 e y'}{a^2 y'^2 + b^2 x'^2 + b^2 x' ae}$$

$$= \frac{a^2 e^2 x' y' + a^3 e y'}{a^2 b^2 + b^2 x' ae} = \frac{eay'}{b^2}.$$

The equation to HP is $y = \frac{y'}{x' - ae}(x - ae)$; hence it may be shewn that the tangent of the angle GPH also $= \frac{eay'}{b^2}$; therefore the angle $SPG =$ the angle HPG.

Hence the angle $SPT' =$ the angle HPT; that is, the tangent at any point is equally inclined to the focal distances of that point.

179. The preceding proposition may also be established thus:

$$CG = e^2 x', \text{ (Art. 176)};$$

therefore $SG = ae + e^2 x'$, and $HG = ae - e^2 x'$.

Also $SP = a + ex'$, $HP = a - ex'$; hence

$$\frac{SG}{HG} = \frac{SP}{HP};$$

therefore by Euclid, VI. 3, PG bisects the angle SPH.

180. *To find the locus of the intersection of the tangent at any point with the perpendicular on it from the focus.*

Let $$y = mx + \sqrt{(b^2 + m^2a^2)} \quad \text{.................... (1)}$$

be the equation to a tangent to the ellipse (Art. 171); then the equation to the perpendicular on it from the focus H is (see the figure to Art. 175)

$$y = -\frac{1}{m}(x - ae) \quad \text{.................... (2).}$$

If we suppose x and y to have respectively the same values in (1) and (2), and eliminate m between the two equations, we shall obtain the required locus.

From (1) $y - mx = \sqrt{(b^2 + m^2a^2)}$; from (2) $my + x = ae$; square and add, then

$$(y^2 + x^2)(1 + m^2) = b^2 + m^2a^2 + a^2e^2 = a^2(1 + m^2);$$

thus $y^2 + x^2 = a^2$ is the equation to the required locus, which is therefore a circle described on the major axis of the ellipse as diameter.

We have supposed the perpendicular drawn from H; we shall arrive at the same result if it be drawn from S; hence if HZ, SZ' be these perpendiculars, CZ and CZ' each $= a$.

181. *To find the length of the perpendicular from the focus on the tangent at any point.*

The equation to the tangent at the point (x', y') is

$$y = -\frac{b^2x'}{a^2y'}x + \frac{b^2}{y'}.$$

The co-ordinates of the focus H are ae, 0. But if p denote the length of the perpendicular from a point (x_1, y_1) on the straight line $y = mx + c$, then by Art. 47

$$p^2 = \frac{(y_1 - mx_1 - c)^2}{1 + m^2}.$$

In the present case $x_1 = ae$, $y_1 = 0$, $m = -\frac{b^2x'}{a^2y'}$, $c = \frac{b^2}{y'}$;

thus $$p^2 = \frac{\left(\frac{b^2x'ae}{a^2y'} - \frac{b^2}{y'}\right)^2}{1 + \frac{b^4x'^2}{a^4y'^2}} = \frac{a^2b^4(a - ex')^2}{a^4y'^2 + b^4x'^2} = \frac{a^2b^4(a - ex')^2}{a^2(a^2b^2 - b^2x'^2) + b^4x'^2}$$

$$=\frac{a^2b^2(a-ex')^2}{a^2(a^2-e^2x'^2)}=\frac{b^2(a-ex')}{a+ex'}=\frac{b^2r}{r'}, \quad \text{(Art. 177).}$$

Since $r'=2a-r$ we have $p^2=\dfrac{b^2r}{2a-r}$.

Similarly if p' be the perpendicular from S on the tangent at (x', y') we shall find $p'^2=\dfrac{b^2r'}{r}$. Therefore $pp'=b^2$.

From the point G in the figure of Art. 175 suppose a perpendicular drawn on PH meeting it at R; then by similar triangles $\dfrac{PR}{PG}=\dfrac{HZ}{HP}$; therefore $PR=PG\times\dfrac{HZ}{HP}$. Substitute the value of HZ just obtained, and the value of PG from Art. 177, and we have $PR=\dfrac{b^2}{a}$. Thus the length PR is constant and equal to half the latus rectum.

182. *From any external point two tangents can be drawn to an ellipse.*

Let the equation to the ellipse be $a^2y^2+b^2x^2=a^2b^2$, and let h, k be the co-ordinates of an external point. Suppose x', y' the co-ordinates of a point on the ellipse, such that the tangent at this point passes through (h, k). The equation to the tangent at (x', y') is $a^2yy'+b^2xx'=a^2b^2$. Since this tangent passes through (h, k)

$$a^2ky'+b^2hx'=a^2b^2 \quad\text{..................(1).}$$

Also since (x', y') is on the ellipse

$$a^2y'^2+b^2x'^2=a^2b^2 \quad\text{....................(2).}$$

Equations (1) and (2) determine the values of x' and y'.

Substitute from (1) in (2), thus $\left(\dfrac{a^2b^2-b^2hx'}{ak}\right)^2+b^2x'^2=a^2b^2$,

or $x'^2(a^2k^2+b^2h^2)-2a^2b^2hx'+a^4(b^2-k^2)=0$. The roots of this quadratic will be found to be both possible since (h, k) is an *external* point and therefore $a^2k^2+b^2h^2$ greater than a^2b^2.

The straight line which passes through the points where these tangents meet the ellipse is called the *chord of contact.*

183. *Tangents are drawn to an ellipse from a given external point: to find the equation to the chord of contact.*

Let h, k be the co-ordinates of the external point; x_1, y_1 the co-ordinates of the point where one of the tangents from (h, k) meets the ellipse; x_2, y_2 the co-ordinates of the point where the other tangent from (h, k) meets the ellipse. The equation to the tangent at (x_1, y_1) is $a^2yy_1 + b^2xx_1 = a^2b^2$; since this tangent passes through (h, k) we have

$$a^2ky_1 + b^2hx_1 = a^2b^2 \quad\text{..................(1)}.$$

Similarly, since the tangent at (x_2, y_2) passes through (h, k)

$$a^2ky_2 + b^2hx_2 = a^2b^2 \quad\text{..................(2)}.$$

Hence it follows that the equation to the *chord of contact* is

$$a^2ky + b^2hx = a^2b^2 \quad\text{..................(3)}.$$

For (3) is obviously the equation to *some straight line;* also this straight line passes through (x_1, y_1) for (3) is satisfied by the values $x = x_1$, $y = y_1$ as we see from (1); similarly from (2) we conclude that this straight line passes through (x_2, y_2). Hence (3) is the required equation.

Thus we may use the following process to draw tangents to an ellipse from a given external point: draw the straight line which is represented by (3); join the points where it meets the ellipse with the given external point, and the straight lines thus obtained are the required tangents.

184. *Through any fixed point chords are drawn to an ellipse, and tangents to the ellipse are drawn at the extremities of each chord: the locus of the intersection of the tangents is a straight line.*

Let h, k be the co-ordinates of the point through which the chords are drawn; let tangents to the ellipse be drawn at the extremities of one of these chords, and let (x_1, y_1) be the point at which they meet. The equation to the corresponding chord of contact is, $a^2yy_1 + b^2xx_1 = a^2b^2$, by Art. 183. But this chord passes through (h, k); therefore $a^2ky_1 + b^2hx_1 = a^2b^2$. Hence the point (x_1, y_1) lies on the straight line $a^2ky + b^2hx = a^2b^2$; that is, the locus of the intersection of the tangents is a straight line.

We will now prove the converse of this proposition.

185. *If from any point in a straight line a pair of tangents be drawn to an ellipse the chords of contact will all pass through a fixed point.*

Let $$Ax + By + C = 0 \text{ ....(1)}$$
be the equation to the straight line; let (x', y') be a point in this straight line from which tangents are drawn to the ellipse; then the equation to the corresponding chord of contact is
$$a^2yy' + b^2xx' = a^2b^2 \text{ ....(2).}$$

Since (x', y') is on (1), we have $Ax' + By + C = 0$; therefore (2) may be written $b^2xx' - \frac{Ax' + C}{B} a^2y = a^2b^2$,

or, $$\left(b^2x - \frac{Aa^2y}{B}\right) x' - \frac{Ca^2y}{B} - a^2b^2 = 0.$$

Now, whatever be the value of x', this straight line passes through the point whose co-ordinates are found by the simultaneous equations $b^2x - \frac{Aa^2y}{B} = 0$, $\frac{Ca^2y}{B} + a^2b^2 = 0$, that is, the point for which $y = -\frac{Bb^2}{C}$, $x = -\frac{Aa^2}{C}$.

The student should observe the different interpretations that can be assigned to the equation $a^2ky + b^2hx = a^2b^2$.

The statements in Art. 103 with respect to the circle may all be applied to the ellipse.

186. Some interesting geometrical investigations relating to tangents to an ellipse from an external point may be noticed.

To draw the two tangents to an ellipse from any external point.

Let O denote the external point, and S either focus. On OS as diameter describe a circle and let it cut the circle described on the major axis as diameter at Z and z. Join OZ and Oz. Then these straight lines, produced if necessary, are the tangents from O by Art. 180 and Euclid, III. 31.

Or we may proceed thus. Let O denote the external point, S the more remote focus. With S as centre and radius equal to the major axis of the ellipse describe a circle. Let H be the other focus. With O as centre and radius equal to OH describe another circle cutting the former at Q and q. Join

SQ and Sq cutting the ellipse at P and p; then OP and Op are the required tangents. For join OS, OH, OP, and OQ.

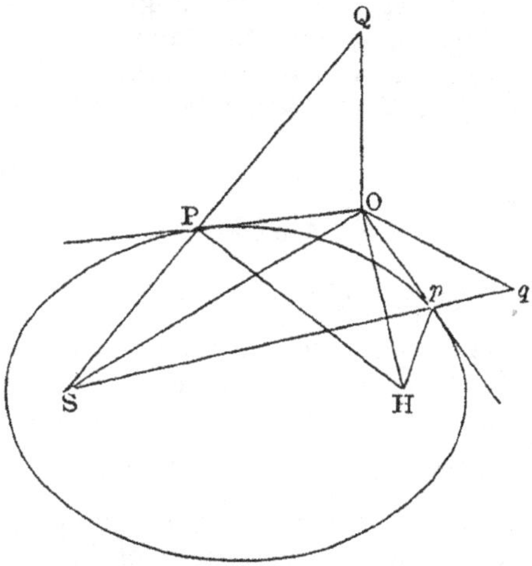

Then in the triangles OPQ and OPH we have $OQ = OH$ by construction, $PQ = 2a - SP = PH$, and OP common. Therefore the angle $OPQ =$ the angle OPH; and OP is the tangent at P by Art. 178.

Similarly Op is the tangent at p.

The two tangents to an ellipse from an external point subtend equal angles at each focus.

Join Hp and Oq. The triangles OSQ and OSq are equal in all respects; thus the tangents OP and Op subtend equal angles at S. Also the angle $OHP =$ the angle OQP, and the angle $OHp =$ the angle Oqp: thus the tangents OP and Op subtend equal angles at H.

The angle between a tangent and a focal distance of the external point is equal to the angle between the other tangent and the other focal distance.

The angle $SOQ =$ the angle SOq; that is,

$$\text{twice the angle } SOP + \text{the angle } SOH$$
$$= \text{twice the angle } HOp + \text{the angle } SOH;$$

therefore the angle $SOP =$ the angle HOp, and also the angle $HOP =$ the angle SOp.

The student should notice the extension which is thus obtained of the result in Art. 178. At any point of the curve the straight line which bisects the angle between the focal distances is at *right angles* to the tangent; at any external point the straight line which bisects the angle between the focal distances *bisects the angle* between the two tangents.

EXAMPLES.

1. Find the excentricity of the ellipse $2x^2 + 3y^2 = c^2$.

2. Find the equation to the tangent at the end of the latus rectum L. (See the figure to Art. 162.) Also find the lengths of the intercepts of this tangent on the axes.

3. Write down the equation to the normal at L.

4. If the normal at L passes through the extremity of the minor axis B', find the excentricity of the ellipse.

5. Find the equation to $A'B$ and to CL. (See the figure to Art. 162.) Find the excentricity of the ellipse if these straight lines are parallel.

6. Find the equation to $B'H$, and determine the abscissa of the point where this straight line cuts the ellipse again.

7. Find the equation to AL, and determine the angle between this straight line and the tangent at L.

8. If from the point P whose abscissa is x', a straight line be drawn through H, determine the abscissa of the point where it meets the ellipse again.

9. Find a point in the ellipse such that the tangent there is equally inclined to the axes.

10. Find a point in the ellipse such that the intercepts made by the tangent on the co-ordinate axes are proportional to the corresponding axes of the ellipse.

11. P is a point on an ellipse, y its ordinate: shew that

$$\tan APA' = -\frac{2b^2}{ae^2y}.$$

12. P is a point on an ellipse, y its ordinate: shew that the tangent of the angle between the focal distance and the tangent at P is $\frac{b^2}{aey}$.

13. If ϕ denote the angle mentioned in the preceding Example, shew that $PC = \sqrt{(a^2 - b^2 \cot^2\phi)}$.

14. From P a point on an ellipse straight lines are drawn to A, A', the extremities of the major axis, and from A, A' straight lines are drawn at right angles to AP, $A'P$: shew that the locus of their intersection will be another ellipse, and find its axes.

15. If any ordinate MP be produced to meet the tangent at L at Q, prove that $QM = PH$. (See the figure to Art. 162.)

16. If a series of ellipses be described having the same major axes the tangents at the ends of their latera recta will pass through one or other of two fixed points.

17. If the focus of an ellipse be the common focus of two parabolas whose vertices are at the ends of the major axis, these parabolas will intersect at right angles, at points whose distance from each other is equal to twice the minor axis.

18. Shew that the length of the longer normal drawn from a point in the minor axis of an ellipse at a distance c from the centre and intercepted between that point and the curve is $\left(a^2 + \frac{c^2}{e^2}\right)^{\frac{1}{2}}$.

19. If any parallel straight lines be drawn from the focus H and the extremity A of the major axis of an ellipse, and if M and N be the points where they meet the minor axis, or the minor axis produced, then the circle whose centre is M and radius NA will either *touch* the ellipse, or fall entirely *outside* of it.

20. A and A' are the extremities of the major axis of an ellipse, T is the point where the tangent at the point P of the curve meets AA' produced; through T a straight line is drawn at right angles to AA' and meeting AP and AP' produced at Q and R respectively: shew that $QT = RT$.

21. If ϕ, ϕ' be the excentric angles of two points, the equation to the chord joining the points is

$$\frac{x}{a}\cos\frac{\phi+\phi'}{2}+\frac{y}{b}\sin\frac{\phi+\phi'}{2}=\cos\frac{\phi-\phi'}{2}.$$

22. Express the equation to the tangent at any point in terms of the excentric angle of that point.

23. Shew that the equation to the normal at the point whose excentric angle is ϕ is $ax\sec\phi - by\operatorname{cosec}\phi = a^2 - b^2$.

24. The locus of the middle point of PG (see Art. 176) is an ellipse of which the excentricity e' is connected with that of the given ellipse by the equation $1-e^2=(1+e^2)^2(1-e'^2)$.

25. Determine the point of intersection of the tangent at L with the straight line HB; find the value of the excentricity of the ellipse when these straight lines are parallel.

26. A tangent at any point P of an ellipse meets the directrix EK at T and $E'K'$ at T': shew that TE varies as the cotangent of PHS, and $T'E'$ varies as the cotangent of PSH. (See the figure to Art. 162.)

27. If the straight line $y=mx+c$ intersect the ellipse $a^2y^2+b^2x^2=a^2b^2$, shew that the length of the chord will be

$$\frac{2ab\sqrt{\{(1+m^2)(m^2a^2+b^2-c^2)\}}}{m^2a^2+b^2}.$$

Hence find the relation between the constants that this straight line may be a tangent to the ellipse.

28. Find the equation to the circle described on HP as diameter, supposing x', y' the co-ordinates of P.

29. Shew that any circle described on HP as diameter, touches the circle described on the major axis as diameter.

30. From a point (h, k) two tangents are drawn to an ellipse: find the sum of the perpendiculars from the foci on the chord of contact.

31. Any ordinate PM of an ellipse is produced to meet the circle on the major axis at Q, and normals to the ellipse and circle at P and Q respectively meet at R: find the locus of R.

32. Two ellipses have a common centre and their axes coincide in direction; also the sum of the squares of the axes is the same in the two ellipses: find the equation to a common tangent.

33. If θ, θ' be the inclinations to the major axis of the ellipse of the two tangents that can be drawn from the point (h, k), shew that

$$\tan\theta + \tan\theta' = -\frac{2hk}{a^2 - h^2}, \qquad \tan\theta\tan\theta' = \frac{b^2 - k^2}{a^2 - h^2}.$$

34. Find the locus of a point such that the two tangents from it to an ellipse are at right angles.

35. Shew that the two tangents which can be drawn to an ellipse through the point (h, k) are represented by

$$(a^2 - h^2)(y - k)^2 + 2(y - k)(x - h)hk + (b^2 - k^2)(x - h)^2 = 0,$$

or by

$$(a^2k^2 + b^2h^2 - a^2b^2)(a^2y^2 + b^2x^2 - a^2b^2) = (a^2ky + b^2hx - a^2b^2)^2.$$

36. Tangents are drawn to an ellipse from the point (h, k): shew that the straight lines drawn from the origin to the points of contact are represented by $\frac{x^2}{a^2} + \frac{y^2}{b^2} = \left(\frac{xh}{a^2} + \frac{ky}{b^2}\right)^2$.

37. Pairs of radii vectores are drawn at right angles to each other from the centre of an ellipse: shew that the tangents at their extremities intersect on the ellipse

$$\frac{x^2}{a^4} + \frac{y^2}{b^4} = \frac{1}{a^2} + \frac{1}{b^2}.$$

38. From an external point T whose co-ordinates are h and k a straight line is drawn to the centre C meeting the ellipse at R: shew that

$$\frac{CT^2}{CR^2} = \frac{a^2k^2 + b^2h^2}{a^2b^2}.$$

39. From an external point (h, k) tangents are drawn: if x_1, x_2 be the abscissæ of the points of contact, shew that

$$x_1 + x_2 = \frac{2ha^2b^2}{a^2k^2 + b^2h^2}, \quad x_1x_2 = \frac{a^4(b^2 - k^2)}{a^2k^2 + b^2h^2}.$$

40. From an external point (h, k) tangents are drawn meeting the ellipse at P and Q: find the value of $HP \,.\, HQ$, H being a focus.

41. From an external point T the straight lines TP, TQ are drawn to touch the ellipse at P and Q. CT cuts the ellipse at R, and RN is drawn parallel to HT to meet the major axis at N: shew that $HP \cdot HQ = RN^2$.

42. Two ellipses of equal excentricity and whose major axes are parallel can only have two points in common: prove this, and shew that if three such ellipses intersect, two and two, at the points P and P', Q and Q', R and R', respectively, the straight lines PP', QQ', RR', meet at a point.

43. Two concentric ellipses which have their axes in the same direction intersect, and four common tangents are drawn so as to form a rhombus, and the points of intersection of the ellipses are joined so as to form a rectangle: prove that the product of the areas of the rhombus and rectangle is equal to half the continued product of the four axes.

44. The ordinate at any point P of an ellipse is produced to meet the circle described on the major axis as diameter at Q: prove that the perpendicular from the focus S on the tangent at Q is equal to SP.

45. Find the equation to the ellipse referred to axes passing through the extremities of the minor axis, and meeting at one extremity of the major axis.

46. If from points of the curve $\frac{a^6}{x^2} + \frac{b^6}{y^2} = (a^2 - b^2)^2$, tangents be drawn to the ellipse $\frac{x^2}{a^2} + \frac{y^2}{b^2} = 1$, the chords of contact will be normal to the ellipse.

47. Prove the proposition in Art. 180 in a manner similar to that used in Art. 138. Also prove the proposition in Art. 138 in a manner similar to that used in Art. 180.

48. Find the equation to the ellipse the origin being the point (h, k) on the ellipse and the axes parallel to the axes of the ellipse.

49. From a point P on an ellipse two chords PQ, PQ' are drawn meeting the ellipse at Q, Q'; if h, k be the co-ordinates of P referred to the centre, and $mx + ny = 1$ the equation to QQ' referred to P as origin, and axes parallel to the axes of the ellipse, shew that with P as origin the straight lines

PQ, PQ' are represented by

$$\frac{x^2}{a^2}+\frac{y^2}{b^2}+\left(\frac{2xh}{a^2}+\frac{2yk}{b^2}\right)(mx+ny)=0.$$

50. Let P be any point on an ellipse; draw PP' parallel to the major axis and cutting the curve at P'; through P draw two chords PQ, PQ', making equal angles with the major axis; join QQ': shew that QQ' is parallel to the tangent at P'.

51. From the equation $y=mx+\sqrt{(m^2a^2+b^2)}$ deduce the equation to the tangent to the parabola.

52. In the figure of Art. 175 suppose GP produced to a point Q such that $GQ=n\,.\,GP$, and find the locus of Q.

53. If PN be any ordinate of a circle, and from the extremity A of the corresponding diameter AB, AQ be drawn meeting PN at Q, so that $AQ=PN$, find the locus of Q and the position of its focus.

54. Express the tangent of the angle between CP and the normal at P in terms of the co-ordinates of P.

55. Find the greatest value of the tangent of the angle between CP and the normal at P.

56. The major axis of an ellipse is equal to twice the minor axis; a straight line of length equal to half the major axis is placed across the major axis with one end on the curve and the other on the minor axis: shew that the middle point of the straight line is on the major axis.

57. A circle is inscribed in the triangle formed by two focal distances and the major axis of an ellipse: find the locus of the centre.

58. If SZ', HZ be perpendiculars on the tangent at the point P of an ellipse, SZ and HZ' will intersect on the normal at P.

59. Shew that the equation to the two straight lines which join the point (h, k) with a focus of the ellipse is

$$(hy-kx)^2-a^2e^2(y-k)^2=0.$$

60. Shew that the straight lines in Examples 35 and 59 have the same bisectors of their angles.

CHAPTER X.

THE ELLIPSE CONTINUED.

Diameters.

187. *To find the length of a straight line drawn from any point in a given direction to meet an ellipse.*

Let x', y' be the co-ordinates of the point from which the straight line is drawn; x, y the co-ordinates of the point to which the straight line is drawn; θ the inclination of the straight line to the axis of x; r the length of the straight line; then (Art. 27)

$$x = x' + r\cos\theta, \qquad y = y' + r\sin\theta.$$

If (x, y) be on the ellipse these values may be substituted in the equation $a^2y^2 + b^2x^2 = a^2b^2$; thus

$$a^2(y' + r\sin\theta)^2 + b^2(x' + r\cos\theta)^2 = a^2b^2;$$

therefore $r^2(a^2\sin^2\theta + b^2\cos^2\theta) + 2r(a^2y'\sin\theta + b^2x'\cos\theta)$

$$+ a^2y'^2 + b^2x'^2 - a^2b^2 = 0.$$

From this quadratic two values of r can be found which are the lengths of the two straight lines that can be drawn from (x', y') in the given direction to the ellipse.

188. *To find the diameter of a given system of parallel chords in an ellipse.* (See Definition, Art. 148.)

Let θ be the inclination of the chords to the major axis of the ellipse; let x', y' be the co-ordinates of the middle point of any one of the chords; the equation which determines the lengths of the straight lines drawn from (x', y') to the curve is (Art. 187)

$$r^2(a^2\sin^2\theta + b^2\cos^2\theta) + 2r(a^2y'\sin\theta + b^2x'\cos\theta)$$
$$+ a^2y'^2 + b^2x'^2 - a^2b^2 = 0 \ldots\ldots\ldots\ldots\ldots(1).$$

Since (x', y') is the *middle* point of the chord, the values of r furnished by this quadratic must be *equal in magnitude and opposite in sign;* hence the coefficient of r must vanish; thus

$$a^2y' \sin\theta + b^2x' \cos\theta = 0, \quad \text{or } y' = -\frac{b^2}{a^2}\cot\theta \,.\, x' \text{..........} (2).$$

Considering x' and y' as variable, this is the equation to a straight line passing through the origin, that is, through the centre of the ellipse.

Hence every diameter passes through the centre.

Also every straight line passing through the centre is a diameter, that is, bisects some system of parallel chords; for by giving to θ a suitable value the equation (2) may be made to represent *any* straight line passing through the centre.

If θ' be the inclination to the axis of x of the diameter which bisects all the chords inclined at an angle θ we have from (2) $\tan\theta' = -\frac{b^2}{a^2}\cot\theta$; therefore

$$\tan\theta \tan\theta' = -\frac{b^2}{a^2}.$$

189. *If one diameter bisect all chords parallel to a second diameter, the second diameter will bisect all chords parallel to the first.*

Let θ_1 and θ_2 be the respective inclinations of the two diameters to the major axis of the ellipse. Since the first bisects all the chords parallel to the second, we have

$$\tan\theta_2 \tan\theta_1 = -\frac{b^2}{a^2}.$$

And this is also the only condition that must hold in order that the second may bisect the chords parallel to the first.

190. *The tangent at either extremity of any diameter is parallel to the chords which that diameter bisects.*

Let h, k be the co-ordinates of either extremity of a diameter; θ the inclination to the major axis of the ellipse of the chords which the diameter bisects. Then the values $x = h$, $y = k$ must satisfy the equation $a^2y \sin\theta + b^2x \cos\theta = 0$; therefore $\tan\theta = -\frac{b^2h}{a^2k}$. But, by Art. 170, the equation to the tangent at (h, k) is $y - k = -\frac{b^2 h}{a^2 k}(x - h)$. Hence the tangent is parallel to the bisected chords.

191. Definition. Two diameters are called *conjugate* when each bisects the chords parallel to the other.

From Art. 190 it follows that each of the conjugate diameters is parallel to the tangent at either extremity of the other.

192. *Given the co-ordinates of one extremity of a diameter to find those of either extremity of the conjugate diameter.*

Let ACA', BCB' be the axes of an ellipse; PCP', DCD' a pair of conjugate diameters.

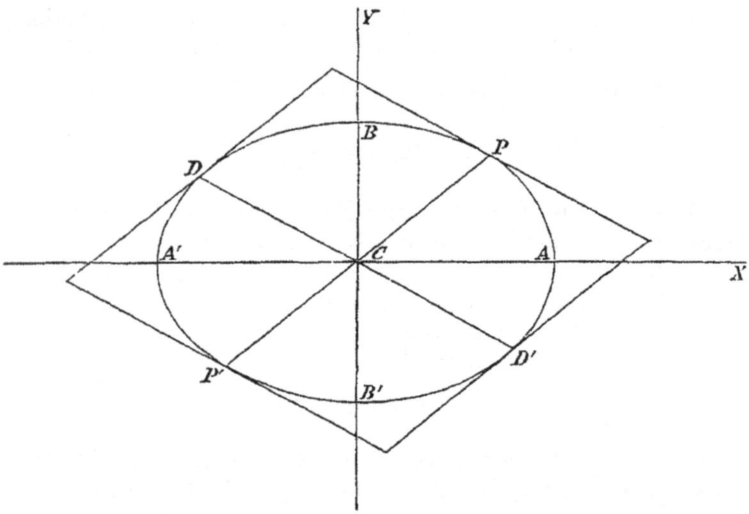

Let x', y' be the given co-ordinates of P; then the equation to CP is

$$y = \frac{y'}{x'}x \;\ldots\ldots\ldots\ldots\ldots\ldots\ldots\ldots (1).$$

Since the conjugate diameter DD' is parallel to the tangent at P, the equation to DD' is

$$y = -\frac{b^2 x'}{a^2 y'}x \;\ldots\ldots\ldots\ldots\ldots\ldots\ldots (2).$$

We must combine (2) with the equation to the ellipse to find the co-ordinates of D and D'. Substitute the value of y from (2) in $a^2y^2 + b^2x^2 = a^2b^2$; then $a^2 \frac{b^4}{a^4} \frac{x'^2}{y'^2} x^2 + b^2x^2 = a^2b^2$; therefore $(b^2x'^2 + a^2y'^2)\, x^2 = a^4y'^2$; therefore $x^2 = \frac{a^4y'^2}{a^2b^2} = \frac{a^2y'^2}{b^2}$; therefore $x = \pm \frac{ay'}{b}$; therefore from (2) $y = \mp \frac{bx'}{a}$.

In the figure the abscissa of D is negative and that of D' positive; hence the upper sign applies to D' and the lower sign to D.

The properties of the ellipse connected with conjugate diameters are numerous and important; we shall now give a few of them.

193. *The sum of the squares of two conjugate semi-diameters is constant.*

Let x', y' be the co-ordinates of P; then by the preceding Article

$$CP^2 + CD^2 = x'^2 + y'^2 + \frac{a^2y'^2}{b^2} + \frac{b^2x'^2}{a^2}$$

$$= \frac{a^2y'^2 + b^2x'^2}{b^2} + \frac{a^2y'^2 + b^2x'^2}{a^2}$$

$$= a^2 + b^2.$$

Thus the sum of the squares of two conjugate semi-diameters is equal to the sum of the squares of the semi-axes.

Moreover

$$CD^2 = a^2 + b^2 - x'^2 - y'^2 = a^2 + b^2 - x'^2 - \frac{b^2}{a^2}(a^2 - x'^2)$$

$$= a^2 - \left(1 - \frac{b^2}{a^2}\right) x'^2 = a^2 - e^2x'^2 = SP\,.\,HP \text{ by Art. 166.}$$

194. *The area of the parallelogram formed by tangents at the ends of conjugate diameters is constant.*

Let PCP', DCD' be the conjugate diameters (see the figure to Art. 192). The area of the parallelogram described so as to touch the ellipse at P, D, P', D', is $4CP\,.\,CD \sin PCD$, or

$4p \, . \, CD$, where p denotes the perpendicular from C on the tangent at P. Let x', y' be the co-ordinates of P; then the equation to the tangent at P is $y = -\frac{b^2 x'}{a^2 y'} x + \frac{b^2}{y'}$.

Hence (Art. 47) $$p = \frac{\frac{b^2}{y'}}{\sqrt{\left(1 + \frac{b^4 x'^2}{a^4 y'^2}\right)}} = \frac{a^2 b^2}{\sqrt{(a^4 y'^2 + b^4 x'^2)}}.$$

And $$CD = \sqrt{\left(\frac{a^2 y'^2}{b^2} + \frac{b^2 x'^2}{a^2}\right)} = \frac{\sqrt{(a^4 y'^2 + b^4 x'^2)}}{ab};$$

therefore $4p \, . \, CD = 4ab$.

Thus the area of any parallelogram which touches the ellipse at the ends of conjugate diameters is equal to the area of the rectangle which touches the ellipse at the ends of the axes.

195. Let a', b' denote the lengths of two conjugate semi-diameters; α the angle between them; by the preceding Article $a'b' \sin \alpha = ab$; therefore

$$\sin^2 \alpha = \frac{a^2 b^2}{a'^2 b'^2} = \frac{4a^2 b^2}{(a'^2 + b'^2)^2 - (a'^2 - b'^2)^2} = \frac{4a^2 b^2}{(a^2 + b^2)^2 - (a'^2 - b'^2)^2}.$$

Hence $\sin^2 \alpha$ has its *least* value when $a' = b'$, and then

$$\sin \alpha = \frac{2ab}{a^2 + b^2}.$$

196. From Art. 194 we have

$$p^2 = \frac{a^2 b^2}{CD^2} = \frac{a^2 b^2}{a^2 + b^2 - CP^2} \quad \text{(Art. 193)}.$$

This gives a relation between p the perpendicular from the centre on the tangent at any point P and the distance CP of that point from the centre.

In Arts. 177 and 193 it is shewn that $PG^2 = \frac{b^2}{a^2} CD^2$. Hence $p \, . \, PG = b^2$. Similarly $p \, . \, PG' = a^2$.

We may also express p in terms of the angle its direction makes with the major axis; for let ψ denote the angle, then the equation to the tangent at (x', y') is $a^2yy' + b^2xx' = a^2b^2$, and this may also be put in the form (Art. 20)

$$x \cos \psi + y \sin \psi = p.$$

Hence $$\frac{p}{\sin \psi} = \frac{b^2}{y'}, \quad \frac{p}{\cos \psi} = \frac{a^2}{x'};$$

therefore $$ay' = \frac{ab^2 \sin \psi}{p}, \quad bx' = \frac{a^2 b \cos \psi}{p};$$

and therefore $$a^2b^2 = \frac{a^2b^2}{p^2}(b^2 \sin^2 \psi + a^2 \cos^2 \psi);$$

therefore $$p^2 = b^2 \sin^2 \psi + a^2 \cos^2 \psi = a^2(1 - e^2 \sin^2 \psi).$$

197. Let ϕ and ϕ' be the excentric angles corresponding to P and D respectively (Art. 168). Then

$$x' = a \cos \phi \ldots\ldots\ldots(1), \qquad y' = b \sin \phi \ldots\ldots\ldots(2),$$

$$-\frac{ay'}{b} = a \cos \phi' \ldots\ldots\ldots(3), \qquad \frac{bx'}{a} = b \sin \phi' \ldots\ldots\ldots(4).$$

From (2) and (3) $\qquad \cos \phi' = -\sin \phi,$

from (1) and (4) $\qquad \sin \phi' = \cos \phi;$

therefore $$\phi' = \frac{\pi}{2} + \phi.$$

198. *To find the equation to the ellipse referred to a pair of conjugate diameters as axes.*

Let CP, CD be two conjugate semi-diameters (see the figure to Art. 192), take CP as the new axis of x, CD as that of y; let $PCA = \alpha$, $DCA = \beta$. Let x, y be the co-ordinates of any point of the ellipse referred to the original axes; x', y' the co-ordinates of the same point referred to the new axes; then (Art. 84)

$$x = x' \cos \alpha + y' \cos \beta, \quad y = x' \sin \alpha + y' \sin \beta.$$

Substitute these values in the equation

$$a^2y^2 + b^2x^2 = a^2b^2;$$

then $$a^2(x' \sin \alpha + y' \sin \beta)^2 + b^2(x' \cos \alpha + y' \cos \beta)^2 = a^2b^2,$$

or $$x'^2(a^2 \sin^2 \alpha + b^2 \cos^2 \alpha) + y'^2(a^2 \sin^2 \beta + b^2 \cos^2 \beta) + 2x'y'(a^2 \sin \alpha \sin \beta + b^2 \cos \alpha \cos \beta) = a^2b^2.$$

But, since CP and CD are conjugate semi-diameters, $\tan\alpha\tan\beta = -\frac{b^2}{a^2}$; hence the coefficient of $x'y'$ vanishes, and the equation becomes

$$x'^2(a^2\sin^2\alpha + b^2\cos^2\alpha) + y'^2(a^2\sin^2\beta + b^2\cos^2\beta) = a^2b^2.$$

In this equation, suppose $x' = 0$, then

$$y'^2 = \frac{a^2b^2}{a^2\sin^2\beta + b^2\cos^2\beta}.$$

This is the value of CD^2, which we shall denote by b'^2; similarly we shall denote CP^2 by a'^2, so that

$$a'^2 = \frac{a^2b^2}{a^2\sin^2\alpha + b^2\cos^2\alpha}.$$

Hence the equation to the ellipse referred to conjugate diameters is

$$\frac{x'^2}{a'^2} + \frac{y'^2}{b'^2} = 1,$$

or, suppressing the accents on the variables,

$$\frac{x^2}{a'^2} + \frac{y^2}{b'^2} = 1.$$

199. A particular case of the preceding is when $a' = b'$; then

$$a^2\sin^2\beta + b^2\cos^2\beta = a^2\sin^2\alpha + b^2\cos^2\alpha;$$

$$\text{therefore } a^2(\sin^2\beta - \sin^2\alpha) = b^2(\cos^2\alpha - \cos^2\beta)$$
$$= b^2(\sin^2\beta - \sin^2\alpha);$$

$$\text{therefore } (a^2 - b^2)(\sin^2\beta - \sin^2\alpha) = 0;$$

$$\text{therefore } \sin^2\beta = \sin^2\alpha;$$

$$\text{therefore } \beta = \pi - \alpha.$$

And since $a'^2 = b'^2$ each of them $= \frac{a^2 + b^2}{2}$, (Art. 193).

Hence from the value of a'^2 in the preceding Article, we have

$$\frac{a^2 + b^2}{2} = \frac{a^2b^2}{a^2\sin^2\alpha + b^2\cos^2\alpha};$$

$$\text{therefore } (a^2+b^2)\{(a^2-b^2)\sin^2\alpha+b^2\}=2a^2b^2;$$

$$\text{therefore } \sin^2\alpha=\frac{a^2b^2-b^4}{(a^2+b^2)(a^2-b^2)}=\frac{b^2}{a^2+b^2}.$$

This shews that the *equal* conjugate diameters are parallel to the straight lines BA and BA'.

200. The equation to the tangent to the ellipse will be of *the same form* whether the axes be rectangular or the oblique system formed by a pair of conjugate diameters; for the investigation of Art. 170 will apply without any change to the equation $a'^2y^2+b'^2x^2=a'^2b'^2$ which represents an ellipse referred to such an oblique system.

201. *Tangents at the extremities of any chord of an ellipse meet on the diameter which bisects that chord.*

Refer the ellipse to the diameter bisecting the chord as the axis of x, and the diameter parallel to the chord as the axis of y; let the equation to the ellipse be $a'^2y^2+b'^2x^2=a'^2b'^2$. Let x', y' be the co-ordinates of one extremity of the chord; then the equation to the tangent at this point is

$$a'^2yy'+b'^2xx'=a'^2b'^2 \quad\text{.....................(1)}.$$

The co-ordinates of the other extremity of the chord are x', $-y'$, and the equation to the tangent there is

$$-a'^2yy'+b'^2xx'=a'^2b'^2 \quad\text{.....................(2)}.$$

The straight lines represented by (1) and (2) meet at the point for which $y=0$, $x=\frac{a'^2}{x'}$: this proves the theorem.

Supplemental chords.

202. DEFINITION. Two straight lines drawn from a point of the ellipse to the extremities of any diameter are called *supplemental* chords. They are called *principal* supplemental chords if that diameter be the major axis.

203. *If a chord and diameter of an ellipse are parallel, the supplemental chord is parallel to the conjugate diameter.*

Let PP' be a diameter of the ellipse; QP, QP' two supplemental chords. Let x', y' be the co-ordinates of P, and therefore $-x'$, $-y'$ the co-ordinates of P'.

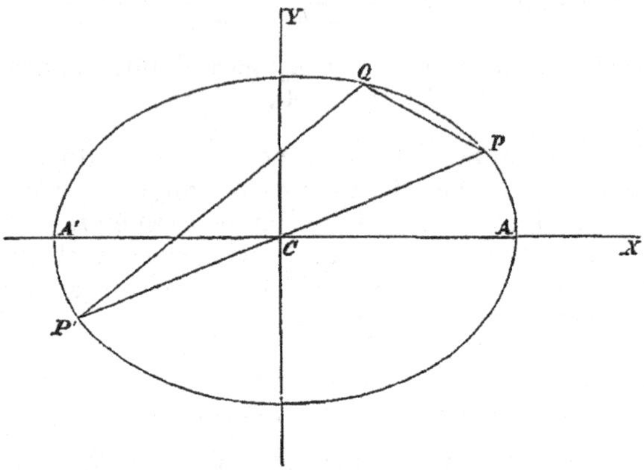

Let the equation to PQ be (Art. 32)

$$y - y' = m(x - x') \qquad (1),$$

and the equation to $P'Q$

$$y + y' = m'(x + x') \qquad (2).$$

The co-ordinates of the point Q satisfy (1) and (2); if then we suppose x, y to denote those co-ordinates, we have from (1) and (2) by multiplication

$$y^2 - y'^2 = mm'(x^2 - x'^2) \qquad (3).$$

But since (x, y) and (x', y') are points on the ellipse

$$a^2y^2 + b^2x^2 = a^2b^2, \quad a^2y'^2 + b^2x'^2 = a^2b^2;$$

$$\text{therefore } a^2(y^2 - y'^2) + b^2(x^2 - x'^2) = 0;$$

$$\text{therefore } y^2 - y'^2 = -\frac{b^2}{a^2}(x^2 - x'^2) \qquad (4).$$

From (3) and (4) we have $mm' = -\frac{b^2}{a^2}$. But we have shewn in Art. 188 that if this relation be satisfied, the two straight lines represented by $y = mx$ and $y = m'x$ are conjugate diameters; this proves the theorem.

Polar Equation.

204. To find the polar equation to the ellipse, the focus being the pole.

Let $SP = r$, $A'SP = \theta$, (see the figure to Art. 158); then $SP = ePN$, by definition; that is, $SP = e(OS + SM)$;

or $$r = a(1 - e^2) + er\cos(\pi - \theta), \text{ (Art. 161)};$$

therefore $$r(1 + e\cos\theta) = a(1 - e^2),$$

and $$r = \frac{a(1 - e^2)}{1 + e\cos\theta}.$$

If we denote the angle ASP by θ, then we have as before $SP = e(OS + SM)$; thus $r = a(1 - e^2) + er\cos\theta$,

and $$r = \frac{a(1 - e^2)}{1 - e\cos\theta}.$$

205. We shall make use of the preceding Article in finding the polar equation to a chord, from which we shall deduce the polar equation to the tangent.

Let P and P' be two points on the ellipse; suppose that $A'SP = \alpha - \beta$, and $A'SP' = \alpha + \beta$, so that $PSP' = 2\beta$; and let l be the semi-latus rectum of the ellipse, so that $l = a(1 - e^2)$: it is required to find the polar equation to the straight line PP'.

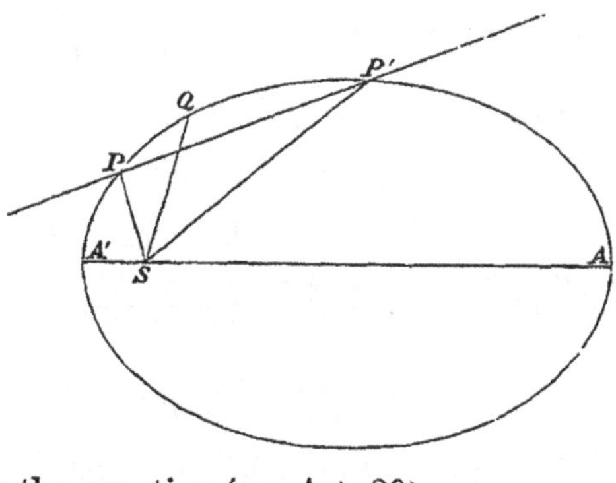

Assume for the equation (see Art. 29)

$$Ar\cos\theta + Br\sin\theta + C = 0 \quad\text{..............(1)}.$$

Since the straight line passes through P, equation (1) must be satisfied by the co-ordinates of P; now $A'SP = \alpha - \beta$, and therefore $SP = \dfrac{l}{1 + e\cos(\alpha - \beta)}$; thus from (1)

$$l\{A\cos(\alpha-\beta)+B\sin(\alpha-\beta)\}+C\{1+e\cos(\alpha-\beta)\}=0\ldots\ldots(2).$$

Similarly, since the straight line passes through P',

$$l\{A\cos(\alpha+\beta)+B\sin(\alpha+\beta)\}+C\{1+e\cos(\alpha+\beta)\}=0\ldots\ldots(3).$$

From (2) and (3), by subtraction,

$$l(A\sin\alpha\sin\beta - B\cos\alpha\sin\beta) + Ce\sin\alpha\sin\beta = 0;$$

$$\text{therefore } l(A\sin\alpha - B\cos\alpha) + Ce\sin\alpha = 0\ldots\ldots\ldots(4).$$

From (2) and (3), by addition,

$$l(A\cos\alpha\cos\beta + B\sin\alpha\cos\beta) + C(1 + e\cos\alpha\cos\beta) = 0;$$

$$\text{therefore } l(A\cos\alpha + B\sin\alpha) + C(\sec\beta + e\cos\alpha) = 0\ldots\ldots(5).$$

From (4) and (5) we find

$$lA + C(\sec\beta\cos\alpha + e) = 0,\quad lB + C\sec\beta\sin\alpha = 0.$$

Substitute the values of A and B in (1) and divide by C;

$$\text{thus } r\left\{(\sec\beta\cos\alpha + e)\cos\theta + \sec\beta\sin\alpha\sin\theta\right\} - l = 0;$$

$$\text{therefore } r = \frac{l}{e\cos\theta + \sec\beta\cos(\alpha - \theta)}\ldots\ldots\ldots(6).$$

If SQ bisect the angle PSP', we have

$$PSQ = \beta,\quad \text{and } A'SQ = \alpha.$$

Now suppose β to diminish indefinitely; then the chord PP' *becomes the tangent at* Q, and we obtain its polar equation by putting $\beta = 0$ in the preceding result; thus we have

$$r = \frac{l}{e\cos\theta + \cos(\alpha - \theta)}.$$

The investigations of this Article will apply to the parabola by supposing $e = 1$.

The investigation of (6) has been put in the following

brief form by Mr F. G. Landon. The equation to SP is $\theta = \alpha - \beta$, and that to SP' is $\theta = \alpha + \beta$; therefore the equation $\theta - \alpha = \pm\beta$ represents the two straight lines SP and SP': this may be written $\cos(\theta - \alpha) = \cos\beta$, or $\cos(\theta - \alpha)\sec\beta = 1$. The equation to the ellipse is $\frac{l}{r} - e\cos\theta = 1$. Combining these equations we get $\frac{l}{r} - e\cos\theta = \cos(\theta - \alpha)\sec\beta$, which must be satisfied at the points P and P'; and as this is the equation to a straight line it is the equation to the straight line PP'.

206. The polar equation to the ellipse referred to the centre is sometimes useful; it may be deduced from the equation $a^2y^2 + b^2x^2 = a^2b^2$, by putting $r\cos\theta$, $r\sin\theta$, for x and y respectively: we thus obtain $r^2(a^2\sin^2\theta + b^2\cos^2\theta) = a^2b^2$.

We add a few miscellaneous propositions on the ellipse.

207. *If tangents be drawn at the extremities of any focal chord of an ellipse, (1) the tangents will intersect on the corresponding directrix, (2) the straight line drawn from the point of intersection of the tangents to the focus will be perpendicular to the focal chord.*

(1) If two tangents to an ellipse meet at the point (h, k) the equation to the chord of contact is, by Art. 183,

$$a^2ky + b^2hx = a^2b^2.$$

Suppose the chord passes through the focus whose co-ordinates are $x = -ae$, $y = 0$; then $-b^2hae = a^2b^2$,

$$\text{therefore } h = -\frac{a}{e};$$

that is, the point of intersection of the tangents is on the directrix corresponding to this focus.

(2) The equation to the straight line through (h, k) and the focus is $y = \frac{k}{h + ae}(x + ae)$. If $h = -\frac{a}{e}$, this becomes $y = -\frac{ke}{a(1 - e^2)}(x + ae) = \frac{ka^2}{hb^2}(x + ae)$, and the straight line is therefore perpendicular to the focal chord of which the equation is $y = -\frac{b^2hx}{a^2k} + \frac{b^2}{k}$.

208. *If through any point within or without an ellipse, two straight lines be drawn parallel to two given straight lines to meet the curve, the rectangles of the segments will be to one another in an invariable ratio.*

Let (x', y') be the given point and suppose α and β respectively the inclinations of the given straight lines to the major axis of the ellipse. By Art. 187 if a straight line be drawn from (x', y') to meet the curve, and be inclined at an angle α to the major axis, the lengths of its segments are given by the equation

$$r^2 (a^2 \sin^2 \alpha + b^2 \cos^2 \alpha) + 2r (a^2 y' \sin \alpha + b^2 x' \cos \alpha) + a^2 y'^2 + b^2 x'^2 - a^2 b^2 = 0;$$

therefore the rectangle of the segments $= \dfrac{a^2 y'^2 + b^2 x'^2 - a^2 b^2}{a^2 \sin^2 \alpha + b^2 \cos^2 \alpha}$.

Similarly the rectangle of the segments of the straight line drawn from (x', y') at an angle $\beta = \dfrac{a^2 y'^2 + b^2 x'^2 - a^2 b^2}{a^2 \sin^2 \beta + b^2 \cos^2 \beta}$.

Hence the ratio of the rectangles $= \dfrac{a^2 \sin^2 \beta + b^2 \cos^2 \beta}{a^2 \sin^2 \alpha + b^2 \cos^2 \alpha}$; and this ratio is constant whatever x' and y' may be.

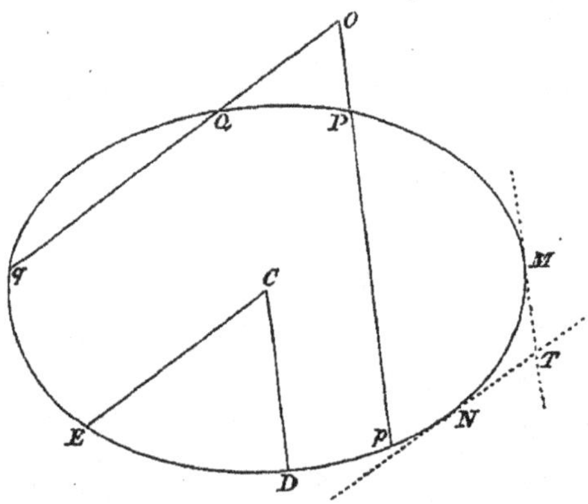

Let O be the point through which the straight lines OPp, OQq, are drawn inclined to the major axis of the ellipse at angles α, β, respectively; then

$$\frac{OP \cdot Op}{OQ \cdot Oq} = \frac{a^2 \sin^2 \beta + b^2 \cos^2 \beta}{a^2 \sin^2 \alpha + b^2 \cos^2 \alpha}.$$

Draw the semi-diameters CD, CE, parallel to Pp, Qq, respectively, then by Art. 206,

$$\frac{CD^2}{CE^2} = \frac{a^2 \sin^2\beta + b^2 \cos^2\beta}{a^2 \sin^2 \alpha + b^2 \cos^2\alpha};$$

$$\text{therefore } \frac{OP \cdot Op}{OQ \cdot Oq} = \frac{CD^2}{CE^2}.$$

Let TM, TN be tangents parallel to Pp, Qq, respectively; then if O coincides with T, the rectangle $OP \cdot Op$ becomes TM^2 and the rectangle $OQ \cdot Oq$ becomes TN^2; therefore

$$\frac{TM^2}{TN^2} = \frac{CD^2}{CE^2}; \quad \text{and } \frac{TM}{TN} = \frac{CD}{CE}.$$

The preceding investigations are very important: we will point out some inferences which may be drawn from them.

Suppose that an ellipse and a circle intersect at four points: denote these points by P, p, Q, q. Then we have seen that

$$\frac{OP \cdot Op}{OQ \cdot Oq} = \frac{CD^2}{CE^2}.$$

But since the four points are on the circle we have $OP \cdot Op = OQ \cdot Oq$ by Euclid, III. 35 and 36, Cor. Therefore $CD^2 = CE^2$. And since CD and CE are equal they make equal angles with the major axis of the ellipse. Thus *if an ellipse and a circle intersect at four points the common chords make equal angles with the major axis of the ellipse.*

Suppose that Q and q coincide so that OQq becomes a common tangent to the ellipse and circle; thus we obtain the following result: *if an ellipse and a circle have a common tangent and a common chord, the tangent and the chord make equal angles with the major axis of the ellipse.*

We may conceive that the three points P, Q, and q move up to coincidence. The circle in this case is called the *circle of curvature* of the ellipse at the point of coincidence. We do not discuss the properties of the circle of curvature in the present work; but we may remark that we have obtained the following result: *the tangent at any point of an ellipse and the chord drawn from the point to the other intersection of the ellipse and the circle of curvature at the point make equal angles with the major axis of the ellipse.*

Similar remarks may be made in connexion with Art. 157.

EXAMPLES.

1. CP and CD are conjugate semi-diameters: given the co-ordinates of $P(x', y')$, find the equation to PD.

2. If straight lines drawn through any point of an ellipse to the extremities of any diameter meet the conjugate CD at the points M, N, prove that $CM \,.\, CN = CD^2$.

3. CP, CD are two conjugate semi-diameters; CP', CD' are two other conjugate semi-diameters: shew that the area of the triangle PCP' is equal to the area of the triangle DCD'.

4. Normals at P and D, the extremities of semi-conjugate diameters, meet at K: find the equation to KC, and shew that KC is perpendicular to PD.

5. In an ellipse the rectangle contained by the perpendicular from the centre upon the tangent, and the part of the corresponding normal intercepted between the axes, is equal to the difference of the squares of the semi-axes.

6. Shew that the locus of the intersection of the perpendicular from the centre on a tangent to the ellipse is the curve which has for its equation $r^2 = a^2 \cos^2\theta + b^2 \sin^2\theta$, the centre being the origin.

7. From A the vertex of an ellipse draw a straight line ARQ to Q the middle point of HP meeting SP at R: shew that the locus of R is an ellipse, and also the locus of Q.

8. Find the polar equation to the ellipse, the vertex being the origin and the major axis the initial line.

9. If any chord AQ meet the minor axis produced at R, and CP be a semi-diameter parallel to AQ, then

$$AQ \,.\, AR = 2CP^2.$$

10. A circle is described on AA' the major axis of an ellipse as diameter; P is any point in the circle; AP, $A'P$

are joined cutting the ellipse at points Q and Q' respectively: shew that

$$\frac{AP}{AQ}+\frac{A'P}{A'Q'}=\frac{a^2+b^2}{b^2}.$$

11. If circles be described on two semi-conjugate diameters of an ellipse as diameters, the locus of their intersection is the curve defined by the equation $2(x^2+y^2)^2=a^2x^2+b^2y^2$.

12. CP, CD are conjugate semi-diameters; CQ is perpendicular to PD: find the locus of Q.

13. Find the points where the ellipse $a(1-e^2)=r+re\cos\theta$ cuts the straight line $a(1-e^2)=r\sin\theta+r(1+e)\cos\theta$.

14. Write down the polar equations to the four tangents at the ends of the latera recta; also the equations to the tangents at the ends of the minor axis: the focus being the pole.

15. Determine the locus of the intersection of tangents drawn at two points P, Q, which are taken so that the sum of the angles ASP, ASQ, is constant.

16. If PSp be a focal chord of an ellipse, and along the straight line SP there be set off SQ a mean proportional between SP and Sp, the locus of Q will be an ellipse having the same excentricity as the original ellipse.

17. Two ellipses have a common focus and their major axes are equal in length and situated in the same straight line: find the polar co-ordinates of the points of intersection.

18. From an external point two tangents are drawn to an ellipse: find between what limits the ratio of the length of one tangent to the length of the other lies.

19. TP, TQ are two tangents to an ellipse, and CP', CQ' are the radii from the centre respectively parallel to these tangents: prove that $P'Q'$ is parallel to PQ.

20. From a point O whose co-ordinates are h, k a straight line is drawn meeting the ellipse at P and p; and CD is the parallel semi-conjugate diameter: shew that

$$\frac{OP\,.\,Op}{CD^2}=\frac{h^2}{a^2}+\frac{k^2}{b^2}-1.$$

21. When the angle between the radius vector from the *focus* and the tangent is least, the radius vector $= a$.

22. When the angle between the radius vector from the *centre* and the tangent is least, the radius vector $= \left(\frac{a^2 + b^2}{2}\right)^{\frac{1}{2}}$.

23. PT, pt are tangents at the extremities of any diameter Pp of an ellipse; any other diameter meets PT at T, and its conjugate meets pt at t; also any tangent meets PT at T' and pt at t': shew that $PT : PT' :: pt' : pt$.

24. From the ends P, D, of conjugate diameters in an ellipse, draw straight lines parallel to any tangent line; and from the centre C draw any straight line cutting these straight lines and the tangent at points p, d, t, respectively: then will

$$Cp^2 + Cd^2 = Ct^2.$$

25. If tangents be drawn from different points of an ellipse of lengths equal to n times the semi-conjugate diameter at each point, then the locus of their extremities will be a concentric ellipse with semi-axes equal to

$$a\sqrt{(n^2+1)} \text{ and } b\sqrt{(n^2+1)}.$$

26. Apply the equation to the tangent in Art. 171 to find the locus of the intersection of tangents at the extremities of conjugate diameters.

27. If from a point (x', y') of an ellipse a chord be drawn parallel to a fixed straight line, shew that the length of this chord varies as $\frac{y' \sin(\alpha - \phi)}{\cos\phi}$, where ϕ is the inclination of the tangent at (x', y') to the axis, and α the inclination of the fixed straight line to the axis.

28. If through any point P of an ellipse two chords PQ, PR be drawn parallel to two fixed straight lines and making angles α and β respectively with the tangent at P, shew that the ratio of $PQ \operatorname{cosec} \alpha$ to $PR \operatorname{cosec} \beta$ is constant.

29. A parabola is touched at the extremities of the latus rectum by an ellipse of given magnitude: find the latus rectum of the parabola.

30. The perpendicular from the centre on a straight line joining the ends of perpendicular diameters of an ellipse is of constant length.

31. Chords are drawn through the end of an axis of an ellipse: find the locus of their middle points.

32. Chords of an ellipse are drawn through any fixed point: find the locus of their middle points.

33. Two focal chords are drawn in an ellipse at right angles to each other: find their position when the rectangle contained by them has respectively its greatest and least value.

34. In an ellipse if PP' and QQ' be focal chords at right angles to each other

$$\frac{1-e^2}{SP \cdot SP'} + \frac{1-e^2}{SQ \cdot SQ'} = \frac{1}{AC^2} + \frac{1}{BC^2}.$$

35. PSp, QSq are focal chords; suppose T the point where the straight lines PQ, pq meet: shew that TS is equally inclined to the focal chords, and that T is on the directrix corresponding to S.

36. If r, θ be the polar co-ordinates of a point P, shew that $\tan HPZ = \dfrac{b}{\sqrt{(2ar - r^2 - b^2)}}$ and $= \dfrac{1 + e\cos\theta}{e\sin\theta}$.

37. Perpendiculars are drawn from P and D the extremities of any pair of conjugate diameters on the diameter $y = x\tan\alpha$: shew that the sum of the squares of the perpendiculars is $a^2\sin^2\alpha + b^2\cos^2\alpha$.

38. The excentric angles of two points P and Q are ϕ and ϕ' respectively: shew that the area of the parallelogram formed by the tangents at the extremities of the diameters through P and Q is $\dfrac{4ab}{\sin(\phi' - \phi)}$; shew also that the area is least when P and Q are the extremities of conjugate diameters.

39. Shew that the equation to the locus of the middle points of all chords of the same length $(2c)$ in an ellipse is

$$c^2\frac{a^2y^2+b^2x^2}{a^4y^2+b^4x^2}+\frac{x^2}{a^2}+\frac{y^2}{b^2}-1=0.$$

40. Chords of an ellipse are drawn at right angles to one another through a point O whose co-ordinates are h, k: if CP, CQ be the radii drawn from the centre parallel to the chords, and E, F the middle points of the chords, shew that

$$\frac{OE^2}{CP^4}+\frac{OF^2}{CQ^4}=\frac{h^2}{a^4}+\frac{k^2}{b^4}.$$

41. Given the co-ordinates of P, find those of the intersection of the tangents at P and D. (See the figure to Art. 192.)

42. Shew that the equation

$$\frac{x^2}{a^2}+\frac{y^2}{b^2}-1=\left\{\frac{x(bx'-ay')}{a^2b}+\frac{y(ay'+bx')}{ab^2}-1\right\}^2$$

represents the tangents at P and D, supposing x', y' the co-ordinates of P. (See the figure to Art. 192.)

43. If CP, CD be any conjugate semi-diameters of an ellipse $APBDA'$, and BP, BD be joined and also AD, $A'P$, these latter intersecting at O, shew that $BDOP$ is a parallelogram.

44. Shew that the area of the parallelogram in the preceding Example $=ay'+bx'-ab$, where x', y' are the co-ordinates of P; and find the greatest value of this area.

45. If a straight line be drawn from the focus of an ellipse to make a given angle α with the tangent, shew that the locus of its intersection with the tangent will be a circle which touches or falls entirely without the ellipse according as $\cos\alpha$ is less or greater than the excentricity of the ellipse.

46. In an ellipse SQ, HQ, drawn perpendicular to a pair of conjugate diameters, intersect at Q: prove that the locus of Q is a concentric ellipse.

47. Two ellipses have their foci coincident; a tangent to one of them intersects at right angles a tangent to the other: shew that the locus of the point of intersection is a circle having the same centre as the ellipses.

48. Find what is represented by the equation $x^2 + y^2 = c^2$ when the axes are oblique.

49. Shew that when the ellipse is referred to any pair of conjugate diameters as axes, the condition that $y = mx$ and $y = m'x$ may represent conjugate diameters is $mm' = -\frac{b'^2}{a'^2}$.

50. The ellipse being referred to equal conjugate diameters, find the equation to the normal at any point.

51. From any point P perpendiculars PM, PN are drawn on the equal conjugate diameters: shew that the normal at P bisects MN.

52. An ellipse intersects the side PQ of a triangle at r and r', the side QR at p and p', and the side RP at q and q': shew that

$$Pr . Pr' . Qp . Qp' . Rq . Rq' = Pq . Pq' . Qr . Qr' . Rp . Rp'.$$

Shew also that a similar result is true for a polygon; and shew what it becomes when the ellipse *touches* the sides.

CHAPTER XI.

THE HYPERBOLA.

209. *To find the equation to the hyperbola.*

The hyperbola is the locus of a point which moves so that its distance from a fixed point bears a constant ratio to its distance from a fixed straight line, the ratio being greater than unity.

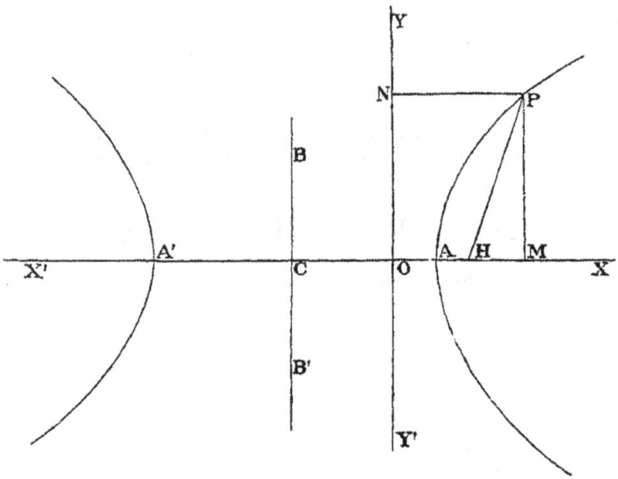

Let H be the fixed point, YY' the fixed straight line. Draw HO perpendicular to YY'; take O as the origin, OH as the direction of the axis of x, OY as that of the axis of y.

Let P be a point on the locus; join HP, draw PM parallel to OY and PN parallel to OX. Let $OH = p$, and let e be the ratio of HP to PN. Let x, y be the co-ordinates of P.

By definition $HP = ePN$; therefore $HP^2 = e^2PN^2$; therefore $PM^2 + HM^2 = e^2PN^2$, that is, $y^2 + (x - p)^2 = e^2x^2$.

This is the equation to the hyperbola with the assumed origin and axes.

210. To find where the hyperbola meets the axis of x we put $y=0$ in the equation to the hyperbola; thus $(x-p)^2=e^2x^2$; therefore $x-p=\pm ex$; therefore $x=\frac{p}{1\mp e}$. Since e is greater than unity, $1-e$ is a negative quantity.

Let $OA'=\frac{p}{e-1}$, $OA=\frac{p}{1+e}$, the former being measured to the *left* of O, then A' and A are points on the hyperbola. A and A' are called the *vertices* of the hyperbola, and C the point midway between A and A' is called the *centre* of the hyperbola.

211. We shall obtain a simpler form of the equation to the hyperbola by transferring the origin to A or C.

I. Suppose the origin at A.

Since $OA=\frac{p}{1+e}$, we put $x=x'+\frac{p}{1+e}$ and substitute this value in the equation $y^2+(x-p)^2=e^2x^2$;

thus
$$y^2+\left(x'+\frac{p}{1+e}-p\right)^2=e^2\left(x'+\frac{p}{1+e}\right)^2,$$

or
$$y^2+\left(x'-\frac{ep}{1+e}\right)^2=e^2\left(x'+\frac{p}{1+e}\right)^2;$$

$$\text{therefore } y^2+x'^2-\frac{2x'ep}{1+e}=e^2\left(x'^2+\frac{2px'}{1+e}\right);$$

$$\text{therefore } y^2=2pex'+(e^2-1)\,x'^2$$
$$=(e^2-1)\left\{\frac{2pex'}{e^2-1}+x'^2\right\}.$$

The distance $A'A=\frac{p}{e-1}+\frac{p}{1+e}=\frac{2ep}{e^2-1}$; we will denote this by $2a$; hence the equation becomes $y^2=(e^2-1)(2ax'+x'^2)$.

We may suppress the accent, if we remember that the origin is at the vertex A, and thus write the equation

$$y^2 = (e^2 - 1)(2ax + x^2) \qquad \text{.................(1).}$$

II. Suppose the origin at C.

Since $CA = a$, we put $x = x' - a$ and substitute this value in (1); thus

$$y^2 = (e^2 - 1)\{2a(x' - a) + (x' - a)^2\} = (e^2 - 1)(x'^2 - a^2).$$

We may suppress the accent, if we remember that the origin is now at the centre C, and thus write the equation

$$y^2 = (e^2 - 1)(x^2 - a^2) \qquad \text{...................(2).}$$

In (2) suppose $x = 0$, then $y^2 = -(e^2 - 1)a^2$; this gives an impossible value to y, and thus the curve does not cut the axis of y. We shall however denote $(e^2 - 1)a^2$ by b^2, and measure off the ordinates CB and CB' each equal to b, as we shall find these ordinates useful hereafter.

Thus (1) may be written

$$y^2 = \frac{b^2}{a^2}(2ax + x^2) \qquad \text{..................(3),}$$

and (2) may be written

$$y^2 = \frac{b^2}{a^2}(x^2 - a^2) \qquad \text{...................(4),}$$

or, more symmetrically,

$$\frac{x^2}{a^2} - \frac{y^2}{b^2} = 1, \qquad \text{or, } a^2y^2 - b^2x^2 = -a^2b^2 \qquad \text{........(5).}$$

212. Since $AH = eOA$ and $OA = \dfrac{p}{1+e}$, we have

$$AH = \frac{ep}{1+e} = \frac{(e-1)ep}{e^2 - 1} = (e-1)a,$$

$$OA = \frac{p}{1+e} = \frac{e-1}{e}a,$$

$$CH = CA + AH = a + (e-1)a = ea,$$

$$CO = CA - OA = a - \frac{e-1}{e}a = \frac{a}{e},$$

and $$OH = p = \frac{a(e^2-1)}{e}.$$

213. We may now ascertain the form of the hyperbola. Take the equation referred to the centre as origin,

$$y^2 = \frac{b^2}{a^2}(x^2 - a^2) \text{ ....(1).}$$

For every value of x less than a, y is impossible. When $x = a$, $y = 0$. For every value of x greater than a there

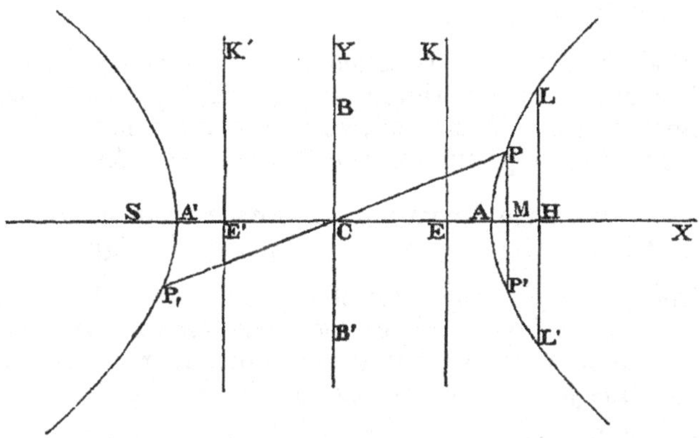

are two values of y equal in magnitude but of opposite sign. Hence if P be a point in the curve on one side of the axis of x, there is a point P' on the other side of the axis, such that $P'M = PM$. Thus the curve is symmetrical with respect to the axis of x, and it extends indefinitely to the right of A.

If we ascribe to x any negative value we obtain for y the same pair of values as when we ascribed to x the corresponding positive value. Hence the portion of the curve to the left of the axis of y is similar to the portion to the right of it.

As the equation (1) may be put in the form

$$x^2 = \frac{a^2}{b^2}(y^2 + b^2) \text{......................(2)},$$

we see that the axis of y also divides the curve symmetrically. Thus the curve consists of two similar branches each extending indefinitely.

The straight line EK is the directrix, H is the corresponding focus. Since the curve is symmetrical with respect to the straight line BCB', it follows that if we take $CS = CH$ and $CE' = CE$, and draw $E'K'$ at right angles to CE', the point S and the straight line $E'K'$ will form respectively a second focus and directrix, by means of which the curve might have been generated.

214. The point C is called the *centre* of the hyperbola, because every chord of the hyperbola which passes through C is bisected at C. This is shewn in the same manner as the corresponding proposition in the ellipse. (See Art. 163.)

215. We have drawn the curve concave towards the axis of x; the following proposition will justify the figure.

The ordinate of any point of the curve which lies between a vertex and a fixed point of the curve on the same branch as the vertex is greater than the corresponding ordinate of the straight line joining that vertex and the fixed point.

Let A be the vertex and take it for the origin; let P be the fixed point; x', y' its co-ordinates. Then the equation to the hyperbola is (Art. 211) $y^2 = \frac{b^2}{a^2}(2ax + x^2)$.

The equation to AP is $y = \frac{y'}{x'}x$, or $y = \frac{b}{a}\sqrt{\left(\frac{2a}{x'} + 1\right)}x$, since (x', y') is on the hyperbola.

Let x denote any abscissa less than x', then since the ordinate of the curve is $\frac{b}{a}\sqrt{(2ax + x^2)}$ or $\frac{b}{a}\sqrt{\left(\frac{2a}{x} + 1\right)}x$, and that of the straight line is $\frac{b}{a}\sqrt{\left(\frac{2a}{x'} + 1\right)}x$, it is obvious that the ordinate of the curve is greater than that of the straight line.

All points may be said to be *outside* the curve for which $\frac{y^2}{b^2} - \frac{x^2}{a^2} + 1$ is *positive;* and all points may be said to be *inside* the curve for which $\frac{y^2}{b^2} - \frac{x^2}{a^2} + 1$ is *negative.* It is easy to see that according to this definition a point is *outside* the curve when no straight line can be drawn from the point to a *focus* without cutting the curve. A very instructive mode of obtaining this result is that exemplified in Art. 54: the expression $\frac{y^2}{b^2} - \frac{x^2}{a^2} + 1$ is negative when the point (x, y) is a focus, vanishes when (x, y) is on the curve, does not vanish in any other case, and is positive when $x = 0$ for all values of y. Hence we infer that the expression is *negative* for every point which can be joined to a focus by a straight line that does not cut the curve, and *positive* in every other case.

Similar remarks might be made in connexion with Art. 127.

216. AA' and BB' are called *axes* of the hyperbola. The axis AA' which if produced passes through the foci, is called the *transverse* axis, and BB' the *conjugate* axis. We do not, as in the case of the ellipse, use the terms *major* and *minor* axis, because since $b = a\sqrt{(e^2 - 1)}$ (Art. 211), and e is greater than unity, b may be greater or less than a.

The ratio which the distance of any point on the hyperbola from the focus bears to the distance of the same point from the corresponding directrix is called the *excentricity* of the hyperbola. We have denoted it by the symbol e.

To find the *latus rectum* (see Art. 128) we put $x = CH$, that is $= ae$, in equation (1) of Art. 213; thus

$$y^2 = \frac{b^2a^2(e^2-1)}{a^2} = \frac{b^4}{a^2};$$

$$\text{therefore } LH = \frac{b^2}{a}, \text{ and the latus rectum} = \frac{2b^2}{a}.$$

Since $b^2 = a^2(e^2 - 1)$; therefore $b^2 + a^2 = a^2e^2$; that is

$$CB^2 + CA^2 = CH^2;$$

$$\text{therefore } AB = CH.$$

217. The equation to the hyperbola may be derived from the equation to the ellipse by writing $-b^2$ for b^2. We shall find that the hyperbola has many properties similar to those which have been proved for the ellipse; and as the demonstrations are similar to those which have been given, we shall in some cases not repeat them for the hyperbola, but refer to the corresponding Articles in the Chapters on the ellipse.

218. *To express the focal distances of any point of the hyperbola in terms of the abscissa of the point.*

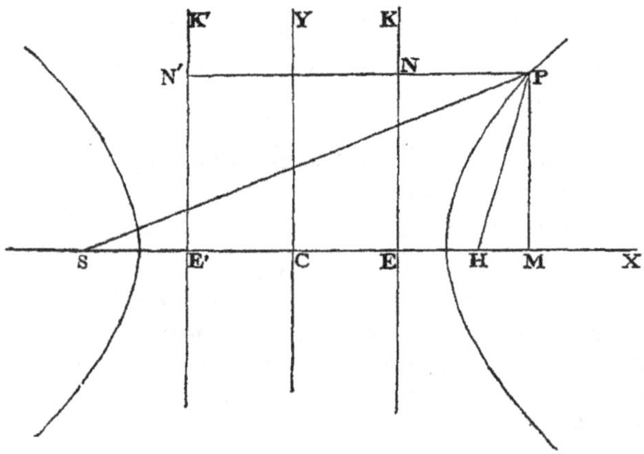

Let S be one focus, $E'K'$ the corresponding directrix; H the other focus, EK the corresponding directrix. Let P be a point on the hyperbola; x, y its co-ordinates, the centre being the origin. Join SP, HP, and draw PNN' parallel to the transverse axis, and PM perpendicular to it. Then

$$SP = ePN' = e(CM + CE') = e\left(x + \frac{a}{e}\right) = ex + a,$$

$$HP = ePN = e(CM - CE) = e\left(x - \frac{a}{e}\right) = ex - a.$$

Hence $SP - HP = 2a$; that is, the *difference* of the focal distances of any point on the hyperbola is equal to the transverse axis.

Let x, y be the co-ordinates of any point Q. Then

$$SQ^2 = (x+ae)^2 + y^2 = (ex+a)^2 + y^2 - (e^2-1)(x^2-a^2)$$
$$= e^2\left(x+\frac{a}{e}\right)^2 + y^2 - \frac{b^2}{a^2}(x^2-a^2).$$

Therefore the focal distance of any point not on the curve bears to the distance of the point from the corresponding directrix a ratio which is greater or less than e according as the point is outside or inside the curve.

Suppose Q a point *outside* the curve; join Q with the nearer focus, which we will denote by H; and let QH cut the curve at P. Let S be the other focus. Join SQ, SP. Then SQ is less than $SP+PQ$ by Euclid, I. 20; therefore $SQ-HQ$ is less than $SP+PQ-HQ$, that is less than $SP-HP$. *Thus the difference of the focal distances of any point outside an hyperbola is less than the transverse axis.* Similarly we may shew that the *difference of the focal distances of any point inside an hyperbola is greater than the transverse axis.*

219. The equation $y^2 = \frac{b^2}{a^2}(x^2-a^2)$ may be written

$$y^2 = \frac{b^2}{a^2}(x-a)(x+a).$$

Hence (see the figure to Art. 213), $\frac{PM^2}{AM \cdot A'M} = \frac{BC^2}{AC^2}$.

Tangent and Normal to an Hyperbola.

220. *To find the equation to the tangent at any point of an hyperbola.*

By a process similar to that in Art. 170, it will be found that the equation to the tangent at the point (x', y') is

$$y - y' = \frac{b^2x'}{a^2y'}(x-x'),$$

or

$$a^2yy' - b^2xx' = -a^2b^2.$$

These equations may be derived from the corresponding equations with respect to the ellipse by writing $-b^2$ for b^2.

221. The equation to the tangent to the hyperbola may also be written in the form $y = mx + \sqrt{(m^2a^2 - b^2)}$; (see Art. 171). Conversely every straight line whose equation is of this form, is a tangent to the hyperbola.

222. It may be shewn as in the case of the circle that a tangent to an hyperbola meets it at only *one* point. Also if a straight line meet an hyperbola at only *one* point, it is in general the tangent to the hyperbola at that point. For suppose the equation to an hyperbola to be $a^2y^2 - b^2x^2 = -a^2b^2$, and the equation to a straight line $y = mx + c$. Then to determine the abscissæ of the points of intersection, we have the equation $a^2(mx+c)^2 - b^2x^2 = -a^2b^2$, or

$$(a^2m^2 - b^2)x^2 + 2a^2mcx + a^2(c^2 + b^2) = 0.$$

This equation has always two roots, except

(1) when $a^4m^2c^2 = (a^2m^2 - b^2)a^2(c^2 + b^2)$, or $c^2 = m^2a^2 - b^2$, and consequently the straight line is a tangent;

(2) when $a^2m^2 - b^2 = 0$; the equation then reduces to one of the first degree, and therefore has but one root. Thus a straight line which meets the hyperbola at only *one* point is the tangent at that point unless the inclination of the straight line to the transverse axis be $\pm \tan^{-1}\dfrac{b}{a}$.

223. The tangents at the vertices A and A' are parallel to the axis of y. (See Art. 172.)

224. *To find the equation to the normal at any point of an hyperbola.* (See Art. 173.)

It will be found that the equation to the normal at (x', y') is $y - y' = -\dfrac{a^2y'}{b^2x'}(x - x')$.

This may also be written in the form

$$y = mx - \frac{(a^2 + b^2)m}{\sqrt{(a^2 - b^2m^2)}}. \quad \text{(See Art. 174.)}$$

225. We shall now deduce some properties of the hyperbola from the preceding Articles.

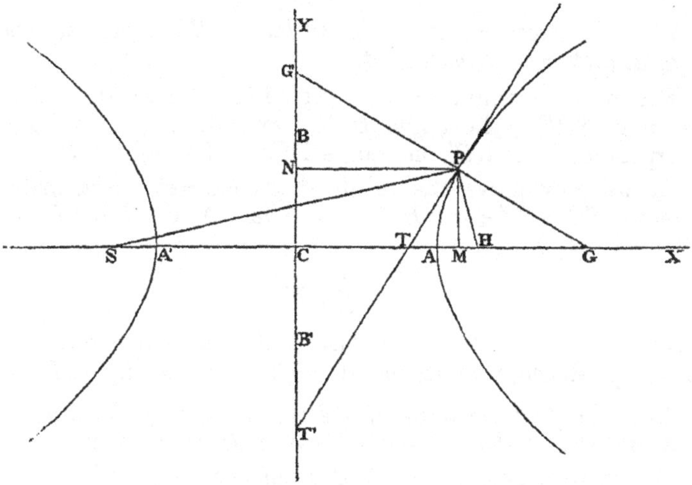

Let x', y' be the co-ordinates of P; let PT be the tangent at P, PG the normal at P; PM, PN perpendiculars on the axes.

The equation to the tangent at P is $a^2yy' - b^2xx' = -a^2b^2$.

Let $y = 0$, then $x = \frac{a^2}{x'}$, hence $CT = \frac{CA^2}{CM}$;

$$\text{therefore } CM \,.\, CT = CA^2.$$

Similarly $$CN \,.\, CT' = CB^2.$$

226. As in Art. 176, we may shew that

$$CG = e^2 CM, \text{ and } CG' = \frac{a^2e^2}{b^2} PM.$$

227. As in Art. 177, we may shew that

$$PG^2 = \frac{b^2rr'}{a^2}, \quad PG'^2 = \frac{a^2rr'}{b^2},$$

where $$SP = r', \quad HP = r.$$

Also the result established in Art. 176 respecting normals at the ends of a focal chord holds for the hyperbola, changing *major* axis into *transverse* axis.

228. *The tangent at any point bisects the angle between the focal distances of that point.*

For in the manner given in Art. 178, we may shew that the angle SPG' = the angle HPG; and therefore since PT is perpendicular to GG', the angle TPS = the angle TPH.

Or we may prove the result thus: $CG = e^2x'$ (Art. 226); therefore $SG = e^2x' + ae$, $HG = e^2x' - ae$. Also $SP = ex' + a$, $HP = ex' - a$; hence

$$\frac{SG}{HG} = \frac{SP}{HP};$$

therefore by Euclid, VI. A, PG bisects the angle between HP and SP produced, that is, the angle SPG' = the angle HPG.

229. *To find the locus of the intersection of the tangent at any point with the perpendicular on it from the focus.*

It may be proved as in Art. 180, that the required locus is the circle described on the transverse axis as diameter.

230. Let p denote the perpendicular from H on the tangent at P, and p' the perpendicular from S; then, as in Art. 181, it may be shewn that $p^2 = \frac{b^2r}{r'}$, and $p'^2 = \frac{b^2r'}{r}$; therefore $pp' = b^2$. Since $r' = 2a + r$, we have $p^2 = \frac{b^2r}{2a+r}$.

The result established in the latter part of Art. 181 holds also for the hyperbola.

231. *From any external point two tangents can be drawn to an hyperbola.*

Let h, k be the co-ordinates of the external point, then as in Art. 182, we shall obtain the following equation for determining the abscissæ of the points of contact of the tangents and hyperbola, $x'^2(a^2k^2 - b^2h^2) + 2a^2b^2hx' - a^4(b^2 + k^2) = 0$.

The roots of this quadratic will be possible if

$$a^4b^4h^2 + a^4(b^2 + k^2)(a^2k^2 - b^2h^2) \text{ is positive};$$

that is, if $k^2a^2 - b^2h^2 + a^2b^2$ is positive.

But if (h, k) be an *external* point the last expression *is* positive, and therefore two tangents can be drawn to the hyperbola from an external point.

The product of the two values of x' given by the above quadratic is $-\frac{a^4(b^2+k^2)}{a^2k^2-b^2h^2}$; this product is therefore positive or negative according as $a^2k^2-b^2h^2$ is negative or positive; that is, the two tangents meet the *same* branch or *different* branches according as $a^2k^2-b^2h^2$ is negative or positive.

The case in which $a^2k^2-b^2h^2=0$ requires to be noticed. Here one root of the quadratic equation becomes infinite, and the other is $\frac{a^4(b^2+k^2)}{2a^2b^2h}$; see *Algebra*, Chapter XXII.

In this case the point (h, k) falls on a certain straight line called an *asymptote*, which we shall consider hereafter; see Art. 255. The asymptote itself may then count as one of the two tangents from the point (h, k). If $h=0$ and $k=0$ the point (h, k) is the origin; in this case the two asymptotes may count as the two tangents from the point (h, k).

232. *Tangents are drawn to an hyperbola from a given external point: to find the equation to the chord of contact.*

Let h, k be the co-ordinates of the external point; then the equation to the chord of contact is

$$a^2ky-b^2hx=-a^2b^2. \quad \text{(See Art. 183.)}$$

233. *Through any fixed point chords are drawn to an hyperbola, and tangents to the hyperbola are drawn at the extremities of each chord: the locus of the intersection of the tangents is a straight line.*

Let h, k be the co-ordinates of the point through which the chords are drawn, then the equation to the locus of the intersection of the tangents is

$$a^2ky-b^2hx=-a^2b^2. \quad \text{(See Art. 184.)}$$

234. *If from any point in a straight line a pair of tangents be drawn to an hyperbola, the chords of contact will all pass through a fixed point.* (See Art. 185.)

The student should observe the different interpretations that can be assigned to the equation $a^2ky - b^2hx = -a^2b^2$.

The statements in Art. 103 with respect to the circle may all be applied to the hyperbola.

235. Some interesting geometrical investigations relating to tangents to an hyperbola from an external point may be noticed.

To draw two tangents to an hyperbola from an external point.

The first method given in Art. 186 may be applied without any change.

In applying the second method we shall have to distinguish between the two cases which present themselves in Art. 231; the distinction between the two cases will be more fully appreciated by the student after he has read the next Chapter. If the external point be between a branch of the curve and the adjacent portions of the asymptotes, the two tangents both touch that branch of the curve: if the external point be so situated that we cannot pass from the point to the curve without crossing an asymptote, the two tangents touch different branches of the curve.

The student can easily construct the figures required in the process we shall now give.

I. Suppose the external point to be between a branch of the curve, and the adjacent portions of the asymptotes. Let O denote the external point, H the nearer focus, S the farther focus. With centre S and radius equal to $2a$ describe a circle; with centre O and radius OH describe another circle cutting the former at Q and q. Join SQ and Sq, and produce these straight lines to meet the curve at P and p. Join OP and Op; these are the required tangents from O.

The demonstration is like that in Art. 186; and we can shew that OP and Op subtend equal angles at H, and also at S.

II. Suppose the external point so situated that we cannot pass from the point to the curve without crossing an asymptote. Let O denote the external point, H the nearer focus, S the farther focus. With centre S and radius equal to $2a$ describe a circle; with centre O and radius OH describe another circle cutting the former at Q and q, the angle HSQ being less than the angle HSq. Join SQ and produce it to meet the curve at P; also join qS and produce it to meet the curve at p. Join OP and Op; these are the required tangents from O.

The demonstration is like that in Art. 186.

From the triangles OSQ and OSq we have the angle OSq equal to the angle OSQ. Thus the angle OSp is the supplement of the angle OSQ; so that the angles subtended at S by the tangent Op and the tangent OP are supplementary.

Also the angle $OHp =$ the angle $OqS =$ the angle OQS $=$ the supplement of the angle $OQP =$ the supplement of the angle OHP; so that the angles subtended at H by the tangent Op and the tangent OP are supplementary.

The straight line which bisects the angle between the focal distances of an external point is equally inclined to the two tangents from that point.

In I. we have

$$\text{angle } SOQ + \text{twice angle } QOP + \text{angle } SOq + \text{twice angle } pOH = 360°;$$

$$\text{therefore angle } SOQ + \text{angle } QOP + \text{angle } pOH = 180°,$$

$$\text{that is, angle } SOP + \text{angle } pOH = 180°;$$

thus the angle pOH is the supplement of the angle SOP, that is, equal to the angle between SO and PO produced.

In II. we have

$$\text{angle } pOq = \text{angle } pOH = \text{angle } pOQ + \text{twice angle } POH,$$

$$\text{angle } SOq = \text{angle } SOQ = \text{angle } SOp + \text{angle } pOQ;$$

therefore by subtraction

$$\text{angle } SOp = \text{twice angle } POH - \text{angle } SOp,$$

$$\text{therefore angle } SOp = \text{angle } POH.$$

The student should observe the extension thus given to the result in Art. 228.

EXAMPLES.

1. Find the equation to an hyperbola of given transverse axis whose vertex bisects the distance between the centre and the focus.

2. If the ordinate MP of an hyperbola be produced to Q so that $MQ = SP$, find the locus of Q.

3. Any chord AP through the vertex of an hyperbola is divided at Q so that $AQ : QP :: AC^2 : BC^2$, and QM is drawn to the foot of the ordinate MP; from Q a straight line is drawn at right angles to QM meeting the transverse axis at O: shew that $AO : A'O :: AC^2 : BC^2$.

4. PQ is a chord of an ellipse at right angles to the major axis AA'; PA, QA' are produced to meet at R: shew that the locus of R is an hyperbola having the same axes as the ellipse.

5. If a circle be described passing through any point P of a given hyperbola and the extremities of the transverse axis, and the ordinate MP be produced to meet the circle at Q, shew that the locus of Q is an hyperbola whose conjugate axis is a third proportional to the conjugate and transverse axes of the original hyperbola.

6. Find the locus of a point such that if from it a pair of tangents be drawn to an ellipse the product of the perpendiculars drawn from the foci on the chord of contact will be constant.

7. If an ellipse and an hyperbola have the same foci their tangents at the points of intersection are at right angles.

8. Shew that the equation

$$x^2 + y^2 = k^2 (Ax + By + C)^2$$

represents an ellipse or an hyperbola according as $k^2(A^2 + B^2)$ is less or greater than unity.

CHAPTER XII.

THE HYPERBOLA CONTINUED.

Diameters.

236. *To find the length of a straight line drawn from any point in a given direction to meet an hyperbola.*

Let x', y' be the co-ordinates of the point from which the straight line is drawn; x, y the co-ordinates of the point to which the straight line is drawn; θ the inclination of the straight line to the axis of x; r the length of the straight line; then (Art. 27) $x = x' + r\cos\theta$, $y = y' + r\sin\theta$.

If (x, y) be on the hyperbola these values may be substituted in the equation $a^2y^2 - b^2x^2 = -a^2b^2$; thus

$$a^2(y' + r\sin\theta)^2 - b^2(x' + r\cos\theta)^2 = -a^2b^2;$$

therefore $r^2(a^2\sin^2\theta - b^2\cos^2\theta) + 2r(a^2y'\sin\theta - b^2x'\cos\theta)$

$$+ a^2y'^2 - b^2x'^2 + a^2b^2 = 0.$$

From this quadratic two values of r can be found which are the lengths of the two straight lines that can be drawn from (x', y') in the given direction to the hyperbola.

237. *To find the diameter of a given system of parallel chords in an hyperbola.* (See Definition in Art. 148.)

Let θ be the inclination of the chords to the transverse axis of the hyperbola; let x', y' be the co-ordinates of the middle point of any one of the chords; the equation which determines the lengths of the straight lines drawn from (x', y') to the curve is (Art. 236)

$$r^2(a^2\sin^2\theta - b^2\cos^2\theta) + 2r(a^2y'\sin\theta - b^2x'\cos\theta)$$
$$+ a^2y'^2 - b^2x'^2 + a^2b^2 = 0 \quad \text{.............(1)}.$$

Since (x', y') is the middle point of the chord, the values of r furnished by this equation must be *equal in magnitude and opposite in sign;* hence the co-efficient of r must vanish; thus

$$a^2y'\sin\theta - b^2x'\cos\theta = 0, \quad \text{or } y' = \frac{b^2}{a^2}\cot\theta \, . \, x' \quad \text{......(2)}.$$

Considering x' and y' as variable this is the equation to a straight line passing through the origin, that is, through the centre of the hyperbola.

Hence every diameter passes through the centre.

Also every straight line passing through the centre is a diameter, that is, bisects some system of parallel chords. For by giving to θ a suitable value the equation (2) may be made to represent *any* straight line passing through the centre. If θ' be the inclination to the axis of x of the diameter which bisects all the chords inclined at an angle θ, we have from (2)

$$\tan\theta' = \frac{b^2}{a^2}\cot\theta;\ \text{therefore}\ \tan\theta\tan\theta' = \frac{b^2}{a^2}.$$

238. *If one diameter bisect all chords parallel to a second diameter, the second diameter will bisect all chords parallel to the first.*

Let θ_1 and θ_2 be the respective inclinations of the two diameters to the transverse axis of the hyperbola. Since the first bisects all the chords parallel to the second, we have $\tan\theta_2\tan\theta_1 = \frac{b^2}{a^2}$. And this is also the only condition that must hold in order that the second may bisect the chords parallel to the first.

The definition in Art. 191 holds for the hyperbola.

239. Every straight line passing through the centre of an ellipse meets that ellipse; this is evident from the figure, or it may be proved analytically. But in the case of an hyperbola this proposition is not true, as we proceed to shew.

240. To find the points of intersection of an hyperbola with a straight line passing through its centre.

Let the equation to the straight line be $y = mx$.

Substitute this value of y in the equation to the hyperbola $a^2y^2 - b^2x^2 = -a^2b^2$; then we have for determining the abscissæ of the points of intersection the equation $(a^2m^2 - b^2)x^2 = -a^2b^2$; therefore $x^2 = \frac{a^2b^2}{b^2 - a^2m^2}$. Hence the values of x are impossible if a^2m^2 is greater than b^2. Thus a straight line drawn through the centre of an hyperbola will not meet the curve if it makes with the transverse axis on either side of it an angle greater than $\tan^{-1}\frac{b}{a}$.

241. It is convenient for the sake of enunciating many properties of the hyperbola to introduce the following important definition.

DEFINITION. The conjugate hyperbola is an hyperbola having for its transverse and conjugate axes the conjugate and transverse axes of the original hyperbola respectively.

242. *To find the equation to the hyperbola conjugate to a given hyperbola.*

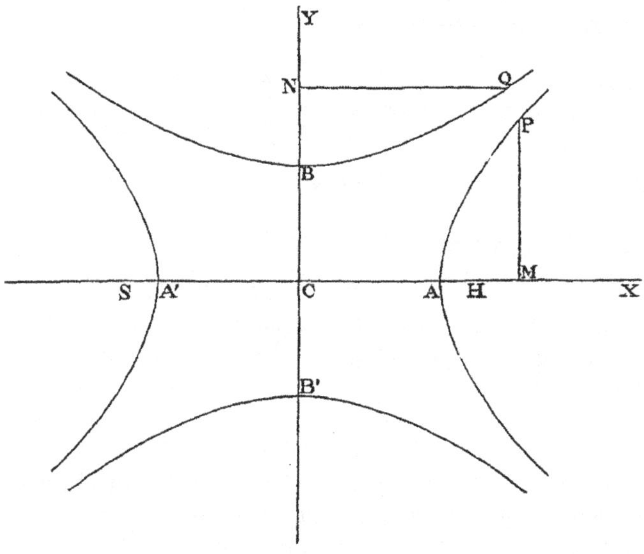

Let AA', BB' be the transverse and conjugate axes respectively of the given hyperbola; then BB' is the transverse axis of the conjugate hyperbola, and AA' is its conjugate axis. Let P be a point in the given hyperbola, Q a point in the conjugate hyperbola. Draw PM, QN perpendicular to

CX, CY respectively. The equation to the given hyperbola is $y^2 = \frac{b^2}{a^2}(x^2 - a^2)$; therefore $PM^2 = \frac{CB^2}{CA^2}(CM^2 - CA^2)$. Hence $QN^2 = \frac{CA^2}{CB^2}(CN^2 - CB^2)$, since Q is a point on an hyperbola having CB, CA for its semi-transverse and semi-conjugate axes respectively. Thus if x, y denote the co-ordinates of Q, we have $x^2 = \frac{a^2}{b^2}(y^2 - b^2)$.

This, therefore, is the equation to the conjugate hyperbola; we observe that it may be deduced from the equation to the given hyperbola by writing $-a^2$ for a^2 and $-b^2$ for b^2.

The foci of the conjugate hyperbola will be on the straight line BCB' at a distance from $C = AB$ (Art. 216); that is, at the same distance from C as S and H.

243. *Every straight line drawn through the centre of an hyperbola meets the hyperbola or the conjugate hyperbola, except the two straight lines inclined to the transverse axis of the hyperbola at an angle* $= \tan^{-1}\frac{b}{a}$.

Let the equation to the straight line be

$$y = mx \quad \text{.........................(1).}$$

To find the abscissæ of the points of intersection of (1) with the given hyperbola, we have, as in Art. 240, the equation

$$x^2 = \frac{a^2b^2}{b^2 - a^2m^2} \quad \text{.....................(2).}$$

Similarly to find the points of intersection of (1) with the conjugate hyperbola, we have the equation

$$x^2 = \frac{a^2b^2}{a^2m^2 - b^2} \quad \text{.....................(3).}$$

If m^2 be less than $\frac{b^2}{a^2}$, (2) gives possible values, and (3) impossible values for x; if m^2 be greater than $\frac{b^2}{a^2}$, (2) gives impossible values, and (3) possible values for x; if $m^2=\frac{b^2}{a^2}$, (2) and (3) make x infinite. Thus the two straight lines that can be drawn at an inclination $\tan^{-1}\frac{b}{a}$ to the transverse axis of the given hyperbola meet neither curve; and every other straight line meets one of the curves.

244. *Of two conjugate diameters one meets the original hyperbola, and the other the conjugate hyperbola.*

Let the equations to the two diameters be

$$y=mx, \qquad y=m'x;$$

then, by Art. 238, $mm'=\frac{b^2}{a^2}$; therefore $m^2m'^2=\frac{b^4}{a^4}$.

Hence if m^2 is less than $\frac{b^2}{a^2}$, m'^2 is greater than $\frac{b^2}{a^2}$; thus the first diameter meets the original hyperbola, and the second meets the conjugate hyperbola. If m^2 is greater than $\frac{b^2}{a^2}$, then m'^2 is less than $\frac{b^2}{a^2}$; thus the first diameter meets the conjugate hyperbola, and the second meets the original hyperbola.

245. We proceed now to some properties connected with conjugate diameters. When we speak of the *extremities* of a diameter we mean the points where that diameter intersects the original hyperbola or the conjugate hyperbola.

We may remark that the original hyperbola bears the same relation to the conjugate hyperbola as the conjugate hyperbola bears to the original hyperbola. Thus the definition may be given as follows: two hyperbolas are called conjugate when each has for its transverse axis the conjugate axis of the other.

Also if a straight line bisect all parallel chords terminated by one of the hyperbolas it bisects all the chords of the same system which are terminated by the other hyperbola. For the equation (Art. 237) $\tan\theta\tan\theta'=\frac{b^2}{a^2}$ remains unchanged when we write $-a^2$ for a^2 and $-b^2$ for b^2, that is, when we pass from the original hyperbola to the conjugate (Art. 242).

Both curves are comprised in the equation

$$(a^2y^2 - b^2x^2)^2 = a^4b^4.$$

246. *The tangent at either extremity of any diameter is parallel to the chords which that diameter bisects.* See Art. 190.

247. *Given the co-ordinates of one extremity of a diameter, to find those of each extremity of the conjugate diameter.*

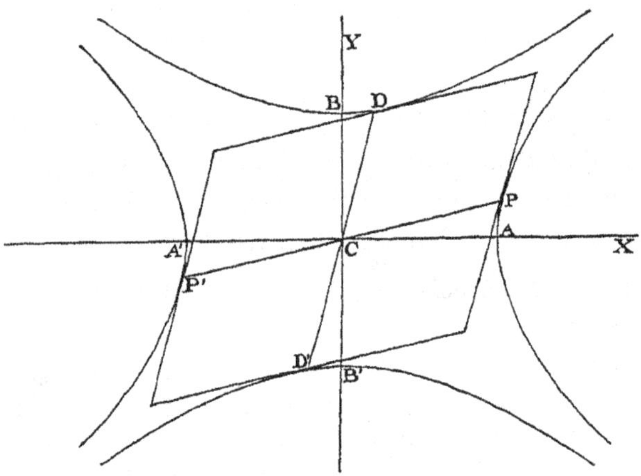

Let ACA', BCB' be the axes of an hyperbola; PCP', DCD' a pair of conjugate diameters. Let x', y' be the given co-ordinates of P; then the equation to CP is

$$y = \frac{y'}{x'}x \text{ .... (1).}$$

Since the conjugate diameter DD' is parallel to the tangent at P, the equation to DD' is

$$y = \frac{b^2x'}{a^2y'}x \text{ .... (2).}$$

We must combine (2) with the equation to the conjugate hyperbola to find the co-ordinates of D and D'. Substitute from (2) in $a^2y^2 - b^2x^2 = a^2b^2$; then $a^2 \frac{b^4}{a^4}\frac{x'^2}{y'^2}x^2 - b^2x^2 = a^2b^2$;

$$\text{therefore } (b^2x'^2 - a^2y'^2)\,x^2 = a^4y'^2;$$

$$\text{therefore } x^2 = \frac{a^4y'^2}{a^2b^2} = \frac{a^2y'^2}{b^2}; \text{ therefore } x = \pm\frac{ay'}{b};$$

$$\text{therefore from (2), } y = \pm\frac{bx'}{a}.$$

In the figure the abscissa of D is positive, and that of D' negative; hence the upper sign applies to D, and the lower sign to D'.

248. *The difference of the squares on two conjugate semi-diameters is constant.*

Let x', y' be the co-ordinates of P; then, by the preceding Article, $CP^2 - CD^2 = x'^2 + y'^2 - \frac{a^2y'^2}{b^2} - \frac{b^2x'^2}{a^2}$

$$= \frac{b^2x'^2 - a^2y'^2}{b^2} + \frac{a^2y'^2 - b^2x'^2}{a^2} = a^2 - b^2.$$

Hence the difference of the squares on two conjugate semi-diameters is equal to the difference of the squares on the semi-axes.

Moreover

$$CD^2 = x'^2 + y'^2 - a^2 + b^2 = x'^2 + \frac{b^2}{a^2}(x'^2 - a^2) - a^2 + b^2$$

$$= x'^2\left(1 + \frac{b^2}{a^2}\right) - a^2 = e^2x'^2 - a^2 = SP\,.\,HP \text{ by Art. 218.}$$

249. *The area of the parallelogram formed by tangents at the ends of conjugate diameters is constant.*

Let PCP', DCD' be the conjugate diameters (see the figure to Art. 247). The area of the parallelogram formed by tangents at P, D, P', D', is $4CP\,.\,CD \sin PCD$, or $4p\,.\,CD$, where p denotes the perpendicular from C on the tangent at P.

Let x', y' be the co-ordinates of P; then the equation to the tangent at P is $y=\frac{b^2x'}{a^2y'}x-\frac{b^2}{y'}$. Hence (Art. 47)

$$p=\frac{\frac{b^2}{y'}}{\sqrt{\left(1+\frac{b^4x'^2}{a^4y'^2}\right)}}=\frac{a^2b^2}{\sqrt{(a^4y'^2+b^4x'^2)}}.$$

$$\text{And } CD=\sqrt{\left(\frac{a^2y'^2}{b^2}+\frac{b^2x'^2}{a^2}\right)}=\frac{\sqrt{(a^4y'^2+b^4x'^2)}}{ab};$$

$$\text{therefore } 4p\,.\,CD=4ab.$$

Hence the area of any parallelogram formed by tangents at the ends of conjugate diameters is equal to the area of the rectangle formed by tangents at the ends of the axes.

250. Let a', b' denote the lengths of two conjugate semi-diameters; α the angle between them; by the preceding Article, $a'b'\sin\alpha=ab$. By making P move along the hyperbola from A we can make CP or a' as great as we please. Also since $a'^2-b'^2$ is constant, b' increases with a'. Thus $\sin\alpha$ can be made as small as we please, that is, CP and CD can be brought as near to coincidence as we please. The limiting position towards which they tend is easily found; for from Art. 237, $mm'=\frac{b^2}{a^2}$; thus the limit to which m and m' approach as CP and CD approach to coincidence is $\pm\frac{b}{a}$.

251. From Art. 249 we have

$$p^2=\frac{a^2b^2}{CD^2}=\frac{a^2b^2}{CP^2-a^2+b^2}. \quad \text{(Art. 248.)}$$

This gives a relation between p the perpendicular from the centre on the tangent at any point P, and the distance CP of that point from the centre.

As in Art. 196 $p\,.\,PG=b^2$, and $p\,.\,PG'=a^2$.

Also if ϕ denote the angle which the perpendicular makes with the transverse axis, we may shew as in Art. 196 that

$$p^2 = a^2(1 - e^2 \sin^2 \phi).$$

252. *To find the equation to the hyperbola referred to a pair of conjugate diameters as axes.*

Let CP, CD be two conjugate semidiameters (see the figure to Art. 247), take CP as the new axis of x, CD as that of y; let $PCA = \alpha$, $DCA = \beta$. Let x, y be the co-ordinates of any point of the hyperbola referred to the original axes; x', y' the co-ordinates of the same point referred to the new axes; then (Art. 84)

$$x = x' \cos\alpha + y' \cos\beta, \quad y = x' \sin\alpha + y' \sin\beta.$$

Substitute these values in the equation $a^2y^2 - b^2x^2 = -a^2b^2$;

then $a^2(x' \sin\alpha + y' \sin\beta)^2 - b^2(x' \cos\alpha + y' \cos\beta)^2 = -a^2b^2$,

or
$$x'^2(a^2 \sin^2\alpha - b^2 \cos^2\alpha) + y'^2(a^2 \sin^2\beta - b^2 \cos^2\beta)$$
$$+ 2x'y'(a^2 \sin\alpha \sin\beta - b^2 \cos\alpha \cos\beta) = -a^2b^2.$$

But since CP and CD are conjugate semidiameters, $\tan\alpha \tan\beta = \frac{b^2}{a^2}$; hence the coefficient of $x'y'$ vanishes, and the equation becomes

$$x'^2(a^2 \sin^2\alpha - b^2 \cos^2\alpha) + y'^2(a^2 \sin^2\beta - b^2 \cos^2\beta) = -a^2b^2.$$

In this equation suppose $y' = 0$, then

$$x'^2 = \frac{-a^2b^2}{a^2 \sin^2\alpha - b^2 \cos^2\alpha} = \frac{a^2b^2}{b^2 \cos^2\alpha - a^2 \sin^2\alpha}.$$

This is the value of CP^2, which we shall denote by a'^2. If we put $x' = 0$ in the above equation, we obtain

$$y'^2 = \frac{-a^2b^2}{a^2 \sin^2\beta - b^2 \cos^2\beta}.$$

Now since we have supposed that the new axis of x meets the curve, we know that the new axis of y will *not* meet the curve (Art. 244), so that $\frac{-a^2b^2}{a^2 \sin^2\beta - b^2 \cos^2\beta}$ is not a *positive* quantity; we shall denote it by $-b'^2$. Hence the equation to the hyperbola referred to conjugate diameters is $\frac{x'^2}{a'^2} - \frac{y'^2}{b'^2} = 1$, or, suppressing the accents on the variables, $\frac{x^2}{a'^2} - \frac{y^2}{b'^2} = 1$.

Also the equation to the conjugate hyperbola referred to the same axes is $\frac{x^2}{a'^2} - \frac{y^2}{b'^2} = -1$.

The equation to the tangent to the hyperbola will be of *the same form* whether the axes be rectangular or the oblique system formed by a pair of conjugate diameters. (See Art. 200.)

253. *Tangents at the extremities of any chord of an hyperbola meet on the diameter which bisects that chord.* (See Art. 201.)

254. *If a chord and diameter of an hyperbola are parallel, the supplemental chord is parallel to the conjugate diameter.* (See Arts. 202, 203.)

Asymptotes.

255. The properties of the hyperbola hitherto given have been similar to those of the ellipse; we have now to consider some properties peculiar to the hyperbola.

Let the equation to the hyperbola be $y^2 = \frac{b^2}{a^2}(x^2 - a^2)$, and let CL be the straight line which has for its equation $y = \frac{bx}{a}$.

Let MPQ be an ordinate meeting the hyperbola at P and the straight line CL at Q; then if CM be denoted by x,

$$PM = \frac{b}{a}\sqrt{(x^2 - a^2)}, \qquad QM = \frac{bx}{a};$$

thus $PQ = \frac{b}{a}\{x - \sqrt{(x^2 - a^2)}\} = \frac{b}{a} \cdot \frac{a^2}{x + \sqrt{(x^2 - a^2)}} = \frac{ab}{x + \sqrt{(x^2 - a^2)}}$.

If then the straight line MPQ be supposed to move parallel to itself from A, the distance PQ continually diminishes, and by taking CM large enough we may make PQ as small as we

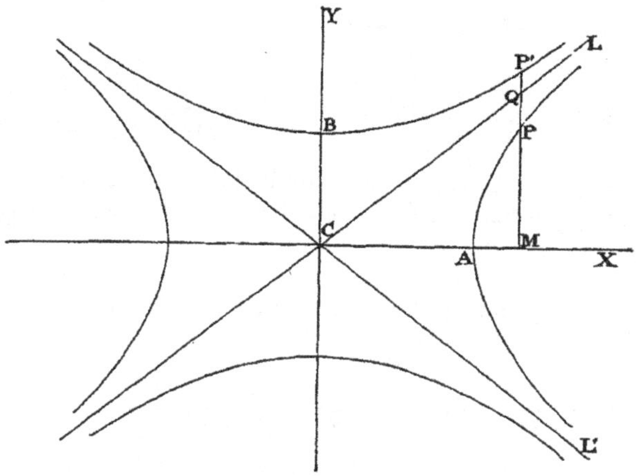

please. The straight line CL is called an *asymptote* of the curve.

Similarly the straight line CL', which has for its equation $y = -\frac{bx}{a}$, is an asymptote.

Thus the equation $\frac{x^2}{a^2} - \frac{y^2}{b^2} = 0$ includes both asymptotes. We may take the following definition.

DEFINITION. An asymptote is a straight line the distance of which from a point of a curve diminishes without limit as the point on the curve moves to an infinite distance from the origin.

The distance of P from CL is $PQ \sin PQC$; and as we have seen that PQ diminishes without limit as P moves away from the origin, CL is an *asymptote* according to the definition here given.

256. In the same manner we may shew that CL is an

asymptote to the conjugate hyperbola. For let MP be produced to meet the conjugate hyperbola at P', then (Art. 242)

$$P'M = \frac{b}{a}\sqrt{(x^2 + a^2)};$$

therefore $$P'Q = \frac{b}{a}\{\sqrt{(x^2 + a^2)} - x\} = \frac{ba}{\sqrt{(x^2 + a^2)} + x}.$$

Hence as CM is increased indefinitely $P'Q$ is diminished indefinitely; therefore CL is an asymptote to the conjugate hyperbola.

257. The equation to the tangent to the hyperbola at the point (x', y') is $a^2yy' - b^2xx' = -a^2b^2$,

therefore $$y = \frac{b^2x'x}{a^2y'} - \frac{b^2}{y'} = \frac{b}{a} \cdot \frac{x'x}{\sqrt{(x'^2 - a^2)}} - \frac{b^2}{y'}$$

$$= \frac{bx}{a\sqrt{\left(1 - \frac{a^2}{x'^2}\right)}} - \frac{b^2}{y'}.$$

If x' and y' are increased indefinitely the limiting form to which the above equation approaches is $y = \frac{bx}{a}$. Thus the tangent to the hyperbola approaches continually to coincidence with an asymptote when the point of contact moves away indefinitely from the origin.

258. It appears from Art. 243 that every straight line drawn through the centre of an hyperbola must meet the hyperbola or its conjugate, unless its direction coincides with that of one of the asymptotes. And from Art. 250 it appears that as conjugate diameters increase indefinitely they approach to coincidence with one of the asymptotes.

259. *The straight line joining the ends of conjugate diameters is parallel to one asymptote and bisected by the other.*

Let x', y' be the co-ordinates of any point P on the hyperbola (see the figure to Art. 247); then the co-ordinates of D, the extremity of the conjugate diameter, are (Art. 247) $\frac{ay'}{b}$ and $\frac{bx'}{a}$. Hence the equation to DP is

$$y - y' = \frac{y' - \frac{bx'}{a}}{x' - \frac{ay'}{b}}(x - x'),$$

that is, $$y - y' = -\frac{b}{a}(x - x');$$

and therefore DP is parallel to the asymptote $y = -\frac{bx}{a}$.

Also the co-ordinates of the middle point of DP are (Art. 10)

$$\frac{1}{2}\left(x' + \frac{ay'}{b}\right) \text{ and } \frac{1}{2}\left(y' + \frac{bx'}{a}\right),$$

that is, $$\frac{ay' + bx'}{2b} \text{ and } \frac{ay' + bx'}{2a}.$$

These co-ordinates satisfy the equation $y = \frac{bx}{a}$; therefore the asymptote $y = \frac{bx}{a}$ bisects PD.

Since the diagonals of a parallelogram bisect each other, and PD is one diagonal of the parallelogram of which CP and CD are adjacent sides, the other diagonal coincides with the asymptote, that is, the tangents at P and D meet on the asymptote.

260. The equation to the hyperbola referred to conjugate diameters as axes is

$$\frac{x^2}{a'^2} - \frac{y^2}{b'^2} = 1 \quad\text{.......................(1).}$$

Hence the equations to the asymptotes referred to these axes are

$$y = \frac{b'x}{a'}, \quad y = -\frac{b'x}{a'} \quad\text{.................(2).}$$

For we may shew as in Art. 243 that the straight lines denoted by (2) are the only straight lines through the centre which meet neither (1) nor its conjugate. Hence these straight lines are the asymptotes by Art. 258.

Or the same conclusion may be obtained thus: the original equation to the hyperbola is $\frac{x^2}{a^2} - \frac{y^2}{b^2} = 1$, and that to the two asymptotes $\frac{x^2}{a^2} - \frac{y^2}{b^2} = 0$. If by substituting for x and y their values in terms of the new co-ordinates x' and y', and suppressing accents on the variables, the former equation is reduced to $\frac{x^2}{a'^2} - \frac{y^2}{b'^2} = 1$, the latter must become, by the same substitution, $\frac{x^2}{a'^2} - \frac{y^2}{b'^2} = 0$.

261. *To find the equation to the hyperbola referred to the asymptotes as axes.*

Let CX, CY be the original axes; CX', CY' the new axes, so that CX' and CY' are inclined to CX on opposite sides of it at an angle α such that $\tan\alpha = \frac{b}{a}$. Let x, y be the co-ordinates of a point P referred to the old axes; x', y' the co-ordinates of the same point referred to the new axes. Draw PM' parallel to CY', and PM and $M'N$ each parallel to CY. Then

$$x = CM = CN + NM = (x' + y')\cos\alpha.$$

So
$$y = PM = (y' - x')\sin\alpha.$$

Also $\cos\alpha = \frac{a}{\sqrt{(a^2+b^2)}}$, and $\sin\alpha = \frac{b}{\sqrt{(a^2+b^2)}}$; substitute these values in the equation $a^2y^2 - b^2x^2 = -a^2b^2$;

then
$$a^2b^2(y' - x')^2 - a^2b^2(y' + x')^2 = -a^2b^2(a^2 + b^2),$$

or
$$x'y' = \frac{a^2 + b^2}{4},$$

or, suppressing the accents, $xy = \frac{a^2 + b^2}{4}$.

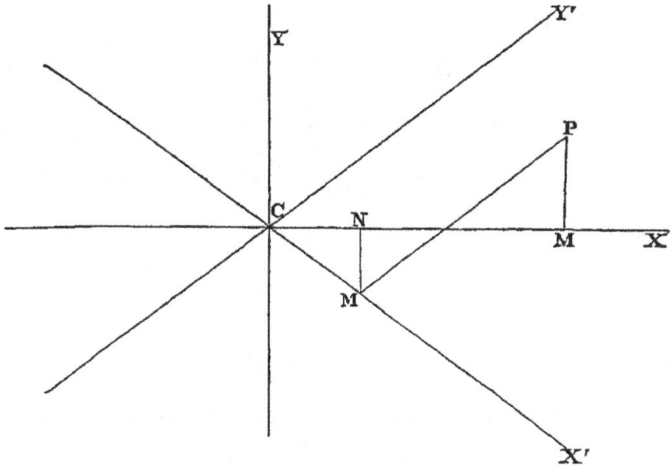

The equation to the conjugate hyperbola referred to the same axes is (Art. 242) $xy = -\frac{a^2 + b^2}{4}$.

262. *To find the equation to the tangent at any point of an hyperbola when the curve is referred to its asymptotes as axes.*

Let x', y' be the co-ordinates of the point; x'', y'' the co-ordinates of an adjacent point on the curve. The equation to the secant through these points is

$$y - y' = \frac{y'' - y'}{x'' - x'}(x - x') \quad \ldots\ldots\ldots\ldots\ldots(1).$$

Since (x', y') and (x'', y'') are points on the hyperbola

$$x'y' = \tfrac{1}{4}(a^2 + b^2), \qquad x''y'' = \tfrac{1}{4}(a^2 + b^2);$$

$$\text{therefore } x''y'' = x'y'.$$

Hence (1) may be written $y - y' = \dfrac{\dfrac{x'y'}{x''} - y'}{x'' - x'}(x - x')$, or $$y - y' = -\frac{y'}{x''}(x - x').$$

Now in the limit $x'' = x'$; hence the equation to the tangent at the point (x', y') is

$$y - y' = -\frac{y'}{x'}(x - x') \dots\dots\dots\dots\dots\dots (2).$$

This equation may be simplified; multiply by x', thus

$$yx' + xy' = 2x'y' = \frac{a^2 + b^2}{2}.$$

263. To find where the tangent at (x', y') meets the axis of x put $y = 0$ in the equation $yx' + xy' = \frac{a^2 + b^2}{2}$;

thus $$x = \frac{a^2 + b^2}{2y'} = \frac{2x'y'}{y'} = 2x'.$$

Similarly to find where the tangent cuts the axis of y put $x = 0$ in the equation; thus $y = \frac{a^2 + b^2}{2x'} = \frac{2x'y'}{x'} = 2y'$.

Thus the product of the intercepts $= 4x'y' = a^2 + b^2$. The area of the triangle contained between the tangent at any point and the asymptotes is equal to the product of the intercepts on the axes into half the sine of the included angle $= \frac{1}{2}(a^2 + b^2) \sin 2\alpha = (a^2 + b^2) \sin \alpha \cos \alpha = ab$, and is therefore constant.

Since the tangent at (x', y') cuts off intercepts $2x'$, $2y'$, from the axes of x and y respectively, the portion of the tangent at any point intercepted between the asymptotes is bisected at the point of contact.

Polar Equation.

264. To find the polar equation to the hyperbola, the focus being the pole.

Let $HP = r$, $AHP = \theta$; (see the figure to Art. 209);

then $$HP = ePN, \text{ by definition;}$$

that is, $$HP = e(OH + HM);$$

or $$r = a(e^2 - 1) + er\cos(\pi - \theta), \text{ (Art. 212);}$$

therefore $$r(1+e\cos\theta)=a(e^2-1),$$

and $$r=\frac{a(e^2-1)}{1+e\cos\theta}\ \ldots\ldots\ldots\ldots(1).$$

If we denote the angle XHP by θ, then we have as before

$$HP=e(OH+HM);$$

thus $$r=a(e^2-1)+er\cos\theta,$$

and $$r=\frac{a(e^2-1)}{1-e\cos\theta}\ \ldots\ldots\ldots\ldots(2).$$

We may also proceed thus: in the figure to Art. 218 suppose $SP=r$ and $PSH=\theta$: then $SP=ePN'$,

that is, $$SP=e(SM-SE');$$

or $$r=er\cos\theta-a(e^2-1);$$

therefore $$r(e\cos\theta-1)=a(e^2-1),$$

and $$r=\frac{a(e^2-1)}{e\cos\theta-1}\ \ldots\ldots\ldots\ldots(3).$$

265. As in Art. 205 it may be shewn that if the equation to the hyperbola be (1) of Art. 264 then the polar equation to a chord subtending at the focus an angle 2β is

$$r=\frac{l}{e\cos\theta+\sec\beta\cos(\alpha-\theta)},$$

$\alpha-\beta$ and $\alpha+\beta$ being respectively the vectorial angles of the straight lines which join the focus to the ends of the chord, and l the semi-latus rectum.

Hence the polar equation to the tangent is

$$r=\frac{l}{e\cos\theta+\cos(\alpha-\theta)}.$$

266. The polar equation to the hyperbola, the centre being the pole, is (Art. 206)

$$r^2(a^2\sin^2\theta-b^2\cos^2\theta)=-a^2b^2.$$

Arts. 207, 208 are applicable to the Hyperbola.

267. It will be a good exercise to trace the form of the hyperbola from any of the polar equations of Art. 264. Take for example the equation (1); suppose $\theta = 0$, then $r = a(e-1)$; we must therefore measure off the length $a(e-1)$ on the initial line from the pole H, and we thus obtain the point A as one of the points of the curve.

As θ increases from 0 to $\frac{\pi}{2}$ we see from (1) that r increases; $\cos\theta$ is negative when θ is greater than $\frac{\pi}{2}$ and r continues to increase. Let α be such an angle that $1 + e\cos\alpha = 0$, that is, $\cos\alpha = -\frac{1}{e}$, then the nearer θ approaches to α the greater r becomes, and by taking θ near enough to α, we may make r as great as we please. Thus as θ increases from 0 to α that portion of the curve is traced out which begins at A and passes on through P to an indefinite distance from the origin.

When θ is greater than α, r is negative, and is at first indefinitely great and diminishes as θ increases from α to π. Since r is negative we measure it in the direction *opposite* to

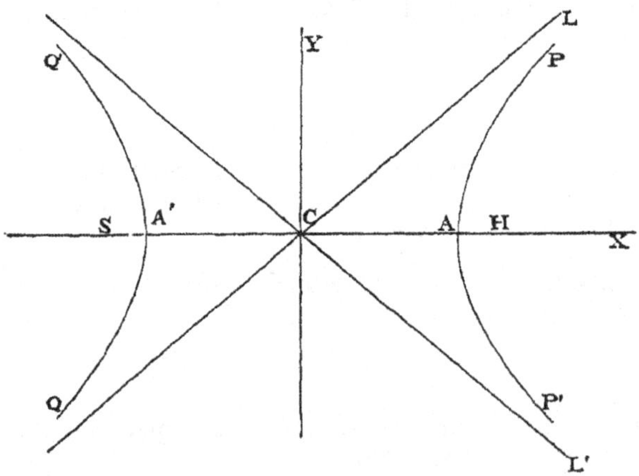

that we should use if it were positive. Thus as θ increases from α to π that portion of the curve is traced out which begins at an indefinite distance from C in the lower left-hand quadrant, and passes on through Q to A'. HA' is found by putting $\theta = \pi$ in (1); then r becomes $-a(e+1)$, therefore HA' is in length $= a(e+1)$.

As θ increases from π to $2\pi - \alpha$, r continues negative and numerically increases, and may be made as great as we please by taking θ sufficiently near to $2\pi - \alpha$. Thus the branch of the curve is traced out which begins at A' and passes on through Q' to an indefinite distance.

As θ increases from $2\pi - \alpha$ to 2π, r is again positive, and is at first indefinitely great and then diminishes. Thus the portion of the curve is traced out which begins at an indefinitely great distance from C in the lower right-hand quadrant and passes on through P' to A.

The asymptotes CL and CL' are inclined to the transverse axis at an angle of which the tangent is $\frac{b}{a}$; hence we have $\cos LCA = \frac{a}{\sqrt{(a^2+b^2)}} = \frac{1}{e}$, and $\cos LCA' = -\frac{1}{e}$; that is, $LCA' = \alpha$. Thus as θ approaches the value α the radius vector approaches to a position parallel to CL. Similarly as θ approaches the value $2\pi - \alpha$ the radius vector approaches to a position parallel to CL'.

Equilateral or Rectangular Hyperbola.

268. If in the equation to the ellipse $a^2y^2 + b^2x^2 = a^2b^2$, we suppose $b = a$, we obtain $x^2 + y^2 = a^2$, which is the equation to a circle; so that the circle may be considered a particular case of the ellipse. If in the equation to the hyperbola $a^2y^2 - b^2x^2 = -a^2b^2$ we suppose $b = a$, we have $y^2 - x^2 = -a^2$. We thus obtain an hyperbola which is called the *equilateral* hyperbola from the equality of the axes. Since the angle between the asymptotes, which $= 2\tan^{-1}\frac{b}{a}$, becomes a right angle when $b = a$, the *equilateral* hyperbola is also called the *rectangular* hyperbola.

The peculiar properties of the rectangular hyperbola can be deduced from those of the ordinary hyperbola by making $b=a$. Thus since $b^2=a^2(e^2-1)$ we have $e^2-1=1$, therefore $e=\sqrt{2}$. The equation to the tangent is (Art. 220)

$$yy'-xx'=-a^2.$$

From Art. 227 $PG=PG'=\sqrt{(rr')}$.

The equation to the conjugate hyperbola is, by Art. 242, $y^2-x^2=a^2$. Thus the conjugate hyperbola is the same curve as the original hyperbola, though differently situated.

By Art. 248, $CP=CD$, and therefore by Art. 259, CP and CD are equally inclined to the asymptotes.

EXAMPLES.

1. The radius of a circle which touches an hyperbola and its asymptotes is equal to that part of the latus rectum which is intercepted between the curve and the asymptote.

2. A straight line drawn through one of the vertices of an hyperbola and terminated by two straight lines drawn through the other vertex parallel to the asymptotes will be bisected at the other point where it cuts the hyperbola.

3. A straight line cuts an hyperbola at P and p, and the asymptotes at Q and q: shew that $PQ=pq$.

4. If a straight line be drawn from the focus of an hyperbola the part intercepted between the curve and the asymptote $=\dfrac{a\sin\alpha}{\sin\alpha+\sin\theta}$, where θ and α are the angles made respectively by the straight line and asymptote with the axis.

5. PQ is one of a series of chords inclined at a constant angle to the diameter AB of a circle: find the locus of the point of intersection of AP and BQ.

6. P is a point in a branch of an hyperbola, P' is a point in a branch of its conjugate, CP, CP', being conjugate semi-diameters. If S, S' be the interior foci of the two branches, prove that the difference of SP and $S'P'$ is equal to the difference of AC and BC.

7. If x, y be co-ordinates of any point of an hyperbola, shew that we may assume $x = a \sec\theta$, $y = b\tan\theta$.

8. A straight line is drawn parallel to the axis of y meeting the hyperbola $\frac{x^2}{a^2} - \frac{y^2}{b^2} = 1$, and its conjugate, at points P, Q: shew that the normals at P and Q intersect each other on the axis of x. Shew also that the tangents at P and Q intersect on the curve whose equation is $y^4(a^2y^2 - b^2x^2) = 4b^6x^2$.

9. Tangents to an hyperbola are drawn from any point in one of the branches of the conjugate: shew that the chord of contact will touch the other branch of the conjugate.

10. Find the equation to the radii from the centre to the points of contact of the two tangents in Example 9, and if these radii are at right angles, shew that the co-ordinates of the point from which the tangents are drawn are

$$a\sqrt{\left(\frac{b^2 - 2a^2}{a^2 + b^2}\right)}, \quad b\sqrt{\left(\frac{2b^2 - a^2}{a^2 + b^2}\right)}.$$

11. Two tangents to a parabola include an angle α: shew that the locus of their point of intersection is an hyperbola with the same focus and directrix.

12. Shew under what limitation the proposition in Example 30 of Chapter X. is true for the hyperbola.

13. The ratio of the sines of the angles made by a diameter of an hyperbola with the asymptotes is equal to the ratio of the sines of the angles made by the conjugate diameter.

14. With two conjugate diameters of an ellipse as asymptotes a pair of conjugate hyperbolas is constructed: prove that if one hyperbola touch the ellipse the other will do so likewise; prove also that the diameters drawn through the points of contact are conjugate to each other.

CHAPTER XIII.

GENERAL EQUATION OF THE SECOND DEGREE.

269. We shall now shew that every locus represented by an equation of the second degree is one of those which we have already discussed, that is, is one of the following: a point, a straight line, two straight lines, a circle, a parabola, an ellipse, or an hyperbola.

The general equation of the second degree may be written

$$ax^2 + bxy + cy^2 + dx + ey + f = 0;$$

we shall suppose the axes rectangular; if the axes were oblique we might transform the equation to one referred to rectangular axes, and as such a transformation cannot affect the degree of the equation (Art. 87), the transformed equation will still be of the form given above.

If the curve passes through the origin $f = 0$; if the curve does *not* pass through the origin f is not $= 0$, we may therefore divide by f and thus the equation will take the form

$$a'x^2 + b'xy + c'y^2 + d'x + e'y + 1 = 0.$$

270. We shall begin by investigating the possibility of removing from the equation the terms involving the *first* power of the variables.

Transfer the origin of co-ordinates to the point (h, k) by putting $x = x' + h$, $y = y' + k$, and substituting these values of x and y in the equation

$$ax^2 + bxy + cy^2 + dx + ey + f = 0 \text{.........}(1);$$

thus we obtain

$$ax'^2 + bx'y' + cy'^2 + (2ah + bk + d)\,x' + (2ck + bh + e)\,y' + f' = 0 \text{.......}(2),$$

where $$f' = ah^2 + bhk + ck^2 + dh + ek + f \text{.........}(3).$$

Now, if possible, let such values be assigned to h and k as will make the coefficients of x' and y' vanish; that is, let

$$2ah + bk + d = 0, \quad \text{and} \quad 2ck + bh + e = 0;$$

thus
$$h = \frac{2cd - be}{b^2 - 4ac}, \quad k = \frac{2ae - bd}{b^2 - 4ac}.$$

It will therefore be possible to assign suitable values to h and k, *provided* $b^2 - 4ac$ *be not* $= 0$.

We shall see that the loci represented by the general equation of the second degree may be separated into two classes, those which have a *centre*, and those which in general have *not* a centre, and that in the former case $b^2 - 4ac$ is not zero, and in the latter case it is zero. We shall first consider the case in which $b^2 - 4ac$ is not zero, and consequently the values found above for h and k are finite.

Equation (2) thus becomes

$$ax'^2 + bx'y' + cy'^2 + f' = 0 \quad \text{.................(4)}.$$

Now if (4) is satisfied by any values x_1, y_1 of the variables, it is also satisfied by the values $-x_1, -y_1$. Hence the new origin of co-ordinates is the *centre* of the locus represented by (1).

Thus if $b^2 - 4ac$ be not $= 0$, the locus represented by (1) *has a centre*, and its co-ordinates are h and k, the values of which are given above.

The value of f' may be found by substituting the values of h and k in (3); the process may be facilitated thus: we have

$$2ah + bk + d = 0, \quad 2ck + bh + e = 0;$$

multiply the first of these equations by h, and the second by k, and add; thus $2ah^2 + 2ck^2 + 2bkh + dh + ek = 0$,

or
$$2f' - dh - ek - 2f = 0;$$

$$\text{therefore } f' = f + \frac{dh + ek}{2} = f + \frac{cd^2 + ae^2 - bed}{b^2 - 4ac}.$$

We shall retain f' for shortness.

271. We may suppress the accents on the variables in equation (4) of the preceding Article and write it

$$ax^2 + bxy + cy^2 + f' = 0 \quad \ldots\ldots\ldots\ldots\ldots (5).$$

This equation we shall further simplify by changing the directions of the axes. (Art. 81.)

Put $x = x' \cos\theta - y' \sin\theta$, $y = x' \sin\theta + y' \cos\theta$, and substitute in (5); thus

$$x'^2 (a\cos^2\theta + c\sin^2\theta + b\sin\theta\cos\theta) \\ + y'^2 (a\sin^2\theta + c\cos^2\theta - b\sin\theta\cos\theta) \\ + x'y' \{2(c-a)\sin\theta\cos\theta + b(\cos^2\theta - \sin^2\theta)\} + f' = 0 \ldots (6).$$

Equate the coefficient of $x'y'$ to zero; thus

$$2(c-a)\sin\theta\cos\theta + b(\cos^2\theta - \sin^2\theta) = 0,$$

or

$$(c-a)\sin 2\theta + b\cos 2\theta = 0;$$

$$\text{therefore } \tan 2\theta = \frac{b}{a-c} \quad \ldots\ldots\ldots\ldots\ldots\ldots\ldots\ldots (7).$$

Since θ can always be found so as to satisfy (7), the term involving $x'y'$ can be removed from (6), and the equation becomes

$$x'^2 (a\cos^2\theta + c\sin^2\theta + b\sin\theta\cos\theta) \\ + y'^2 (a\sin^2\theta + c\cos^2\theta - b\sin\theta\cos\theta) + f' = 0,$$

or

$$Ax'^2 + By'^2 + f' = 0 \quad \ldots\ldots\ldots\ldots\ldots\ldots\ldots (8),$$

where

$$A = \tfrac{1}{2}\{a + c + (a-c)\cos 2\theta + b\sin 2\theta\},$$
$$B = \tfrac{1}{2}\{a + c - (a-c)\cos 2\theta - b\sin 2\theta\}.$$

Since

$$\tan 2\theta = \frac{b}{a-c},$$

$$\cos 2\theta = \frac{a-c}{\sqrt{\{b^2 + (a-c)^2\}}}, \text{ and } \sin 2\theta = \frac{b}{\sqrt{\{b^2 + (a-c)^2\}}}.$$

Hence

$$A = \tfrac{1}{2}[a + c + \sqrt{\{b^2 + (a-c)^2\}}],$$
$$B = \tfrac{1}{2}[a + c - \sqrt{\{b^2 + (a-c)^2\}}].$$

We may suppress the accents on the variables in (8) and write it $-\frac{A}{f'}x^2 - \frac{B}{f'}y^2 = 1$.

(1) If A, B, and f' have the same sign, the locus is impossible.

(2) If A and B have the same sign and f' have the contrary sign, the locus is an ellipse of which the semi-axes are respectively

$$\sqrt{\left(-\frac{f'}{A}\right)}, \text{ and } \sqrt{\left(-\frac{f'}{B}\right)}. \quad \text{(Art. 160.)}$$

The locus is of course a circle if $A = B$.

(3) If A and B have different signs, the locus is an hyperbola. (Art. 211.)

We have supposed in these three cases that f' is not $= 0$; if $f' = 0$, and A and B have the *same* sign the locus is the origin; if $f' = 0$, and A and B have *different* signs the locus consists of two straight lines represented by

$$y = \pm\sqrt{\left(-\frac{A}{B}\right)}x.$$

From the values of A and B we see that

$$AB = \frac{(a+c)^2 - b^2 - (a-c)^2}{4} = \frac{4ac - b^2}{4}.$$

Hence A and B have the same sign or different signs according as $b^2 - 4ac$ is negative or positive.

272. Hence we have the following summary of the results of the preceding Articles of this Chapter. The equation

$$ax^2 + bxy + cy^2 + dx + ey + f = 0$$

represents an ellipse if $b^2 - 4ac$ be *negative*, subject to three exceptions in which it represents respectively a circle, a point, and an impossible locus. If $b^2 - 4ac$ be *positive*, the equation represents an hyperbola subject to one exception when it represents two intersecting straight lines.

273. We may notice that the equation found in Art. 271,

$\tan 2\theta = \frac{b}{a-c}$, has an infinite number of solutions; for if 2α be *one* value of 2θ which satisfies the equation, then if $2\theta = 2\alpha + n\pi$, where n is any integer, the equation will be satisfied. But these different solutions will all give the same position for the axes. For the values of θ are comprised in the expression $\alpha + \frac{n\pi}{2}$, and by ascribing different values to n we obtain a series of angles each differing from α by a multiple of $\frac{\pi}{2}$, and the only changes that will arise from selecting different values of n are that the axes of x and y in one case may occupy respectively the positions of the axes of y and x in another, or the positive and negative directions of the axes may be interchanged.

The denominator in the value of $\cos 2\theta$ and of $\sin 2\theta$ in Art. 271 may have *either* sign; but the sign must be the same in both in order that the relation $\tan 2\theta = \frac{b}{a-c}$ may hold.

274. It appears from the former part of Art. 271, that by turning the axes through an angle θ the equation

$$ax^2 + bxy + cy^2 + f' = 0$$

becomes
$$a'x'^2 + b'x'y' + c'y'^2 + f' = 0,$$

where
$$a' = \tfrac{1}{2}\{a + c + (a-c)\cos 2\theta + b\sin 2\theta\},$$
$$b' = (c-a)\sin 2\theta + b\cos 2\theta,$$
$$c' = \tfrac{1}{2}\{a + c - (a-c)\cos 2\theta - b\sin 2\theta\}.$$

Hence
$$a' + c' = a + c; \quad \text{and}$$
$$b'^2 - 4a'c' = \{(c-a)\sin 2\theta + b\cos 2\theta\}^2 - (a+c)^2 + \{(a-c)\cos 2\theta + b\sin 2\theta\}^2$$
$$= (a-c)^2 + b^2 - (a+c)^2 = b^2 - 4ac.$$

Thus the expression $b^2 - 4ac$ has the same value whether it be formed from the coefficients of the general equation of the second degree *before* or *after* the axes have been shifted.

The same remark applies to the expression $a + c$.

Hence we conclude that if the curve represented by the general equation $ax^2 + bxy + cy^2 + dx + ey + f = 0$ be a rectangular hyperbola, $a + c = 0$; for if the curve were referred to its transverse and conjugate diameters as axes this relation would hold, and therefore, as we have just seen, it must always hold whatever be the axes.

275. We have next to consider the case in which $b^2 - 4ac$ is zero. We cannot now as in Art. 270 remove the terms involving the first power of the variables from the general equation, but we can still simplify the equation as in Art. 271, by changing the direction of the axes.

Let the equation be

$$ax^2 + bxy + cy^2 + dx + ey + f = 0 \quad \text{..........(1)};$$

put $\quad x = x' \cos\theta - y' \sin\theta, \quad y = x' \sin\theta + y' \cos\theta,$

then (1) becomes

$$\begin{aligned} x'^2 (a \cos^2\theta + c \sin^2\theta + b \sin\theta \cos\theta) \\ + y'^2 (a \sin^2\theta + c \cos^2\theta - b \sin\theta \cos\theta) \\ + x'y' \{2 (c - a) \sin\theta \cos\theta + b (\cos^2\theta - \sin^2\theta)\} \\ + x' (d \cos\theta + e \sin\theta) + y' (e \cos\theta - d \sin\theta) + f = 0 \quad \text{...... (2)}. \end{aligned}$$

Now let $\tan 2\theta = \dfrac{b}{a - c}$, then the coefficient of $x'y'$ in (2) vanishes, and as in Art. 271 the coefficients of x'^2 and y'^2 are $\frac{1}{2}[a + c \pm \sqrt{\{(a - c)^2 + b^2\}}]$. One of these coefficients must therefore vanish since their product is $\dfrac{4ac - b^2}{4}$, which, by hypothesis, $= 0$; suppose the coefficient of $x'^2 = 0$, thus, by suppressing accents on the variables, (2) may be written

$$Cy^2 + Dx + Ey + f = 0 \quad \text{.............. (3)}.$$

If D be *not* $= 0$, this may be written

$$C\left(y + \frac{E}{2C}\right)^2 = -D\left(x - \frac{E^2}{4CD} + \frac{f}{D}\right),$$

and thus the locus is a parabola. (Art. 125.)

If $D = 0$, then (3) represents two parallel straight lines, or one straight line, or an impossible locus, according as E^2 is greater, equal to, or less than $4Cf$.

Hence if $b^2 - 4ac = 0$ the equation

$$ax^2 + bxy + cy^2 + dx + ey + f = 0$$

represents a parabola subject to three exceptions, in which it represents respectively two parallel straight lines, one straight line, and an impossible locus.

By combining this result with those stated in Art. 272, we have a complete account of the general equation of the second degree.

276. We have shewn in Art. 270, that when $b^2 - 4ac$ is not $= 0$, the general equation of the second degree represents a *central* curve; we shall now prove that when $b^2 - 4ac = 0$ the curve has *not* a centre *except when the locus consists of two parallel straight lines.*

If a curve of the second degree have the origin of co-ordinates for its centre, no term involving the first power of either of the variables alone can exist in the equation.

For if possible suppose that the origin of co-ordinates is the centre of the curve

$$ax^2 + bxy + cy^2 + dx + ey + f = 0 \quad\text{...........}\quad (1),$$

and let x_1, y_1 be the co-ordinates of a point on the curve, and therefore $-x_1$, $-y_1$ co-ordinates of another point on the curve; substitute successively in (1), then

$$ax_1^2 + bx_1y_1 + cy_1^2 + dx_1 + ey_1 + f = 0,$$
$$ax_1^2 + bx_1y_1 + cy_1^2 - dx_1 - ey_1 + f = 0;$$

therefore, by subtraction,

$$2(dx_1 + ey_1) = 0 \quad\text{...................}\quad (2).$$

Now unless d and e both vanish, (2) can only be true when (x_1, y_1) lies on the straight line $dx + ey = 0$. But the centre of a curve is a point which bisects *every* chord passing through it; hence the origin of co-ordinates cannot be the centre of the curve (1) unless both d and e vanish.

277. Suppose then that we have an equation

$$ax^2 + bxy + cy^2 + dx + ey + f = 0 \quad\text{........}\quad (1),$$

in which $b^2 - 4ac = 0$. Here a and c cannot both be zero, for then b would also be zero, and (1) would not be an equation of the second degree; we shall suppose that a is not zero. Now if the curve denoted by (1) had a centre, and we took that centre as the origin of co-ordinates, the terms involving the first power of x and y would vanish by Art. 276. But from Arts. 270 and 274 it follows that when $b^2 - 4ac = 0$, we cannot *in general* make these terms vanish by changing the origin or the axes. The *only exception* that can arise is when the numerators in the values of h and k in Art. 270 vanish, so that the values of h and k become indeterminate, and the two equations for determining them reduce to one; see *Algebra*, Chapter XV. We have then $2ae - bd = 0$, so that $e = \frac{bd}{2a}$. Hence, by substituting for c and e, the equation (1) becomes

$$ax^2 + bxy + \frac{b^2y^2}{4a} + dx + \frac{bd}{2a}y + f = 0,$$

that is,
$$a\left(x + \frac{by}{2a}\right)^2 + d\left(x + \frac{by}{2a}\right) + f = 0 \ldots\ldots\ldots\ldots(2).$$

Equation (2) will furnish two values of $x + \frac{by}{2a}$, so that if these values are possible the locus consists of two parallel straight lines. In this case any point on the straight line which is parallel to these two and midway between them will be a centre.

Thus the result enunciated in the beginning of Art. 276 is demonstrated.

278. We may observe that relations similar to those obtained in Art. 274 hold when the axes of co-ordinates are *oblique*. For suppose the equation $ax^2 + bxy + cy^2 + f' = 0$ to be referred to rectangular axes, and let the axes be transformed into an oblique system inclined at an angle ω; suppose moreover that the new axis of x coincides with the old axis of x. We have then to put (Art. 84)

$$x = x' + y' \cos\omega, \qquad y = y' \sin\omega;$$

substitute these values in the above equation and it becomes

$$a'x'^2 + b'x'y' + c'y'^2 + f' = 0,$$

where $a' = a$,

$b' = 2a\cos\omega + b\sin\omega$,

$c' = a\cos^2\omega + b\sin\omega\cos\omega + c\sin^2\omega$;

thus $$b'^2 - 4a'c' = (b^2 - 4ac)\sin^2\omega,$$

and $$a' + c' - b'\cos\omega = (a + c)\sin^2\omega;$$

so that $$\frac{b'^2 - 4a'c'}{\sin^2\omega} = b^2 - 4ac,$$

and $$\frac{a' + c' - b'\cos\omega}{\sin^2\omega} = a + c.$$

Therefore, by means of Art. 274, we conclude that for any system of axes, rectangular or oblique, the expressions $\frac{b'^2 - 4a'c'}{\sin^2\omega}$ and $\frac{a' + c' - b'\cos\omega}{\sin^2\omega}$ remain unchanged when the axes are changed.

These results are very important, because as we have seen, the curve will in general be an ellipse, parabola, or hyperbola according as the former expression is negative, zero, or positive; and a rectangular hyperbola if the latter expression be zero.

These results may be obtained by another method, which will be found instructive. Suppose that the axes of x and y are inclined at an angle λ; and let us determine the points of intersection of the curve

$$ax^2 + bxy + cy^2 = g \text{.....................(1)},$$

and the circle

$$x^2 + 2xy\cos\lambda + y^2 = r^2 \text{..................(2)}.$$

Combining (1) and (2) we obtain

$$g(x^2 + 2xy\cos\lambda + y^2) = r^2(ax^2 + bxy + cy^2);$$

that is $(g - r^2a)x^2 + (2g\cos\lambda - r^2b)xy + (g - r^2c)y^2 = 0.$

This is a quadratic equation for finding $\frac{y}{x}$. Solving the quadratic we find that the expression under the radical sign is

$$r^4(b^2 - 4ac) - 4r^2g(b\cos\lambda - a - c) - 4g^2\sin^2\lambda.$$

If this expression vanishes the two values of $\frac{y}{x}$ are equal; this indicates that the circle (2) *touches* the curve (1): and hence we may draw the important inference that the squares of the semi-axes of the curve (1) are numerically equal to the values of r^2 given by the equation

$$r^4 \frac{b^2 - 4ac}{\sin^2 \lambda} - 4r^2 g \frac{b \cos \lambda - a - c}{\sin^2 \lambda} - 4g^2 = 0 \text{ ....(4).}$$

Now suppose the axes of co-ordinates transformed into another system inclined at the angle λ', and let (1) become

$$a'x'^2 + b'x'y' + c'y'^2 = g;$$

then the quadratic equation

$$r^4 \frac{b'^2 - 4a'c'}{\sin^2 \lambda'} - 4r^2 g \frac{b' \cos \lambda' - a' - c'}{\sin^2 \lambda'} - 4g^2 = 0 \text{ ....(5)}$$

has the same geometrical meaning as (4), and the roots will therefore be the same. Hence (4) and (5) must coincide, and therefore

$$\frac{b^2 - 4ac}{\sin^2 \lambda} = \frac{b'^2 - 4a'c'}{\sin^2 \lambda'} \text{ ....(6),}$$

and

$$\frac{b \cos \lambda - a - c}{\sin^2 \lambda} = \frac{b' \cos \lambda' - a' - c'}{\sin^2 \lambda'} \text{ ....(7).}$$

In fact if we divide $-4g^2$ by either member of (6) we obtain the numerical value of the product of the squares of the semi-axes of the curve. Similarly if we divide $4g$ times either member of (7) by the corresponding member of (6) we obtain the numerical value of the sum or of the difference of the squares of the semi-axes, according as the curve is an ellipse or an hyperbola.

279. We shall now shew how to trace a curve of the second degree from its equation without transformation of co-ordinates; the axes may be supposed oblique or rectangular.

Let the equation be

$$ax^2 + bxy + cy^2 + dx + ey + f = 0 \text{ ....(1).}$$

Solve the equation with respect to y; thus

$$y=-\frac{bx+e}{2c}\pm\frac{1}{2c}\{(bx+e)^2-4c(ax^2+dx+f)\}^{\frac{1}{2}}$$

$$=-\frac{bx+e}{2c}\pm\frac{1}{2c}\{(b^2-4ac)x^2+2(be-2cd)x+e^2-4cf\}^{\frac{1}{2}}\ldots(2)$$

$$=\alpha x+\beta\pm\left\{\frac{b^2-4ac}{4c^2}(x^2+2px+q)\right\}^{\frac{1}{2}}\ldots\ldots\ldots\ldots\ldots\ldots\ldots(3);$$

where $\alpha=-\dfrac{b}{2c}$, $\beta=-\dfrac{e}{2c}$, $p=\dfrac{be-2cd}{b^2-4ac}$, $q=\dfrac{e^2-4cf}{b^2-4ac}$.

I. Suppose b^2-4ac negative, and write $-\mu$ for $\dfrac{b^2-4ac}{4c^2}$; thus (3) becomes

$$y=\alpha x+\beta\pm\{-\mu(x^2+2px+q)\}^{\frac{1}{2}}\ldots\ldots\ldots(4).$$

Now $x^2+2px+q=(x+p)^2+q-p^2$; if then $q-p^2$ be positive, the quantity under the radical is negative and the locus impossible; if $q-p^2=0$, the locus is the point determined by $x=-p$, $y=\alpha x+\beta$; if $q-p^2$ be negative, we may put $(x+p)^2+q-p^2=\{x+p+\sqrt{(p^2-q)}\}\{x+p-\sqrt{(p^2-q)}\}$ $=(x-\gamma)(x-\delta)$ suppose; and thus (4) may be written

$$y=\alpha x+\beta\pm\{-\mu(x-\gamma)(x-\delta)\}^{\frac{1}{2}}\ldots\ldots\ldots\ldots(5).$$

Since $(x-\gamma)(x-\delta)$ is positive, except when x lies between γ and δ, the values of y in (5) are real only so long as x lies between γ and δ. Moreover y is always *finite*, and thus the curve represented by (5) is limited in every direction.

Since we know from our previous investigations that (5) must represent one of the curves enumerated in Art. 269, it follows that it must represent an *ellipse*.

From the form of equation (5) we see that the chords parallel to the axis of y are bisected by the straight line

$$y=\alpha x+\beta\ldots\ldots\ldots\ldots\ldots\ldots\ldots\ldots\ldots(6).$$

For let there be two points on the curve (5) having the common abscissa x_1, and the ordinates y', y'', respectively; and let y_1 be the corresponding ordinate of (6), then $$y_1 = \alpha x_1 + \beta,$$

$$y' = \alpha x_1 + \beta + \{-\mu(x_1 - \gamma)(x_1 - \delta)\}^{\frac{1}{2}},$$

$$y'' = \alpha x_1 + \beta - \{-\mu(x_1 - \gamma)(x_1 - \delta)\}^{\frac{1}{2}}.$$

Thus $y_1 = \frac{1}{2}(y' + y'')$; and therefore the point (x_1, y_1) lies midway between the points (x_1, y') and (x_1, y'').

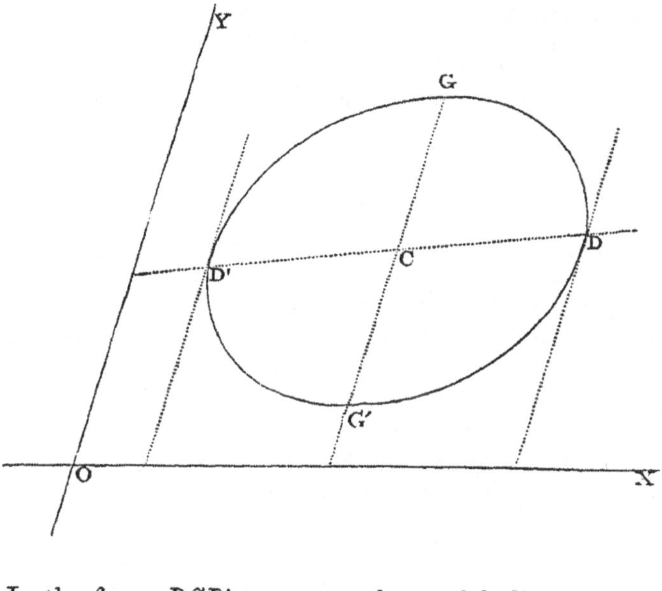

In the figure DCD' represents the straight line $y = \alpha x + \beta$; the abscissæ of D' and D are γ and δ respectively; supposing δ greater than γ. The centre C is midway between D' and D; its abscissa is therefore $\frac{1}{2}(\gamma + \delta)$. The equation to the curve will give the ordinates of D', D, G', G. Since GG' is parallel to the chords which $D'D$ bisects, DD' and GG' are conjugate diameters. GG' is a known quantity since the ordinates of G and G' are known. DD' is also a known quantity since the abscissæ and ordinates of D and D' are known. The angle between GG' and DD' is known from the equation to DD'; the axes of the ellipse may therefore be found (Arts. 193, 195).

II. Suppose b^2-4ac positive; put μ for $\dfrac{b^2-4ac}{4c^2}$; thus equation (3) becomes

$$y=\alpha x+\beta \pm \{\mu (x^2+2px+q)\}^{\frac{1}{2}} \quad \text{...........(7).}$$

Now $x^2+2px+q=(x+p)^2+q-p^2$; if then $q-p^2$ be positive, the quantity under the radical is always positive, whatever positive or negative value be assigned to x. The curve therefore extends to infinity. Also it may be shewn as before, that the straight line $y=\alpha x+\beta$ is a diameter of the curve; but it never *meets* the curve, because the quantity $x^2+2px+q$ or $(x+p)^2+q-p^2$ cannot vanish. Hence the curve consists of *two* unconnected branches extending to infinity, and is therefore an hyperbola.

If $q-p^2=0$, (7) becomes $y=\alpha x+\beta \pm \sqrt{\mu}\,(x+p)$.

The locus now consists of two intersecting straight lines.

If $q-p^2$ be negative we may as before write (7) in the form $y=\alpha x+\beta \pm \{\mu\,(x-\gamma)\,(x-\delta)\}^{\frac{1}{2}}$. Hence x may have any value, positive or negative, except those between γ and δ; thus the curve consists of two unconnected branches extending to infinity, and is therefore an hyperbola.

We shall be assisted in drawing an example of this case by ascertaining the position of the asymptotes.

The equation to the curve is

$$y=\alpha x+\beta \pm \{\mu (x^2+2px+q)\}^{\frac{1}{2}};$$

therefore
$$y=\alpha x+\beta \pm x\sqrt{\mu}\left(1+\frac{2p}{x}+\frac{q}{x^2}\right)^{\frac{1}{2}}.$$

Expand by the Binomial Theorem; thus

$$y=\alpha x+\beta \pm x\sqrt{\mu}\left\{1+\tfrac{1}{2}\left(\frac{2p}{x}+\frac{q}{x^2}\right)+\text{\&c.}\right\}$$

$$=\alpha x+\beta \pm \sqrt{\mu}\,(x+p)+\text{\&c.}$$

The terms included in the &c. involve negative powers of x, and may therefore be made as small as we please by sufficiently increasing x; hence from the nature of an asymptote the required equations to the asymptotes are

$$y = ax + \beta + \sqrt{\mu}\,(x + p), \quad \text{and } y = ax + \beta - \sqrt{\mu}\,(x + p).$$

Hence we can draw the asymptotes, and therefore the axes, for they bisect the angles between the asymptotes. The intersection of the asymptotes is the centre, and thus the situation and form of the hyperbola are known.

We may observe that the tangent of the angle between the asymptotes is, by Art. 41,

$$\frac{\alpha + \sqrt{\mu} - (\alpha - \sqrt{\mu})}{1 + \alpha^2 - \mu}, \text{ that is } \frac{2\sqrt{\mu}}{1 + \alpha^2 - \mu}:$$

substitute for α and μ their values and we obtain $\dfrac{\sqrt{(b^2 - 4ac)}}{a + c}$.

The expression $q - p^2 = \dfrac{(e^2 - 4cf)(b^2 - 4ac) - (be - 2cd)^2}{(b^2 - 4ac)^2}$; this vanishes when $(e^2 - 4cf)(b^2 - 4ac) - (be - 2cd)^2 = 0$, and therefore when $(b^2 - 4ac)f + ae^2 + cd^2 - bed = 0$; so that if this relation holds the locus represented by (1) consists of two intersecting straight lines.

We have hitherto supposed that c is not zero, and as $b^2 - 4ac$ cannot be *negative* if c be zero, it was not necessary to advert to the possibility of c being zero while considering the first case. But as c may be zero consistently with $b^2 - 4ac$ being *positive*, we must now examine the consequences of supposing c zero.

The equation (1) may be solved with respect to x instead of with respect to y. Hence it will be found on investigation that the results hitherto obtained, when $b^2 - 4ac$ is positive, are certainly true provided that a and c are not *both* zero; the latter case requires further examination. Suppose then $a = 0$ and $c = 0$; thus (1) becomes $bxy + dx + ey + f = 0$; by changing the origin this can be put in the form $bx'y' + f' = 0$, where $f' = \dfrac{bf - de}{b}$; the curve is therefore an hyperbola with the new axes for its asymptotes, except when $bf - de = 0$, and then it becomes two intersecting straight lines. When $a = 0$

and $c=0$, the expression $(b^2-4ac)f+ae^2+cd^2-bed$ reduces to $b(bf-de)$; thus we conclude that when b^2-4ac is positive the equation (1) always represents an hyperbola, except when $(b^2-4ac)f+ae^2+cd^2-bed=0$, and then it represents two intersecting straight lines.

III. Suppose $b^2-4ac=0$, then (2) becomes

$$y=-\frac{bx+e}{2c}\pm\frac{1}{2c}\{2(be-2cd)x+e^2-4cf\}^{\frac{1}{2}},$$

which may be written $y=\alpha x+\beta\pm\frac{1}{2c}(p'x+q')^{\frac{1}{2}}$,

where $$\alpha=-\frac{b}{2c},\quad \beta=-\frac{e}{2c},$$

$$p'=2(be-2cd),\quad q'=e^2-4cf.$$

If p' be positive, the expression under the radical is positive or negative, according as x is algebraically greater or less than $-\frac{q'}{p'}$; if p' be negative, the statement must be reversed. In both cases the curve extends to infinity in *one* direction only and is therefore a *parabola*.

The straight line $y=\alpha x+\beta$ is a diameter, bisecting all ordinates parallel to the axis of y, and meeting the parabola at the point for which $x=-\frac{q'}{p'}$.

If $p'=0$, the equation becomes $y=\alpha x+\beta\pm\frac{\sqrt{q'}}{2c}$; this equation represents two parallel straight lines if q' is positive, and one straight line if $q'=0$; if q' is negative, the locus is impossible.

We have hitherto supposed in considering the third case that c is not zero; if $c=0$, then $b=0$, since $b^2-4ac=0$; hence a and c cannot *both* be zero, for the equation (1) is supposed to be of the *second* degree. As before, we may solve equation (1) with respect to x, and thus determine the peculiarities which occur when $c=0$. We have found for example when c is not zero, that the locus will consist of two parallel straight lines, when $be-2cd=0$, and e^2-4cf is positive; in like manner, if a be not zero, we can shew that the locus will consist of two parallel straight lines when $bd-2ae=0$, and d^2-4af is positive. By means of the relation $b^2-4ac=0$, it is easily shewn that the second form of the conditions coincides with the first when a and c are both different from zero. When $a=0$ the first is the necessary form of the conditions, but we see that the second form will then also hold. When $c=0$ the second is the necessary form, though the first will then also hold. Hence we shall include every case by stating that *both* forms of the conditions must hold.

Similarly the conditions under which the locus will consist of one straight line, or will be impossible, may be investigated.

280. We will recapitulate the results of the present Chapter with respect to the locus of the equation

$$ax^2+bxy+cy^2+dx+ey+f=0.$$

I. If b^2-4ac be negative, the locus is an ellipse admitting of the following varieties:

(1) $c=a$, and $\frac{b}{2a}=$ cosine of the angle between the axes; locus a circle (Art. 104).

(2) $(e^2-4cf)(b^2-4ac)-(be-2cd)^2$ positive; locus impossible.

(3) $(e^2-4cf)(b^2-4ac)-(be-2cd)^2=0$; locus a point.

II. If b^2-4ac be positive, the locus is an hyperbola, except when $(b^2-4ac)f+ae^2+cd^2-bde=0$, and then it consists of two intersecting straight lines.

III. If $b^2-4ac=0$, the locus is a parabola, except when $be-2cd=0$, and $bd-2ae=0$; and then it consists of two parallel straight lines, or of one straight line, or is impossible, according as e^2-4cf and d^2-4af are positive, zero, or negative.

EXAMPLES.

1. Find the centre of the curve

$$x^2 - 4xy + 4y^2 - 2ax + 4ay = 0.$$

2. Find the centre of the ellipse

$$by\left(1 - \frac{y}{c}\right) + cx\left(1 - \frac{x}{b}\right) = xy.$$

3. Find what is represented by $ax^2 + 2bxy + cy^2 = 1$, when $b^2 = ac$.

4. Find the locus of the centre of a circle inscribed in a sector of a given circle, one of the bounding radii of the sector remaining fixed.

5. In the side AB of a triangle ABC, any point P is taken, and PQ is drawn perpendicular to AC: find the locus of the point of intersection of the straight lines BQ and CP.

6. DE is any chord parallel to the major axis AA' of an ellipse whose centre is C; and AD and CE intersect at P: shew that the locus of P is an hyperbola, and find the direction of its asymptotes.

7. Tangents to two concentric ellipses, the directions of whose axes coincide, are drawn from a point P, and the chords of contact intersect at Q: if the point P always lies on a straight line, shew that the locus of Q will be a rectangular hyperbola.

8. Find what form the result in the preceding Example takes when two of the axes whose directions are coincident are equal.

9. Prove that an hyperbola may be described by the intersection of two straight lines which move parallel to themselves while the product of their distances from a fixed point remains constant.

10. Two straight lines are drawn from the focus of an ellipse including a constant angle; tangents are drawn to the ellipse at the points where the straight lines meet the ellipse: find the locus of the intersection of the tangents.

11. Find the latus rectum of the parabola $(y-x)^2 = ax$.

12. Shew that the product of the semi-axes of the ellipse $y^2 - 4xy + 5x^2 = 2$ is 2.

13. Find the angle between the asymptotes of the hyperbola $xy = bx^2 + c$.

14. Find the equation to a parabola which touches the axis of x at a distance α, and cuts the axis of y at distances β, β' from the origin.

15. If two points be taken in each of two rectangular axes, so as to satisfy the condition that a rectangular hyperbola may pass through all the four, shew that the position of the hyperbola is indeterminate, and that its centre describes a circle which passes through the origin and bisects all the straight lines which join the points two and two.

16. Two straight lines of given lengths coincide with and move along two fixed axes in such a manner that a circle may always be drawn through their extremities: find the locus of the centre of the circle, and shew that it is an equilateral hyperbola.

17. A variable ellipse always touches a given ellipse, and has a common focus with it: find the locus of its other focus, (1) when the major axis is given, (2) when the minor axis is given.

18. Draw the curve $y^2 - 5xy + 6x^2 - 14x + 5y + 4 = 0$.

19. Draw the curve $x^2 + y^2 - 3(x+y) - xy = 0$.

20. Find the nature and position of the curve

$$y^2 - 8xy + 25x^2 + 6cy - 42cx + 9c^2 = 0.$$

21. The equation to a conic section is $ax^2 + 2bxy + cy^2 = 1$: shew that the equation to its axes is $xy(a-c) = b(x^2 - y^2)$.

22. The locus of the vertices of all similar triangles whose bases are parallel chords of a parabola will in general be another parabola; but if any one of the triangles *touch* the parabola with its sides, the locus becomes a straight line.

23. A series of circles pass through a given point O, have their centres in a straight line OA, and meet another straight line BC. Let M be the point at which one of the circles meets the straight line OA again, and let N be either of the points at which this circle meets BC. From M and N straight lines are drawn parallel to BC and OA respectively, intersecting at P. Shew that the locus of P is an hyperbola which becomes a parabola when the two straight lines are at right angles.

24. The chord of contact of two tangents to a parabola subtends an angle β at the vertex: shew that the locus of their point of intersection is an hyperbola whose asymptotes are inclined to the axis of the parabola at an angle ϕ such that $\tan\phi = \frac{1}{2}\tan\beta$.

25. Determine the locus of the middle points of the chords of the curve $ax^2 + 2bxy + cy^2 + 2ex + 2fy + g = 0$, which are parallel to the straight line $x\sin\theta - y\cos\theta = 0$; and hence find the position of the principal axes of the curve.

26. Shew that the equation $(x^2 - a^2)^2 + (y^2 - a^2)^2 = a^4$ represents two ellipses.

27. AB and AC are given in position, and BC is of constant length: shew that if PB and PC be drawn making any constant angle with AB and AC the locus of P is an ellipse.

28. A number of parabolas whose axes are parallel have a common tangent at a given point: shew that if parallel tangents be drawn to all the parabolas the points of contact will lie on a straight line passing through the given point.

29. If on one of the longer sides of a rectangle as major axis an ellipse be described which passes through the intersection of the diagonals, and straight lines be drawn from

any point of that part of the ellipse which is external to the rectangle to the extremities of the remote side, they will divide the major axis into segments which are in geometrical progression.

30. A series of ellipses have their equal conjugate diameters of the same magnitude, one of them being common to all while the other varies in position: shew that tangents drawn from any point in the fixed diameter produced will touch the ellipses at points situated on a circle.

31. TP, TQ are tangents to a central conic section, and the chord PQ is produced to meet the directrices at R and R': shew that

$$RP.R'P : RQ.R'Q :: TP^2 : TQ^2.$$

32. In any conic section if PQ, PR make equal angles with a fixed chord PK, and QR be joined, shew that QR will pass through a fixed point for all positions of PQ, PR.

CHAPTER XIV.

MISCELLANEOUS PROPOSITIONS.

281. WE shall give in this Chapter some miscellaneous propositions for the most part applicable to all the conic sections.

To find the equation to a conic section, the origin and axes being unrestricted in position.

Let a, b be the co-ordinates of the focus; and let the equation to the directrix be $Ax + By + C = 0$. The distance of any point (x, y) from the focus is $\{(x-a)^2 + (y-b)^2\}^{\frac{1}{2}}$, and the distance of the same point from the directrix is

$$\frac{Ax + By + C}{\sqrt{(A^2 + B^2)}}.$$

Let e be the *excentricity* of the conic section; then if (x, y) be a point on the curve, we have, by definition,

$$\{(x-a)^2 + (y-b)^2\}^{\frac{1}{2}} = \frac{e\,(Ax + By + C)}{\sqrt{(A^2 + B^2)}} \quad \text{.......(1)};$$

therefore $$(x-a)^2 + (y-b)^2 = \frac{e^2\,(Ax + By + C)^2}{A^2 + B^2} \quad \text{......(2)}.$$

We see from (1) that the distance of any point on a conic section from the focus can be expressed in terms of the *first* power of the co-ordinates of that point whatever be the origin and axes. This is usually expressed by saying *the distance of any point from the focus is a linear function of the co-ordinates of the point.*

282. It will be seen by examining the equations to the conic sections given in the preceding Chapters that any conic section may be represented by the equation $y^2 = mx + nx^2$. The origin is a vertex of the curve and the axis of x an

axis of the curve; m is the latus rectum; in the parabola $n=0$; n is negative in the ellipse and positive in the hyperbola. In the circle m is the diameter of the circle and $n=-1$.

283. *To find the equation to the tangent at any point of a curve of the second degree.*

Let the equation to the curve be

$$ax^2+bxy+cy^2+dx+ey+f=0 \quad\text{..........}\quad (1),$$

the axes being oblique or rectangular.

Let x', y' be the co-ordinates of the point,

x'', y'' the co-ordinates of an adjacent point on the curve.

The equation to the secant through these points is

$$y-y'=\frac{y''-y'}{x''-x'}(x-x') \quad\text{..................}\quad (2).$$

Since (x', y') and (x'', y'') are on the curve,

$$ax'^2+bx'y'+cy'^2+dx'+ey'+f=0,$$
$$ax''^2+bx''y''+cy''^2+dx''+ey''+f=0;$$

therefore $a(x''^2-x'^2)+b(x''y''-x'y')+c(y''^2-y'^2)$
$$+d(x''-x')+e(y''-y')=0,$$

or $(x''-x')\{a(x''+x')+by''+d\}$
$$+(y''-y')\{c(y''+y')+bx'+e\}=0;$$

$$\text{therefore } \frac{y''-y'}{x''-x'}=-\frac{a(x''+x')+by''+d}{c(y''+y)+bx'+e}.$$

Hence (2) may be written

$$y-y'=-\frac{a(x''+x')+by''+d}{c(y''+y')+bx'+e}(x-x').$$

Now in the limit $x''=x'$ and $y''=y'$; hence the equation to the tangent at the point (x', y') is

$$y-y'=-\frac{2ax'+by'+d}{2cy'+bx'+e}(x-x').$$

This equation may be simplified; we have by reduction

$$y(2cy' + bx' + e) + x(2ax' + by' + d)$$
$$= y'(2cy' + bx' + e) + x'(2ax' + by' + d)$$
$$= 2(ax'^2 + bx'y' + cy'^2 + dx' + ey' + f) - dx' - ey' - 2f;$$

therefore

$$y(2cy' + bx' + e) + x(2ax' + by' + d) + dx' + ey' + 2f = 0.$$

If $f = 0$, the curve passes through the origin, and the equation to the tangent at that point becomes $y = -\frac{d}{e}x$, which we see does not involve the coefficients of x^2, y^2, or xy, in the equation to the curve.

The equation to the tangent at any point of (1) may also be found in the following manner:

Let x', y' be the co-ordinates of one point on the curve; and x'', y'' the co-ordinates of another point on the curve.

The equation to the secant through these points may be written

$$a(x - x')(x - x'') + b(x - x')(y - y'') + c(y - y')(y - y'')$$
$$= ax^2 + bxy + cy^2 + dx + ey + f.$$

For it is obvious that this equation is really of the *first degree* in x and y, and therefore represents some straight line. Moreover the equation is satisfied when $x = x'$, and $y = y'$; and also when $x = x''$, and $y = y''$. Therefore the equation represents the straight line passing through the points (x', y') and (x'', y'').

Now suppose $x'' = x'$, and $y'' = y'$; then the secant becomes the tangent at the point (x', y'), and the equation becomes

$$a(x - x')^2 + b(x - x')(y - y') + c(y - y')^2$$
$$= ax^2 + bxy + cy^2 + dx + ey + f:$$

and by simplifying we obtain the same form as before.

284. The equation to the normal at the point (x', y')

when the curve is expressed by equation (1) of the preceding Article and the axes are rectangular, will be

$$y - y' = \frac{2cy' + bx' + e}{2ax' + by' + d}(x - x').$$

285. It may be shewn as in Art. 183, that if from a point (h, k) two tangents be drawn to the curve expressed by equation (1) of Art. 283, the equation to the *chord of contact* is $y(2ck + bh + e) + x(2ah + bk + d) + dh + ek + 2f = 0$.

286. *All chords of a conic section which subtend a right angle at a given point of the curve intersect on the normal at that point.*

Take the given point of the curve as the origin of a system of rectangular axes, and let the equation to the curve be

$$ax^2 + bxy + cy^2 + dx + ey = 0 \quad \text{.............} \quad (1).$$

The axis of x meets the curve at the points found by making $y = 0$ in the above equation, that is, at the points $x = 0$, and $x = -\frac{d}{a}$. Similarly the axis of y meets the curve at the origin and also at the point for which $y = -\frac{e}{c}$.

Hence the equation $\dfrac{x}{-\frac{d}{a}} + \dfrac{y}{-\frac{e}{c}} = 1$,

or

$$\frac{xa}{d} + \frac{yc}{e} + 1 = 0 \quad \text{........................} \quad (2)$$

represents the chord joining the points of intersection of the axes and curve.

Also the equation to the normal to the curve at the origin is by Art. 284,

$$y = \frac{e}{d}x \quad \text{.........................} \quad (3).$$

Hence (2) and (3) meet at the point whose co-ordinates are $\frac{-d}{a+c}$, $\frac{-e}{a+c}$, and whose distance from the origin is therefore $\frac{\sqrt{(d^2 + e^2)}}{a+c}$.

Now change the *directions* of the axes preserving the same origin; the equation (1) will then become

$$a'x'^2 + b'x'y' + c'y'^2 + d'x' + e'y' = 0.$$

Also it appears from Arts. 274 and 275, that

$$a' + c' = a + c, \text{ and } d'^2 + e'^2 = d^2 + e^2.$$

Hence the normal at the origin will meet the new chord at the same distance from the origin as it met the original chord, that is, will meet it *at the same point.* Since this is true whatever be the directions of the axes, it follows that all the chords intersect at the same point.

287. By comparing Arts. 154, 204, and 264, we see that the polar equation to any conic section, the focus being the pole and the initial line the axis, is $r = \frac{l}{1 + e\cos\theta}$, where l = half the latus rectum.

We shall use this in proving the following proposition:

The semi-latus rectum of any conic section is an harmonic mean between the segments made by the focus of any focal chord of that conic section.

Let $A'SP = \theta$, see the figure to Art. 158;

$$\text{therefore } SP = \frac{l}{1 + e\cos\theta}.$$

Suppose PS produced to meet the curve again at P';

$$\text{therefore } SP' = \frac{l}{1 + e\cos(\pi + \theta)};$$

$$\text{therefore } \frac{1}{SP} + \frac{1}{SP'} = \frac{1 + e\cos\theta}{l} + \frac{1 - e\cos\theta}{l} = \frac{2}{l},$$

which proves the proposition.

288. The polar equation to the tangent to a conic section, the focus being the pole and the initial line the axis, is (Arts. 205, 265)

$$\frac{l}{r} = e\cos\theta + \cos(\alpha - \theta) \quad \text{.................} \quad (1),$$

where α is the angular co-ordinate of the point of contact.

Similarly the polar equation to the tangent at the point whose angular co-ordinate is β, is

$$\frac{l}{r} = e\cos\theta + \cos(\beta - \theta) \quad\text{.................(2).}$$

At the point where these tangents meet, we have

$$\cos(\alpha - \theta) = \cos(\beta - \theta).$$

Now we cannot have $\alpha - \theta = \beta - \theta$, since α and β are by supposition different; we therefore take $\alpha - \theta = \theta - \beta$, therefore $\theta = \frac{\alpha + \beta}{2}$.

Thus the two tangents (1) and (2) meet at the point whose angular co-ordinate is $\frac{\alpha + \beta}{2}$.

For example, suppose the conic section an ellipse; let $ASP = \alpha$, $ASQ = \beta$, and let the tangents at P and Q meet at T;

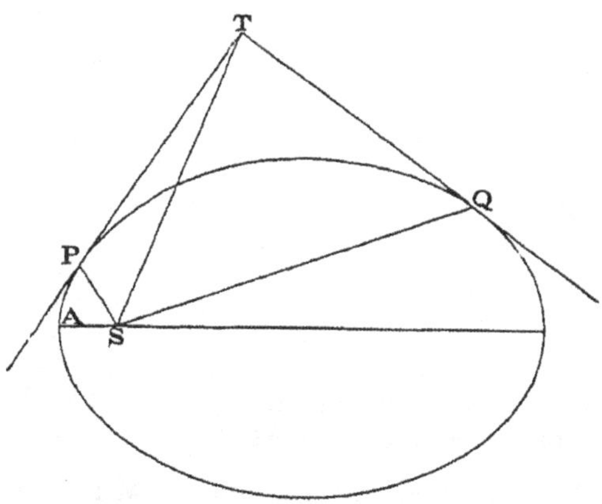

then $$AST = \frac{\alpha + \beta}{2}; \text{ therefore } PST = \frac{\beta - \alpha}{2} = QST;$$

that is, the *two tangents drawn from any point to an ellipse subtend equal angles at either focus.*

Similarly the two tangents drawn from any point to a parabola subtend equal angles at the focus.

With respect to the hyperbola we have to distinguish two cases. We have shewn in Art. 231, that from any point included between the asymptotes and the curve, two tangents can be drawn both meeting the *same* branch of the curve, but from any point included within the supplemental angles of the asymptotes two tangents can be drawn meeting *different* branches of the curve.

If now the two tangents from a point meet the *same* branch of an hyperbola, it may be shewn as in the case of the ellipse, that they subtend equal angles at either focus. We will consider the case in which the tangents meet *different* branches.

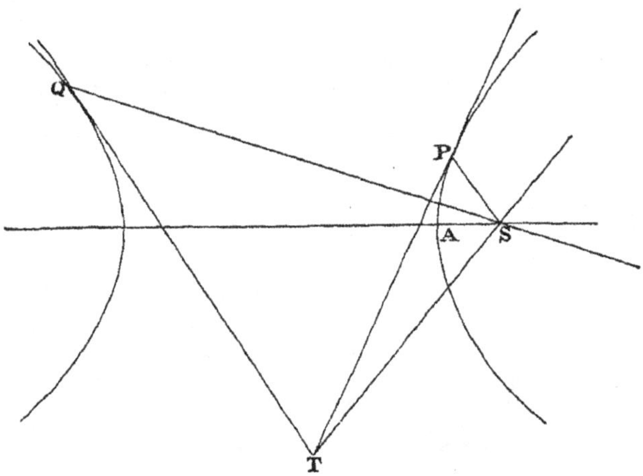

Let T be a point from which tangents TP, TQ are drawn to different branches of an hyperbola.

Let $ASP = \alpha$; and let the angle which QS produced through S makes with AS be β; then β is an angle greater than π, and $ASQ = \beta - \pi$.

Thus the equations to TP and TQ will be respectively

$$\frac{l}{r} = e \cos \theta + \cos (\alpha - \theta), \quad \frac{l}{r} = e \cos \theta + \cos (\beta - \theta).$$

At the point T where the tangents meet, we have

$$\cos(\alpha-\theta)=\cos(\beta-\theta).$$

We may therefore take $\theta=\frac{\alpha+\beta}{2}$, that is, we have $\frac{\alpha+\beta}{2}$ as the angle which TS produced makes with AS; thus

$$AST=\pi-\frac{\alpha+\beta}{2},$$

$$\text{therefore } TSP=\pi-\frac{\beta-\alpha}{2}, \qquad TSQ=\frac{\beta-\alpha}{2};$$

$$\text{therefore } TSP+TSQ=\pi;$$

that is, the angle which one tangent subtends at either focus is the supplement of the angle which the other tangent subtends at the same focus.

289. We have given in Art. 120 the definitions of a pole and polar with respect to a given circle. The same definitions are used generally substituting *conic section* for *circle.* If then the equation to the curve be

$$ax^2+bxy+cy^2+dx+ey+f=0,$$

the equation to the *polar* of (x', y') is (Art. 283)

$$x(2ax'+by'+d)+y(2cy'+bx'+e)+dx'+ey'+2f=0.$$

The equation just given always represents a straight line at a finite distance from the origin except when both

$$2ax'+by'+d=0, \text{ and } 2cy'+bx'+e=0.$$

But if x' and y' satisfy these relations they are the co-ordinates of the *centre* of the curve; see Arts. 270 and 276. Hence strictly speaking there is no polar corresponding to the centre of a conic section; this fact is frequently expressed by saying that *the polar of the centre is the straight line at infinity.* See page 74.

290. *If one straight line pass through the pole of another straight line, the second straight line will pass through the pole of the first straight line.*

Let (x', y') be the pole of the *first* straight line, and therefore the equation to the *first* straight line

$$x(2ax' + by' + d) + y(2cy' + bx' + e) + dx' + ey' + 2f = 0 \ldots (1).$$

Let (x'', y'') be the pole of the *second* straight line, and therefore the equation to the *second* straight line

$$x(2ax'' + by'' + d) + y(2cy'' + bx'' + e) + dx'' + ey'' + 2f = 0 \ldots (2).$$

Since (1) passes through (x'', y'') we have

$$x''(2ax' + by' + d) + y''(2cy' + bx' + e) + dx' + ey' + 2f = 0,$$

that is,

$$x'(2ax'' + by'' + d) + y'(2cy'' + bx'' + e) + dx'' + ey'' + 2f = 0;$$

hence (2) passes through (x', y').

291. *The intersection of two straight lines is the pole of the straight line which joins the poles of those straight lines.* See Art. 122.

292. *If a quadrilateral* ABCD *be inscribed in a conic section, of the three points* E, F, G, *each is the pole of the straight line joining the other two.*

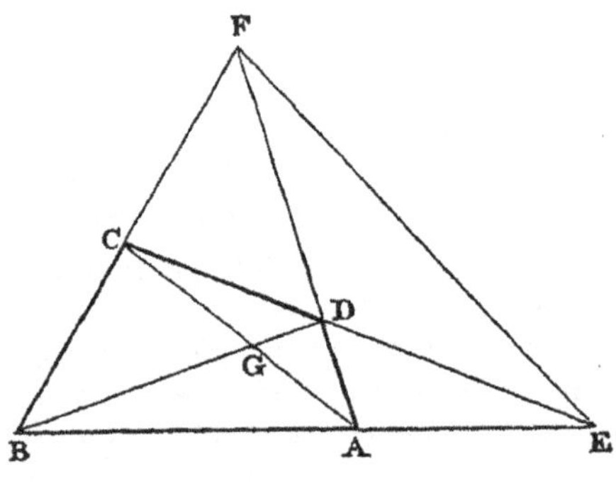

Let E be the origin; EA, ED the directions of the axes of x and y; and let the equation to the conic section be

$$ax^2 + bxy + cy^2 + dx + ey + f = 0 \ldots\ldots\ldots\ldots (1).$$

Also suppose

$$EA = h, \qquad EB = h',$$
$$ED = k, \qquad EC = k'.$$

The equation to AC is $\frac{x}{h} + \frac{y}{k'} = 1$ .... (2);

the equation to BD is $\frac{x}{h'} + \frac{y}{k} = 1$ .... (3);

the equation to AD is $\frac{x}{h} + \frac{y}{k} = 1$ .... (4);

the equation to CB is $\frac{x}{h'} + \frac{y}{k'} = 1$ .... (5).

From (2) and (3) it follows that the equation

$$x\left(\frac{1}{h} + \frac{1}{h'}\right) + y\left(\frac{1}{k} + \frac{1}{k'}\right) = 2 \text{ .... (6)}$$

represents *some* straight line passing through G. But from (4) and (5) it follows that (6) represents *some* straight line passing through F. Hence (6) must be the equation to FG.

Suppose in (1) that $y = 0$; then we have the quadratic $ax^2 + dx + f = 0$; and the roots of this equation are h and h'; hence $h + h' = -\frac{d}{a}$, $hh' = \frac{f}{a}$; therefore $\frac{1}{h} + \frac{1}{h'} = -\frac{d}{f}$. Similarly, $\frac{1}{k} + \frac{1}{k'} = -\frac{e}{f}$.

Hence (6) becomes $dx + ey + 2f = 0$.

But this, by Art. 289, is the equation to the polar of the origin; therefore FG is the polar of E. Similarly EG is the polar of F. Hence, by Art. 291, G is the pole of EF.

293. *To determine the form of the general equation to a conic section when the axes are tangents.*

Let $\qquad ax^2 + bxy + cy^2 + dx + ey + f = 0$ ....(1)

be the equation to the conic section.

To find where the curve meets the axis of x, put $y = 0$ in the above equation; thus $ax^2 + dx + f = 0$.

If the axis of x is a *tangent* to the curve it must meet the curve at only one point (see Art. 171); hence the roots of the above quadratic must be equal; therefore

$$d^2 = 4af \quad\ldots\ldots\ldots\ldots\ldots\ldots\ldots\ldots (2).$$

Similarly that the axis of y may be a tangent to (1) we must have

$$e^2 = 4cf \quad\ldots\ldots\ldots\ldots\ldots\ldots\ldots\ldots (3).$$

Substitute the values of a and c from (2) and (3), then (1) becomes $d^2x^2 + 4dfx + e^2y^2 + 4efy + 4bfxy + 4f^2 = 0$,

or $$(dx + ey + 2f)^2 + (4bf - 2de)\,xy = 0,$$

or $$\left(\frac{d}{2f}x + \frac{e}{2f}y + 1\right)^2 + \frac{2bf - de}{2f^2}\,xy = 0.$$

Put $$\frac{d}{2f} = -\frac{1}{h}, \quad \frac{e}{2f} = -\frac{1}{k}, \quad \frac{2bf - de}{2f^2} = \mu;$$

thus we obtain for the required equation

$$\left(\frac{x}{h} + \frac{y}{k} - 1\right)^2 + \mu xy = 0.$$

By putting successively x and $y = 0$, we see that h is the distance from the origin to the point where the curve meets the axis of x, and k is the distance from the origin to the point where the curve meets the axis of y.

If it be required to determine a conic section which touches two given straight lines at given points, and also passes through another given point, we may assume the last written equation to represent it, so that the straight lines to be touched are taken as the axes of x and y; then by putting the co-ordinates of the additional given point in the equation we find a single value for μ. Thus there is only one conic section satisfying the data.

294. Suppose the equation

$$\left(\frac{x}{h} + \frac{y}{k} - 1\right)^2 + \mu xy = 0 \quad\ldots\ldots\ldots\ldots\ldots\ldots (1)$$

to represent a parabola. Then, by Art. 280,

$$\left(\frac{2}{hk} + \mu\right)^2 = \frac{4}{h^2k^2};$$

$$\text{therefore } \mu = 0, \text{ or } \mu = -\frac{4}{hk}.$$

If $\mu = 0$, (1) becomes $\frac{x}{h} + \frac{y}{k} - 1 = 0$; this equation represents the straight line joining the points of contact of (1) with the axes.

If $\mu = -\frac{4}{hk}$, we have from (1),

$$\left(\frac{x}{h} + \frac{y}{k} - 1\right)^2 = \frac{4xy}{hk} \quad \text{.........................} \quad (2);$$

$$\text{therefore } \frac{x}{h} + \frac{y}{k} - 1 = \pm 2\sqrt{\left(\frac{xy}{hk}\right)};$$

$$\text{therefore } \frac{x}{h} \mp 2\sqrt{\left(\frac{xy}{hk}\right)} + \frac{y}{k} = 1;$$

$$\text{therefore } \sqrt{\frac{x}{h}} \mp \sqrt{\frac{y}{k}} = \pm 1.$$

We may write this

$$\sqrt{\frac{x}{h}} + \sqrt{\frac{y}{k}} = 1 \quad \text{.......................} \quad (3),$$

remembering that the radicals may be positive or negative. Thus (3) is the equation to a parabola referred to two tangents as axes.

295. We may notice the form of the equation to the tangent to the parabola

$$\sqrt{\frac{x}{h}} + \sqrt{\frac{y}{k}} = 1 \quad \text{.......................} \quad (1).$$

The equation to the secant through (x', y') and (x'', y'') is

$$y - y' = \frac{y'' - y'}{x'' - x'}(x - x').$$

Since (x', y') and (x'', y'') are on the parabola, we have

$$\sqrt{\frac{x'}{h}} + \sqrt{\frac{y'}{k}} = 1, \text{ and}$$

$$\sqrt{\frac{x''}{h}} + \sqrt{\frac{y''}{k}} = 1;$$

$$\text{therefore } \frac{\sqrt{x''}-\sqrt{x'}}{\sqrt{h}} = -\frac{\sqrt{y''}-\sqrt{y'}}{\sqrt{k}};$$

and $$\frac{y''-y'}{x''-x'} = \frac{\sqrt{y''}-\sqrt{y'}}{\sqrt{x''}-\sqrt{x'}} \cdot \frac{\sqrt{y''}+\sqrt{y'}}{\sqrt{x''}+\sqrt{x'}} = -\frac{\sqrt{k}}{\sqrt{h}} \cdot \frac{\sqrt{y''}+\sqrt{y'}}{\sqrt{x''}+\sqrt{x'}}.$$

Hence the equation to the secant may be written

$$y - y' = -\frac{\sqrt{k}}{\sqrt{h}} \cdot \frac{\sqrt{y''}+\sqrt{y'}}{\sqrt{x''}+\sqrt{x'}} (x - x').$$

Hence we have for the equation to the tangent at (x', y')

$$y - y' = -\frac{\sqrt{(ky')}}{\sqrt{(hx')}} (x - x'),$$

or $$\frac{y}{\sqrt{(ky')}} + \frac{x}{\sqrt{(hx')}} = \frac{y'}{\sqrt{(ky')}} + \frac{x'}{\sqrt{(hx')}} = 1.$$

Similar Curves.

296. DEFINITION. Two curves are said to be *similar* and *similarly* situated when a radius vector drawn from some fixed point in any direction to the first curve bears a constant ratio to the radius vector drawn from some fixed point in a parallel direction to the second curve.

Two curves are said to be *similar* when a radius vector drawn from some fixed point in any direction to the first curve bears a constant ratio to the radius vector drawn from some fixed point to the second curve in a direction inclined at a constant angle to the former.

The two fixed points are called *centres of similarity.*

297. If two curves are similar, so that a pair of *centres of similarity* exists, then an infinite number of pairs of centres of similarity can be found.

For, suppose O, O' to denote one pair of *centres of similarity;* and let OP, OQ be radii vectores of the first curve, and $O'P'$, $O'Q'$ the corresponding radii vectores of the second curve, so that the angle $POQ =$ the angle $P'O'Q'$, and $\frac{OP}{O'P'} = \frac{OQ}{O'Q'}$. Suppose any point S taken and joined to O;

then make the angle $P'O'S'$ = the angle POS, the angles being measured in the same direction, and take $O'S'$ so that $\frac{O'S'}{OS}=\frac{O'P'}{OP}$: then S and S' shall be centres of similarity.

For join SP, SQ, $S'P'$, $S'Q'$; then the triangles SOP, $S'O'P'$ are similar; and so also are the triangles SOQ, $S'O'Q'$. Hence it easily follows that the angle $QSP=Q'S'P'$; and that $\frac{SP}{S'P'}=\frac{SQ}{S'Q'}$; and thus the proposition is established.

298. *All parabolas are similar curves.*

Let $4a$ be the latus rectum of a parabola, and $4a'$ the latus rectum of a second parabola. The polar equations of these curves, the foci being the respective poles, are

$$r=\frac{2a}{1+\cos\theta}, \quad r'=\frac{2a'}{1+\cos\theta'}.$$

Hence, if $\theta=\theta'$, we have $\frac{r}{r'}=\frac{a}{a'}$. Thus any two parabolas are similar, and the foci are centres of similarity.

299. To find the conditions which must hold in order that the curves

$$ax^2+bxy+cy^2+dx+ey+f=0 \ldots\ldots\ldots\ldots(1),$$

$$a'x^2+b'xy+c'y^2+d'x+e'y+f'=0 \ldots\ldots\ldots\ldots(2),$$

may be similar and similarly situated.

Suppose (h, k), (h', k') the respective centres of similarity; for x and y in (1) put $h+r\cos\theta$, and $k+r\sin\theta$ respectively; we shall thus obtain a quadratic in r which may be written

$$Lr^2+Mr+N=0 \ldots\ldots\ldots\ldots\ldots\ldots\ldots\ldots(3).$$

For x and y in (2) put $h'+r'\cos\theta$, and $k'+r'\sin\theta$ respectively; we shall thus obtain a quadratic in r' which may be written

$$L'r'^2+M'r'+N'=0 \ldots\ldots\ldots\ldots\ldots\ldots\ldots\ldots(4).$$

Now that the curves may be similar and similarly situated, we must always have $r'=\lambda r$, where λ is some constant quantity; thus (4) becomes

$$\lambda^2L'r^2+\lambda M'r+N'=0 \ldots\ldots\ldots\ldots\ldots\ldots(5).$$

Since (3) or (5) will give the values of r, these equations must be *identical;* thus

$$\frac{L}{\lambda^2 L'} = \frac{M}{\lambda M'} = \frac{N}{N'} \text{ ....(6).}$$

Since neither N nor N' involves θ, we deduce as a necessary condition that $\frac{L}{L'}$ must be constant whatever θ may be. Put for L and L' their values; then

$$\frac{a\cos^2\theta + b\sin\theta\cos\theta + c\sin^2\theta}{a'\cos^2\theta + b'\sin\theta\cos\theta + c'\sin^2\theta} = \text{a constant} = \mu \text{ say.....(7)};$$

therefore $(a-\mu a')\cos^2\theta + (b-\mu b')\sin\theta\cos\theta + (c-\mu c')\sin^2\theta = 0.$

Since this is to be true whatever θ may be, it follows that

$$\frac{a}{a'} = \frac{b}{b'} = \frac{c}{c'} \text{ ....(8).}$$

Hence we have arrived at (8) as *necessary* conditions, in order that (1) and (2) may be similar and similarly situated. We have still to ascertain whether these are *sufficient* to ensure the similarity. The direct method would be to examine if h, k, h', k' can be so chosen as to make (6) hold; but the following method is more simple. The equations (1) and (2), by means of (8), may be written

$$ax^2 + bxy + cy^2 + dx + ey + f = 0,$$
$$ax^2 + bxy + cy^2 + \mu(d'x + e'y + f') = 0.$$

I. Suppose $b^2 - 4ac = 0$; then each curve is in general a parabola, and therefore the curves are similar; also their diameters are parallel so that the curves are similarly situated. See Art. 279. This conclusion is subject to the exceptions that may arise when either locus instead of a parabola, becomes one or two straight lines, or impossible.

II. Suppose $b^2 - 4ac$ not $= 0$. We may then by changing the origin of co-ordinates for each curve reduce the equations to the form

$$ax^2 + bxy + cy^2 + f_1 = 0,$$
$$ax^2 + bxy + cy^2 + f_2 = 0.$$

By expressing these equations in polar co-ordinates, they give

$$r^2 = \frac{-f_1}{a\cos^2\theta + b\sin\theta\cos\theta + c\sin^2\theta},$$

$$r'^2 = \frac{-f_2}{a\cos^2\theta' + b\sin\theta'\cos\theta' + c\sin^2\theta'}.$$

Thus, if $\theta = \theta'$, we have $\frac{r}{r'} = \text{constant}$. Hence the curves are in general similar and similarly situated. This conclusion is subject to the exceptions that may arise when either locus instead of a curve becomes two straight lines, or a point, or impossible.

300. Next, suppose we require the curves (1) and (2) of Art. 299 to be similar *without the limitation of being similarly situated.* For x and y in (1) we put respectively

$$h + r\cos\theta, \qquad k + r\sin\theta.$$

For x and y in (2) we put respectively

$$h' + r'\cos(\theta + \alpha), \qquad k' + r'\sin(\theta + \alpha),$$

where α is some constant angle at present undetermined. Proceed as in Article 299; instead of equation (7) we shall now have

$$\frac{a\cos^2\theta + b\sin\theta\cos\theta + c\sin^2\theta}{a'\cos^2(\theta+\alpha) + b'\sin(\theta+\alpha)\cos(\theta+\alpha) + c'\sin^2(\theta+\alpha)}$$
$$= \text{a constant} = \mu \text{ say.}$$

This may be written

$$\frac{a\cos^2\theta + b\sin\theta\cos\theta + c\sin^2\theta}{A\cos^2\theta + B\sin\theta\cos\theta + C\sin^2\theta} = \mu,$$

where

$$A = a'\cos^2\alpha + c'\sin^2\alpha + b'\sin\alpha\cos\alpha,$$

$$B = 2(c' - a')\sin\alpha\cos\alpha + b'(\cos^2\alpha - \sin^2\alpha),$$

$$C = a'\sin^2\alpha + c'\cos^2\alpha - b'\sin\alpha\cos\alpha.$$

That the curves may be similar we must have

$$\frac{A}{a} = \frac{B}{b} = \frac{C}{c}.$$

Hence each of these ratios must equal $\frac{A+C}{a+c}$;

$$\text{therefore } \frac{B^2}{b^2} = \frac{(A+C)^2}{(a+c)^2};$$

$$\text{therefore } \frac{B^2}{(A+C)^2} = \frac{b^2}{(a+c)^2}.$$

And $$\frac{AC}{ac} = \frac{(A+C)^2}{(a+c)^2};$$

therefore $$\frac{AC}{(A+C)^2} = \frac{ac}{(a+c)^2}.$$

Hence, $$\frac{B^2-4AC}{(A+C)^2} = \frac{b^2-4ac}{(a+c)^2}.$$

But $$A+C = a'+c',$$

and $$B^2-4AC = b'^2-4a'c', \text{ (Art. 274)};$$

therefore $$\frac{b'^2-4a'c'}{(a'+c')^2} = \frac{b^2-4ac}{(a+c)^2}.$$

This relation must therefore hold, in order that the given curves may be similar.

From the results obtained in Art. 278 it is easy to derive an instructive verification of the condition of similarity just demonstrated. It will be seen that similar conic sections have the same excentricity.

Area of a Polygon.

301. In Art. 11 we have given an expression for the area of a *triangle* in terms of the co-ordinates of its angular points: we shall now investigate the corresponding expression for the area of any *polygon.*

Let the angular points of the polygon taken in order be (x_1, y_1), (x_2, y_2), ...(x_n, y_n); take any point (x, y) within the polygon and draw straight lines to the angular points of the polygon, thus dividing the polygon into triangles having a common vertex at (x, y). Then by Art. 11 the *numerical* values of the areas of these triangles are respectively

$$\frac{1}{2}\left\{x(y_2-y_1)+x_1(y-y_2)+x_2(y_1-y)\right\},$$

$$\frac{1}{2}\left\{x(y_3-y_2)+x_2(y-y_3)+x_3(y_2-y)\right\},$$

....

$$\frac{1}{2}\left\{x(y_n-y_{n-1})+x_{n-1}(y-y_n)+x_n(y_{n-1}-y)\right\},$$

$$\frac{1}{2}\left\{x(y_1-y_n)+x_n(y-y_1)+x_1(y_n-y)\right\}.$$

Let us *assume*, for the present, that the *sum* of these expressions will give the area. By addition x and y disappear, and we obtain

$$\frac{1}{2}\left\{x_1(y_n-y_2)+x_2(y_1-y_3)+x_3(y_2-y_4)+\ldots\right.$$
$$\left.+x_{n-1}(y_{n-2}-y_n)+x_n(y_{n-1}-y_1)\right\}.$$

By multiplying out this expression may be written thus:

$$\frac{1}{2}\left\{x_2y_1-x_1y_2+x_3y_2-x_2y_3+\ldots\right.$$
$$\left.+x_ny_{n-1}-x_{n-1}y_n+x_1y_n-x_ny_1\right\}.$$

The expression may also be written thus:

$$\frac{1}{2}\left\{y_1(x_2-x_n)+y_2(x_3-x_1)+y_3(x_4-x_2)+\ldots\right.$$
$$\left.+y_{n-1}(x_n-x_{n-2})+y_n(x_1-x_{n-1})\right\}.$$

302. We now proceed to examine the admissibility of the assumption made in the preceding Article. Suppose that the polygon has no *re-entrant* angle. We must then shew that the expressions for the areas of the triangles used in the preceding Article are *all of the same sign;* for unless this is the case we do not obtain a correct numerical value of the area of the polygon by adding these expressions. The required result may be obtained by the aid of a principle which we have already applied; see Arts. 54 and 215.

Consider the expression given for the area of the first triangle in the preceding Article. The expression will retain the same sign for all positions of (x, y) which are on the same side of the straight line passing through (x_1, y_1) and (x_2, y_2). Similarly the expression given for the area of the second triangle in the preceding Article will retain the same sign for all positions of (x, y) which are on the same side of the straight line passing through (x_2, y_2) and (x_3, y_3). Thus if the two expressions have the same sign for one position of (x, y) within the polygon, they will have the same sign for all such positions. But by trial we can ascertain that the two expressions *have* the same sign when $x = \frac{1}{2}(x_1 + x_3)$ and $y = \frac{1}{2}(y_1 + y_3)$: the two expressions will in fact be found then to coincide. Thus the two expressions have the same sign for all positions of (x, y) within the polygon. Similarly the expressions for the areas of the second and third triangles have the same sign. And so on. Thus the assumption made in the preceding Article is justified.

303. We will now briefly illustrate the method by which it may be shewn that the expressions obtained in Art. 301 for the area of a polygon hold even when the polygon has re-entrant angles.

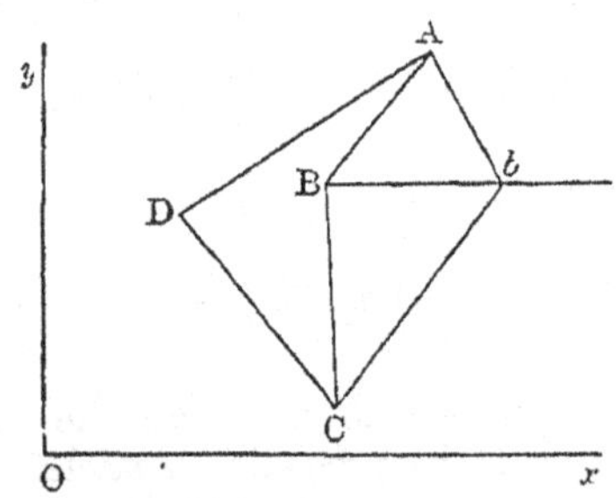

Suppose, for example, we have a quadrilateral figure $ABCD$, with a re-entrant angle at B. Through B draw a straight line parallel to the axis of x, and take a point b on this straight line, such that $AbCD$ is a quadrilateral figure without a re-entrant angle.

Let the co-ordinates of A be x_1, y_1; let those of B be

x_2, y_2; and so on. Let the abscissa of b be x. Then we know that the area of $AbCD$ is numerically expressed by

$$\frac{1}{2}\left\{x_1(y_4 - y_2) + x(y_1 - y_3) + x_3(y_2 - y_4) + x_4(y_3 - y_1)\right\}.$$

Now as x increases this expression becomes algebraically greater since $y_1 - y_3$ is positive; and as x increases we see from the figure that the area increases: hence it follows that the expression is *positive.* Put $x = x_2 + h$, so that $h = Bb$. The expression then becomes

$$\frac{1}{2}\left\{x_1(y_4 - y_2) + x_2(y_1 - y_3) + x_3(y_2 - y_4) + x_4(y_3 - y_1)\right\} + \frac{1}{2}h(y_1 - y_3);$$

and as $\frac{1}{2}h(y_1 - y_3)$ is obviously equal to the area of $ABCb$, it follows that the other part of the expression is equal to the area of $ABCD$.

304. Although the results given in Art. 301 are not of great importance, yet the reasoning in Arts. 302 and 303 is very instructive. The method of Art. 303 may be applied whatever be the form of the figure, with slight modifications which do not affect the principle.

Homologous Triangles.

305. In Art. 76 we have spoken of *homologous triangles;* we will here give another property relating to such triangles.

Suppose ABC, $A'B'C'$, $A''B''C''$ three triangles such that any two of them are homologous; and suppose moreover that AB, $A'B'$, $A''B''$ meet at a point: then the three centres of homology will lie on a straight line.

For consider the triangles $AA'A''$ and $BB'B''$. By supposition AB, $A'B'$, and $A''B''$ meet at a point: therefore, by Art. 76, the intersections of corresponding sides of the triangles lie on a straight line; that is the intersection of AA' and BB', of $A'A''$ and $B'B''$, and of $A''A$ and $B''B$ lie on a straight line.

And conversely if the three centres of homology lie on a straight line the sides AB, $A'B'$, $A''B''$ meet at a point; so also do BC, $B'C'$, $B''C''$; and CA, $C'A'$, $C''A''$. This also follows from Art. 76.

306. It may be easily shewn that if we take the equations to the sides of two triangles as in Art. 76, then the equations

$$l''u + mv + nw = 0, \quad lu + m''v + nw = 0, \quad lu + mv + n''w = 0$$

will determine a third triangle such that any two of the triangles are homologous, and that any three corresponding sides meet at a point.

EXAMPLES.

1. Straight lines are drawn through a fixed point: shew that the locus of the middle points of the portions of them intercepted between two fixed straight lines is an hyperbola whose asymptotes are parallel to those fixed straight lines.

2. Through any point P of an ellipse QPQ' is drawn parallel to the major axis, and PQ and PQ' each made equal to the focal distance SP: find the loci of Q and Q'.

3. In the given straight lines AP, AQ are taken variable points p, q, such that $Ap : pP :: Qq : qA$; shew that the locus of the point of intersection of Pq and Qp is an ellipse which touches the given straight lines at the points P, Q.

4. TP, TQ are two tangents to a parabola, P, Q being the points of contact; a third tangent cuts these at p, q respectively: shew that $\dfrac{Tp}{TP} + \dfrac{Tq}{TQ} = 1$.

5. TP, TQ are *equal* tangents to a parabola, P, Q being the points of contact: if PT, QT be both cut by a third tangent, shew that their alternate segments will be equal.

6. From a point O are drawn two straight lines to touch a parabola at the points P and Q; another straight line touches the parabola at R and intersects OP, OQ at S and T:

if V be the intersection of the straight lines joining PT, QS, crosswise, O, R, V are on the same straight line.

7. From an external point two tangents are drawn to an ellipse: shew that an ellipse similar and similarly situated will pass through the external point, the points of contact, and the centre of the given ellipse.

8. A and B are two similar, similarly situated, and concentric ellipses; C is a third ellipse similar to A and B, its centre being on the circumference of B, and its axes parallel to those of A or B: shew that the chord of intersection of A and C is parallel to the tangent to B at the centre of C.

9. The straight line joining any point with the intersection of the polar of that point with a directrix subtends a right angle at the corresponding focus.

10. If normals be drawn to an ellipse from a given point, the points where they cut the curve will lie on a rectangular hyperbola which passes through the given point and has its asymptotes parallel to the axes of the ellipse.

11. If CM, MP are the abscissa and ordinate of any point P, on the circumference of a circle, and MQ is taken equal to MP and inclined to it at a constant angle, the locus of the point Q is an ellipse.

12. Having given the equation to a conic section

$$ax^2 + 2bxy + y^2 + f = 0,$$

find the locus of the intersection of normals drawn at the extremities of each pair of ordinates to the same abscissa.

13. Any two points P, Q are taken in two fixed straight lines in one plane such that the straight line PQ is always parallel to a given straight line; P, Q are severally joined with two fixed points H, R: find the locus of the intersection of PH and QR.

14. The tangent at any point P of a circle meets the tangent at a fixed point A at T, and T is joined with B the extremity of the diameter passing through A: shew that the locus of the point of intersection of AP and BT is an ellipse.

15. The polar equation to a conic section from the focus being $\frac{1}{r} - c\cos\theta = b$, shew that the equation to a straight line which cuts it at the points for which $\theta = \alpha$ and β respectively, is $\frac{1}{r} - c\cos\theta = b\cos\left(\theta - \frac{\alpha+\beta}{2}\right)\sec\frac{\alpha-\beta}{2}$.

16. Chords are drawn in a conic section so as to subtend a constant angle at the focus: prove that the locus of the foot of the perpendicular drawn from the focus on the chord is a circle, except in a particular case when it becomes a straight line.

17. If SP, SQ be focal distances of a conic section including a constant angle, shew that PQ touches a confocal conic.

18. Having given two fixed points through which a conic section is to pass, and the directrix, find the locus of the corresponding focus.

19. The focus and the directrix of an ellipse are given; through the former a straight line is drawn making with the latter an angle whose sine is the excentricity of the ellipse. Find the locus of the points where this straight line meets the curve, the excentricity being variable.

20. A series of conic sections is described having a common focus and directrix, and in each curve a point is taken whose distance from the focus varies inversely as the latus rectum: find the locus of these points.

21. Two conic sections have a common focus S through which any radius vector is drawn meeting the curves at P, Q, respectively. Shew that the locus of the point of intersection of the tangents at P, Q, is a straight line.

Shew that this straight line passes through the intersection of the directrices of the conic sections, and that the sines of the angles which it makes with these straight lines are inversely proportional to the corresponding excentricities.

22. A straight line is drawn cutting an ellipse at the points P, p; let Q be either of the points at which the same straight line meets a similar, similarly situated, and concentric

ellipse: shew that if the straight line moves parallel to itself, $PQ \cdot Qp$ is constant.

23. In two straight lines OX, OY, which intersect at O, take $OA = a$, $OB = b$: shew that the centres of all the conic sections which touch the straight lines at A and B lie on the straight line $ay = bx$.

24. About two equal ellipses whose centres coincide, and whose major axes are inclined to each other at a given angle an ellipse is circumscribed: if A and B be the semi-axes of the circumscribing ellipse, a and b the semi-axes of the equal ellipses, and 2α the inclination of their major axes, then will

$$a^2b^2 + A^2B^2 = (A^2b^2 + B^2a^2)\cos^2\alpha + (A^2a^2 + B^2b^2)\sin^2\alpha.$$

Hence shew that about the two equal ellipses a *similar* ellipse may be circumscribed.

25. Two similar ellipses have a common centre and touch each other; if n be the ratio of their linear magnitudes, m the ratio of the major to the minor axis in either, and α the inclination of their major axes, prove that

$$\sin\alpha = \left(n - \frac{1}{n}\right) \div \left(m - \frac{1}{m}\right).$$

26. Two tangents (a, b) to a parabola intersect at P at an angle ω, and a circle is described between these tangents and the curve: shew that the distance of its centre from P is

$$\frac{ab}{(a+b)\sec\frac{\omega}{2} + 2\sqrt{(ab)}\tan\frac{\omega}{2}}.$$

27. If two chords at right angles be drawn through a fixed point to meet a curve of the second degree, shew that $\dfrac{1}{Rr} + \dfrac{1}{R'r'}$ is constant; where R and r are the segments of one chord made by the fixed point, and R' and r' those of the other.

28. The equation to the locus of the foci of all parabolas whose chords of contact with axes inclined at an angle α cut off a triangle of constant area is $r = k\sqrt{\{\sin\theta\sin(\alpha-\theta)\}}$.

29. A parabola slides between two rectangular axes, find the curve traced out by the focus.

30. A parabola slides between two rectangular axes, find the curve traced out by the vertex.

31. Successive circles are drawn each touching the preceding one externally and each having double contact with a given parabola: shew that their radii form an arithmetical progression whose common difference is the latus rectum.

32. A system of ellipses is represented by the equation in rectangular co-ordinates $ax^2 + 2cxy + by^2 = n(a+b)$, where a, b, c are variable and n constant: shew that every parallelogram constructed on a pair of perpendicular diameters as diagonals will circumscribe a certain fixed circle.

33. If from any point in the tangent to a conic section a perpendicular be drawn on the straight line joining the focus and the point of contact, prove that the distance of the point in the tangent from the directrix is to the distance of the foot of the perpendicular from the focus as 1 is to e.

34. Upon a given straight line as latus rectum, let any number of conic sections be drawn, and from the focus let two straight lines be drawn intersecting them all: then the chords of all the intercepted arcs will, if produced, pass through a single point.

35. A straight line of constant length moves so that its ends always lie on two given straight lines: find the locus traced out by a point in the straight line which divides it in a given ratio.

36. In any conic section if r and r' be focal distances at right angles to each other, and l be half the latus rectum, then $\left(\frac{1}{r} - \frac{1}{l}\right)^2 + \left(\frac{1}{r'} - \frac{1}{l}\right)^2$ is constant.

37. Two conic sections equal in every respect are placed with their axes at right angles and with a common focus S; SP, SQ being radii vectores of the one and the other at right

angles to each other, find the locus of the intersection of the tangents at P and Q. Also find the locus when SPQ is a straight line.

38. S and H are the foci of an ellipse, and round S, H, as focus and centre, another ellipse is described, having its minor axis equal to the latus rectum of the former; through any point P in the first draw SPQ to meet the second: it is required to find the locus of the intersection of HP and the ordinate QM.

39. A and B are the centres of two equal circles; AP, BQ, radii of these circles at right angles. If $AB^2 = 2AP^2$, the straight line PQ always passes through one of the points of intersection of the circles.

40. Tangents are drawn to a conic section at the points P, R; another tangent is drawn at an intermediate point Q, and meets the other tangents at M, N: shew that the angle MSN is half the angle PSR, S being a focus.

41. Tangents are drawn from the point (h, k) to the parabola $y^2 = 4ax$: shew that the straight lines from the focus to the points of contact are determined by

$$y^2\{(h+a)^2 - k^2\} - 2ky(x-a)(h-a) + 4ah(x-a)^2 = 0.$$

42. If two equal ellipses have the same centre, shew that their points of intersection are at the extremities of diameters at right angles to one another.

43. Given a focus and two tangents to a conic section, shew that the chord of contact passes through a fixed point.

44. A circle is described on the minor axis of an ellipse as diameter: find the locus of the pole with respect to the ellipse of a tangent to the circle.

45. In a parabola two focal chords PSp, QSq, are drawn: shew that a focal chord parallel to PQ will meet pq produced on the tangent at the vertex.

46. If from the vertex of a parabola a pair of chords be drawn at right angles to each other, and on them a rectangle be completed, prove that the locus of the further angle is another parabola.

47. From a point P in the circumference of an ellipse chords PQ, PR are drawn at right angles: express the co-ordinates of the point of intersection of QR with the normal at P in terms of the co-ordinates of P. Shew that as P moves along the ellipse this point of intersection will describe the ellipse $\frac{x^2}{a^2}+\frac{y^2}{b^2}=\left(\frac{a^2-b^2}{a^2+b^2}\right)^2$.

48. Shew that the locus of the centre of an equilateral hyperbola described about a given equilateral triangle is the circle inscribed in the triangle.

49. Two equal parabolas have the same axis and vertex, but are turned in opposite directions; chords of one parabola are tangents to the other: shew that the locus of the middle points of the chords is a parabola whose latus rectum is one-third of that of the given parabolas.

50. The co-ordinates of the focus of the parabola whose equation when referred to two tangents inclined at an angle ω is $\sqrt{\left(\frac{x}{a}\right)}+\sqrt{\left(\frac{y}{b}\right)}=1$, are

$$\frac{ab^2}{a^2+b^2+2ab\cos\omega}, \text{ and } \frac{a^2b}{a^2+b^2+2ab\cos\omega}.$$

51. If $ax^2+2bxy+cy^2+2a'x+2c'y+d=0$ be the equation to a parabola, the axis of the parabola will be given by the equation $(a+b)\left(x+\frac{a'}{a+c}\right)+(b+c)\left(y+\frac{c'}{a+c}\right)=0$.

52. Two equal parabolas have the same focus and their axes are at right angles to each other, and a normal to one of them is perpendicular to a normal to the other: prove that the locus of the intersection of such normals is a parabola.

53. Find the locus of the intersection of two normals in an ellipse which are at right angles.

54. Normals are drawn at the extremities of the conjugate diameters of an ellipse, and by their intersections form a parallelogram. If ϕ denote the excentric angle of an ex-

tremity of one of the conjugate diameters, shew that the area of the parallelogram is $\frac{4(a^2-b^2)^2}{ab}\sin^2\phi\cos^2\phi$.

55. Through the four angular points of a given square a circle is drawn, and also a series of curves of the second order, and common tangents to the circle and each curve are drawn. Find the locus of the points of contact of each curve with its tangent.

56. From any point T outside an ellipse two tangents TP and TQ are drawn to the ellipse: shew that a circle can be described with T as centre so as to touch SP, HP, SQ, HQ, or these straight lines produced.

57. If x and y are the co-ordinates of T, in Example 56, shew that the radius of the circle is $\frac{\sqrt{(a^2y^2+b^2x^2-a^2b^2)}}{a}$.

58. If from a point three radii vectores are drawn to a circle, and from the same point in the same directions three radii vectores are drawn to another circle, and the corresponding radii are in a constant ratio, that point is a centre of similitude of the circles.

59. Tangents are drawn to the parabola $r(1+\cos\theta)=l$ at three points for which θ is equal to α, β, γ respectively: shew that the equation to the circle which passes round the triangle formed by the tangents is

$$r\cos\frac{\alpha}{2}\cos\frac{\beta}{2}\cos\frac{\gamma}{2}=\frac{l}{2}\cos\left(\theta-\frac{\alpha+\beta+\gamma}{2}\right).$$

Hence shew that the circle which passes through the intersections of three tangents to a parabola will pass through the focus.

60. Let u stand for $ax^2+bxy+cy^2+dx+ey+f$, and let u_1 denote what u becomes when h and k are put for x and y respectively: then the two straight lines which can be drawn from the point (h, k) to touch the curve $u=0$ are determined by

$$4uu_1=\{x(2ah+bk+d)+y(2ck+bh+e)+dh+ek+2f\}^2.$$

CHAPTER XV.

ABRIDGED NOTATION.

307. *Through five points, no three of which are in one straight line, one conic section and only one can be drawn.*

Let the axis of x pass through two of the five points, and the axis of y through two of the remaining three points. Let the distances of the first two points from the origin be h_1, h_2, respectively, and those of the second two points k_1, k_2, respectively; also let h, k be the co-ordinates of the remaining point. Suppose (Art. 269)

$$ax^2 + bxy + cy^2 + dx + ey + 1 = 0 \quad\ldots\ldots\ldots\ldots(1)$$

to be the equation to a conic section passing through the five points. Since the curve passes through the points $(h_1, 0)$ $(h_2, 0)$, we have from (1)

$$ah_1^2 + dh_1 + 1 = 0 \quad\ldots\ldots\ldots\ldots(2),$$

$$ah_2^2 + dh_2 + 1 = 0 \quad\ldots\ldots\ldots\ldots(3).$$

Similarly, since the curve passes through $(0, k_1)$, $(0, k_2)$, we have

$$ck_1^2 + ek_1 + 1 = 0 \quad\ldots\ldots\ldots\ldots(4),$$

$$ck_2^2 + ek_2 + 1 = 0 \quad\ldots\ldots\ldots\ldots(5).$$

Lastly, since the curve passes through (h, k), we have

$$ah^2 + bhk + ck^2 + dh + ek + 1 = 0 \quad\ldots\ldots\ldots\ldots(6).$$

From (2) and (3) we find $a = \dfrac{1}{h_1 h_2}$, $d = -\dfrac{h_1 + h_2}{h_1 h_2}$.

From (4) and (5) we find $c = \dfrac{1}{k_1 k_2}$, $e = -\dfrac{k_1 + k_2}{k_1 k_2}$;

then from (6) we can determine the value of b. Since no three of the five given points are in the same straight line, none of the quantities h_1, h_2, k_1, k_2, h, k, can be zero; hence the values of the coefficients a, b, c, d, e are all finite. If we substitute these values in (1), we obtain the equation to a conic section passing through the five given points. As each of the quantities a, b, c, d, e, has only *one* value, only *one* conic section can be made to pass through the five given points.

308. The investigation of the preceding Article may still be applied when *three* of the given points are on one straight line; the point (h, k) for instance may be supposed to lie on the straight line joining $(0, k_1)$ and $(h_1, 0)$; the conic section in this case cannot be an ellipse, parabola, or hyperbola, since these curves cannot be cut by a straight line in more than two points; the conic section must therefore reduce to two straight lines, namely the straight line joining the three points already specified, and the straight line joining the other two points. If, however, *four* of the given points are on one straight line, the method of the preceding Article is inapplicable; it is obvious that more than one pair of straight lines can then be made to pass through the five points.

309. We shall now give some useful forms of the equations to conic sections passing through the angular points of a triangle or touching its sides.

Let $u = 0$, $v = 0$, $w = 0$ be the equations to three straight lines which meet and form a triangle; the equation

$$lvw + mwu + nuv = 0 \quad\quad (1),$$

where l, m, n are constants, will represent a conic section described round the triangle; also by giving suitable values to l, m, n, the above equation may be made to represent *any* conic section described round the triangle. This we proceed to demonstrate.

I. The equation (1) is of the *second degree* in the variables x and y, which occur in the expressions u, v, w; hence (1) must represent a conic section.

II. The equation (1) is satisfied by the values of x and

y, which make simultaneously $v=0$, $w=0$; the conic section therefore passes through the intersection of the straight lines represented by $v=0$ and $w=0$. Similarly the conic section passes through the intersection of $w=0$ and $u=0$, and also through the intersection of $u=0$ and $v=0$. Hence the conic section represented by (1) is described round the triangle formed by the intersection of the straight lines represented by $u=0$, $v=0$, $w=0$.

III. By giving suitable values to l, m, n, the equation (1) will represent any conic section described round the triangle. For let S denote a given conic section described round the triangle; take two points on S; suppose h_1, k_1 the co-ordinates of one of these points, and h_2, k_2 those of the other. If we first substitute h_1 and k_1 for x and y respectively in (1), and then substitute h_2 and k_2, we have two equations from which we can find the values of $\frac{m}{l}$ and $\frac{n}{l}$; suppose $\frac{m}{l}=p$ and $\frac{n}{l}=q$. Substitute these values in (1), which becomes

$$vw+pwu+quv=0 \quad\text{.................... (2);}$$

this is therefore the equation to a conic section which has *five points* in common with S, namely, the three angular points of the triangle and the points (h_1, k_1), (h_2, k_2). The conic section (2) must therefore coincide with S by Art. 307. Hence the assertion is proved.

We might replace one of the constants in (1) by unity, but we retain the more symmetrical form; (1) may be written $\frac{l}{u}+\frac{m}{v}+\frac{n}{w}=0$.

310. Equation (1) of the preceding Article may be written

$$w(lv+mu)+nuv=0 \quad\text{.................. (1);}$$

we will now determine where (1) meets the straight line represented by

$$lv+mu=0 \quad\text{.......................... (2).}$$

By combining (2) with (1) we deduce $nuv=0$; therefore either $u=0$, or $v=0$; but by taking either of these suppositions and making use of (2), we see that the other supposition must also hold; hence the straight line (2) meets the curve (1) at only *one* point, namely, the point of intersection of $u=0$ and $v=0$.

Hence (2) is the *tangent* to (1) at this point. Similarly $mw+nv=0$ is the tangent to (1) at the point of intersection of $w=0$ and $v=0$, and $nu+lw=0$ is the tangent to (1) at the point of intersection of $u=0$ and $w=0$.

311. The demonstration of the preceding Article is imperfect, because we know from Arts. 132, 222, that a straight line parallel to the axis of a parabola or to either asymptote of an hyperbola meets the curve at only one point, but is not the tangent at that point. The proposition may however be established in the following manner. Take the axis of x coincident with the straight line $u=0$, so that u becomes qy, where q is some constant; also take the axis of y coincident with the straight line $v=0$, so that v becomes px, where p is some constant. Suppose $w=Ax+By+C$. Then (1) of the preceding Article becomes $(Ax+By+C)(lpx+mqy)+npqxy=0$. By Art. 283 the equation to the tangent at the origin, that is at the intersection of $x=0$ and $y=0$, is $lpx+mqy=0$, or $lv+mu=0$; which was to be proved.

312. Let each of the three tangents in Art. 310 be produced to meet the opposite side of the triangle formed by the straight lines $u=0$, $v=0$, $w=0$; then it may be shewn that the three points of intersection lie on the straight line

$$\frac{u}{l}+\frac{v}{m}+\frac{w}{n}=0.$$

The straight lines joining the angular points of the triangle formed by the tangents with the angular points of the original triangle respectively opposite to them, are represented by the equations $\frac{u}{l}-\frac{v}{m}=0$, $\frac{v}{m}-\frac{w}{n}=0$, $\frac{w}{n}-\frac{u}{l}=0$; these three straight lines meet at a point. Thus when a triangle is inscribed in a conic section the straight lines joining each point with the pole of the opposite side meet at a point.

313. Let $u=0$, $v=0$, $w=0$ be the equations to three straight lines, then the equation

$$Au^2+Bv^2+Cw^2+2A'vw+2B'wu+2C'uv=0$$

will generally represent any assigned conic section, if the constants A, B, C, A', B', C' are properly determined.

For suppose we divide the equation by one of the constants as C', there are then five independent constants left. Now let S denote any assigned conic section; take five points on S and substitute the co-ordinates of the five points successively in the above equation; we shall thus have five equations for determining the five constants. Suppose a, b, c, a', b' the values thus determined, then the equation

$$au^2 + bv^2 + cw^2 + 2a'vw + 2b'wu + 2uv = 0$$

represents a conic section which has five points in common with S, and which therefore coincides with S. (Art. 307.)

314. The method of the preceding Article, although important and instructive, is not satisfactory, because we have not shewn that the five equations from which the constants are to be determined are *consistent* and *independent*. There may be exceptions to the theorem, and we therefore use the word *generally* in the enunciation. If the three straight lines *meet at a point*, then the curve denoted by the equation always passes through that point, and the equation in this case will *not* represent *any assigned conic section*. If the three straight lines are parallel, u, v, w take the forms

$$lx + my + p, \qquad lx + my + p', \qquad lx + my + p'',$$

and the equation takes the form

$$\lambda (lx + my)^2 + \mu (lx + my) + \nu = 0,$$

which represents two parallel straight lines, and thus will *not* represent *any assigned conic section*. With these exceptions, however, the theorem is universally true, as we shall now shew by another demonstration.

Since the straight lines are not all parallel, two of them at least will meet; suppose $u = 0$ and $v = 0$ to be these two, and take their directions for the axes of y and x respectively; then $u = 0$ becomes $x = 0$, and $v = 0$ becomes $y = 0$; also $w = 0$ may be written $lx + my + n = 0$. We have then to shew that the equation

$$Ax^2 + By^2 + C(lx + my + n)^2 + 2A'y(lx + my + n) \\ + 2B'x(lx + my + n) + 2C'xy = 0 \quad \text{.........(1)}$$

will represent any assigned conic section by properly determining the constants A, B, ... Suppose

$$ax^2 + 2bxy + cy^2 + 2dx + 2ey + f = 0 \text{..........(2)}$$

to be the equation to the assigned conic section. Arrange the terms in (1) and equate the coefficients of the corresponding terms in (1) and (2); thus

$$Cn^2 = f,\ \ A'n + Cmn = e,\ \ B'n + Cln = d,\ \ B + Cm^2 + 2A'm = c,$$
$$Clm + A'l + B'm + C' = b,\ \ A + Cl^2 + 2B'l = a.$$

These equations determine successively C, A', B', B, C', A. As the given straight lines do not meet at a point, n is not zero; hence the values found for $C, A', \ldots$ are all finite and determinate. Thus (1) is shewn to coincide with (2), and the required theorem is demonstrated.

315. We will now investigate the equation to the tangent at any point of the curve represented by

$$Au^2 + Bv^2 + Cw^2 + 2A'vw + 2B'wu + 2C'uv = 0.$$

Let u', v', w' be the values of u, v, w respectively at one point of the curve, and u'', v'', w'' their values at another point of the curve. Then the equation to the straight line joining these two points may be put in the form

$$A(u-u')(u-u'') + B(v-v')(v-v'') + C(w-w')(w-w'')$$
$$+ 2A'(v-v')(w-w'') + 2B'(w-w')(u-u'') + 2C'(u-u')(v-v'')$$
$$= Au^2 + Bv^2 + Cw^2 + 2A'vw + 2B'wu + 2C'uv.$$

For this equation is really of the first degree in the variables u, v, and w, and therefore represents *some* straight line; moreover the equation is satisfied at the point (u', v', w'), and also at the point (u'', v'', w''), and therefore it represents the straight line which passes through these two points.

Now suppose the point (u'', v'', w'') to move along the curve until it coincides with the point (u', v', w'). Then the secant becomes ultimately the tangent at (u', v', w'), and the equation to this tangent is

$$Auu' + Bvv' + Cww' + A'(vw' + wv') + B'(wu' + uw')$$
$$+ C'(uv' + vu') = 0.$$

316. As a particular case of the preceding Article suppose that A', B', and C' are zero. Then the equation to the curve is

$$Au^2 + Bv^2 + Cw^2 = 0 \qquad (1);$$

and the equation to the tangent at (u', v', w') is

$$Auu' + Bvv' + Cww' = 0 \qquad (2).$$

Hence we can find the condition which must hold in order that a proposed straight line may *touch* the curve denoted by (1). Let the equation to the proposed straight line be

$$\lambda u + \mu v + \nu w = 0 \qquad (3).$$

If (3) denotes the equation to the tangent at (u', v', w'), we find by comparing (3) with (2) that

$$\frac{Au'}{\lambda} = \frac{Bv'}{\mu} = \frac{Cw'}{\nu}.$$

Let r denote the value of each of these fractions; then

$$u' = \frac{\lambda r}{A}, \quad v' = \frac{\mu r}{B}, \quad w' = \frac{\nu r}{C}.$$

These values must satisfy (1) since (u', v', w') is a point on the curve; thus

$$\frac{\lambda^2}{A} + \frac{\mu^2}{B} + \frac{\nu^2}{C} = 0:$$

this is therefore the required condition.

317. The investigation of Art. 315 may be modified in special cases by using a different form for the equation to the secant. For example suppose that A, B, and C are zero. Then the equation to the curve is

$$A'vw + B'wu + C'uv = 0,$$

which may be also put in the form

$$\frac{A'}{u} + \frac{B'}{v} + \frac{C'}{w} = 0 \qquad (1).$$

The equation to the straight line which passes through

the points (u', v', w') and (u'', v'', w'') on the curve may be put in the form

$$\frac{A'u}{u'u''}+\frac{B'v}{v'v''}+\frac{C'w}{w'w''}=0.$$

For this equation is of the first degree in the variables u, v, w, and therefore represents some straight line; moreover the equation is satisfied at the point (u, v', w') and also at the point (u'', v'', w''), and therefore it represents the straight line which passes through these points.

Therefore the equation to the tangent at (u', v', w') is

$$\frac{A'u}{u'^2}+\frac{B'v}{v'^2}+\frac{C'w}{w'^2}=0 \text{..................(2)}.$$

Hence we can find the condition which must hold in order that a proposed straight line may *touch* the curve denoted by (1). Let the equation to the proposed straight line be

$$\lambda u+\mu v+\nu w=0 \text{....................(3)}.$$

If (3) denotes the equation to the tangent at (u', v', w'), we find by comparing (3) with (2) that

$$\frac{A'}{\lambda u'^2}=\frac{B'}{\mu v'^2}=\frac{C'}{\nu w'^2}.$$

From these relations and (1) we obtain as the required condition

$$\sqrt{(A'\lambda)}+\sqrt{(B'\mu)}+\sqrt{(C'\nu)}=0.$$

318. *To express the equation to a conic section which touches the sides of a triangle.*

Let $u=0$, $v=0$, $w=0$ be the equations to the sides of a triangle; then any conic section may be represented by the equation

$$Au^2+Bv^2+Cw^2+2A'vw+2B'wu+2C'uv=0 \text{......(1)}.$$

To find where this conic section meets the straight line $u=0$, we must put $u=0$; thus (1) becomes

$$Bv^2+Cw^2+2A'vw=0 \text{..................(2)}.$$

Now from (2) we obtain by solution *two* values of $\frac{v}{w}$, say $\frac{v}{w}=\mu_1$, and $\frac{v}{w}=\mu_2$. The equation $v=\mu_1 w$ represents some straight line passing through the intersection of $v=0$, and $w=0$. Hence since (1) is satisfied by those values of x and y which make simultaneously $u=0$ and $v-\mu_1 w=0$, the intersection of the straight lines $u=0$ and $v-\mu_1 w=0$ is a point on (1). Similarly the intersection of $u=0$ and $v-\mu_2 w=0$ is a point on (1). Hence the straight line $u=0$ will meet (1) at *two* points, and therefore will not be a tangent to it, unless the straight lines $v-\mu_1 w=0$, and $v-\mu_2 w=0$, *coincide.* Hence that $u=0$ may *touch* (1) we must have $\mu_1=\mu_2$, and therefore $A'^2=BC$.

Similarly that $v=0$ may touch (1) we must have $B'^2=CA$; and that $w=0$ may touch (1) we must have $C'^2=AB$. From these three relations we see that A, B, and C must have the *same* sign, because the product of each two is positive. Also the sign of A, B, and C may be supposed positive, because if each of them were negative we could change the sign of every term in (1), and thus make the coefficients of u^2, v^2, and w^2 positive. We may therefore put

$$A=l^2, \quad B=m^2, \quad C=n^2;$$

thus

$$A'=\pm mn, \quad B'=\pm nl, \quad C'=\pm lm.$$

Hence (1) becomes

$$l^2u^2+m^2v^2+n^2w^2\pm 2mnvw\pm 2nlwu\pm 2lmuv=0 \ldots\ldots(3).$$

We shall now examine the ambiguity of signs that appears in this expression.

I. Suppose all the upper signs to be taken. The equation may then be written

$$(lu+mv+nw)^2=0.$$

This is the equation to a straight line, or rather to two coincident straight lines.

II. Suppose the lower sign to be taken twice and the upper sign once; we have then three cases,

$$(lu+mv-nw)^2=0, \quad \text{or} \quad (lu-mv+nw)^2=0,$$

$$\text{or} \quad (-lu+mv+nw)^2=0.$$

Each equation represents two coincident straight lines.

III. Since then the forms in I. and II. represent straight lines, we see by excluding these cases from (3), that if *a curve* of the second degree touch the straight lines

$$u=0, \quad v=0, \quad w=0,$$

its equation must take one of the forms

$$l^2u^2+m^2v^2+n^2w^2-2mnvw-2nlwu-2lmuv=0 \ldots (4),$$

$$l^2u^2+m^2v^2+n^2w^2-2mnvw+2nlwu+2lmuv=0 \ldots (5),$$

$$l^2u^2+m^2v^2+n^2w^2+2mnvw-2nlwu+2lmuv=0 \ldots (6),$$

$$l^2u^2+m^2v^2+n^2w^2+2mnvw+2nlwu-2lmuv=0 \ldots (7).$$

These four forms may also be written

$$\sqrt{(lu)}+\sqrt{(mv)}+\sqrt{(nw)}=0 \;\ldots\ldots\; (8) \text{ from } (4),$$

$$\sqrt{(-lu)}+\sqrt{(mv)}+\sqrt{(nw)}=0 \;\ldots\ldots\; (9) \text{ from } (5),$$

$$\sqrt{(lu)}+\sqrt{(-mv)}+\sqrt{(nw)}=0 \;\ldots\ldots\; (10) \text{ from } (6),$$

$$\sqrt{(lu)}+\sqrt{(mv)}+\sqrt{(-nw)}=0 \;\ldots\ldots\; (11) \text{ from } (7),$$

which may be verified by transposing and squaring, so as to put the equations in a rational form.

319. It is easy to verify the proposition that the curve represented by the equation

$$\sqrt{(lu)}+\sqrt{(mv)}+\sqrt{(nw)}=0$$

cannot cut the straight lines $u=0$, $v=0$, $w=0$. For suppose the above equation satisfied by the co-ordinates of a point; then these co-ordinates must make lu, mv, and nw, *all positive, or all negative*. Suppose lu is positive; then for any point on the other side of $u=0$, the expression lu becomes negative, and thus the co-ordinates of such a point will *not* satisfy the equation unless both mv and nw are also negative. But if the curve cuts the straight line $u=0$, there will be points on both sides of $u=0$ lying on the curve, and it will be possible to change the sign of u without changing the signs of v and w. Hence the curve cannot cut the straight line $u=0$. Similarly it cannot cut the straight lines $v=0$, $w=0$.

The same mode of proof will shew that the curves represented by equations (9), (10), and (11), of the preceding Article cannot cut the straight lines $u=0$, $v=0$, $w=0$.

320. The forms in equations (5), (6), and (7) of Art. 318 may be derived from (4) by changing the sign of one of the constants. Thus, for example, (5) may be derived from (4) by changing the sign of l. In the following Article we shall use (4) as the equation to a conic section touching the sides of a triangle; it will be found that we might have used (5), (6), or (7). We shall see in a subsequent Article, a case in which it is necessary to distinguish the forms. See Arts. 324, 325.

321. Equation (4) of Art. 318 may be written

$$(lu - mv)^2 + nw(nw - 2mv - 2lu) = 0 \quad \text{.........(1)}.$$

If we combine this with $w=0$, we deduce that

$$lu - mv = 0 \quad \text{..........................(2)};$$

hence we can interpret the last equation; it represents a straight line passing through the intersection of $u=0$ and $v=0$, and also through the point where the straight line $w=0$ meets the curve (1). It may be shewn as in Art. 310, that

$$nw - 2mv - 2lu = 0 \quad \text{.................(3)}$$

represents the tangent to (1) at the other point where (2) meets it.

Similarly we can interpret

$$mv - nw = 0 \quad \text{....................(4)},$$

$$lu - 2nw - 2mv = 0 \quad \text{....................(5)},$$

$$nw - lu = 0 \quad \text{....................(6)},$$

$$mv - 2lu - 2nw = 0 \quad \text{....................(7)}.$$

The intersection of (3) with $w=0$, of (5) with $u=0$, and of (7) with $v=0$ will lie on the straight line

$$lu + mv + nw = 0.$$

The straight line $lu + mv = 0$ passes through the intersection of $u=0$, and $v=0$, and also through the intersection of (3) with $w=0$; hence its position is known.

Similarly the equations $mv + nw = 0$, and $nw + lu = 0$, can be interpreted.

322. We will now investigate the equation to the tangent at any point of the curve represented by

$$A\sqrt{u} + B\sqrt{v} + C\sqrt{w} = 0 \quad\text{..............(1).}$$

We might clear this equation of radicals and so obtain the form already considered in Art. 315, and then express the equation to the tangent at any point. Or we may proceed thus:

The equation to the straight line passing through the points (u', v', w') and (u'', v'', w'') on the curve may be put in the form

$$\frac{A(u-u')}{\sqrt{u'}+\sqrt{u''}} + \frac{B(v-v')}{\sqrt{v'}+\sqrt{v''}} + \frac{C(w-w')}{\sqrt{w'}+\sqrt{w''}} = 0.$$

For this equation is of the first degree in the variables u, v, w, and therefore represents some straight line; moreover the equation is satisfied at the point (u', v', w') and also at the point (u'', v'', w''), and therefore it represents the straight line passing through these two points.

Now suppose the point (u'', v'', w'') to move along the curve until it coincides with the point (u', v', w'). Then the secant becomes ultimately the tangent at (u', v', w'); and the equation to this tangent is

$$\frac{A(u-u')}{\sqrt{u'}} + \frac{B(v-v')}{\sqrt{v'}} + \frac{C(w-w')}{\sqrt{w'}} = 0,$$

that is

$$\frac{Au}{\sqrt{u'}} + \frac{Bv}{\sqrt{v'}} + \frac{Cw}{\sqrt{w'}} = 0 \quad\text{........(2).}$$

Hence we can find the condition which must hold in order that a proposed straight line may *touch* the curve denoted by (1). Let the equation to the proposed straight line be

$$\lambda u + \mu v + \nu w = 0 \quad\text{......................(3).}$$

If (3) denotes the equation to the tangent at (u', v', w') we find by comparing (3) with (2) that

$$\frac{A}{\lambda\sqrt{u'}} = \frac{B}{\mu\sqrt{v'}} = \frac{C}{\nu\sqrt{w'}}.$$

From these relations and (1) we obtain as the required condition

$$\frac{A^2}{\lambda}+\frac{B^2}{\mu}+\frac{C^2}{\nu}=0.$$

323. *To find the equation to the circle described round a triangle.*

It will be convenient in this and the two following Articles to use the form $x\cos\alpha+y\sin\alpha-p=0$ as the type of the equation to a straight line; we shall therefore put α, β, γ for u, v, w respectively (Art. 71).

Let $\alpha=0$, $\beta=0$, $\gamma=0$ be the equations to the sides of a triangle; then, by Art. 309,

$$l\beta\gamma+m\gamma\alpha+n\alpha\beta=0 \qquad (1)$$

will represent *any* conic section described round the triangle; hence by giving proper values to l, m, n, this equation may be made to represent the circle which we know by geometry can be described round the triangle. We might proceed thus: in (1) write for α, β, γ the expressions which they represent, then equate the coefficient of xy to zero, and the coefficient of x^2 to that of y^2; we shall thus have two equations for determining $\frac{n}{l}$ and $\frac{m}{l}$; and with the values thus obtained (1) will represent the required circle. We leave this as an exercise for the student, and adopt another method. The equation to the tangent to (1) at the intersection of $\alpha=0$, and $\beta=0$, is, by Art. 310,

$$l\beta+m\alpha=0 \qquad (2).$$

Let A, B, C denote the angles of the triangle opposite the sides $\alpha=0, \beta=0, \gamma=0$, respectively; by Euclid, III. 32, the tangent denoted by (2) must make an angle A with the straight line $\alpha=0$, and an angle B with the straight line $\beta=0$. Suppose the origin of co-ordinates *within* the triangle, then the equation to the straight line passing through the intersection of $\alpha=0$ and $\beta=0$, and making angles A and B respectively with these straight lines, is

$$\alpha\sin B+\beta\sin A=0 \qquad (3).$$

Thus (2) must coincide with (3); therefore we have $\frac{l}{m}=\frac{\sin A}{\sin B}$. Similarly, $\frac{m}{n}=\frac{\sin B}{\sin C}$.

Thus the equation to the circle described round the triangle is

$$\beta\gamma \sin A + \gamma\alpha \sin B + \alpha\beta \sin C = 0.$$

324. *To find the equation to the circle inscribed in a triangle.*

Suppose the origin of co-ordinates *within* the triangle; then for all points on the circle α, β, γ are *negative* quantities (see Art. 54). Now the equation to the circle must be of one of the forms (8), (9), (10), (11) given in Art. 318; the first is the only form applicable, namely,

$$\sqrt{(l\alpha)} + \sqrt{(m\beta)} + \sqrt{(n\gamma)} = 0 \ldots\ldots\ldots\ldots\ldots (1),$$

which is equivalent to

$$\sqrt{(-l\alpha)} + \sqrt{(-m\beta)} + \sqrt{(-n\gamma)} = 0 \ldots\ldots\ldots\ldots (2).$$

The other forms are inapplicable, because they would introduce impossible expressions. We have then to determine the values of l, m, and n. If we put $\alpha = 0$ in (1), we obtain $\frac{\beta}{\gamma}=\frac{n}{m}$; thus $\frac{n}{m}$ is the ratio of the perpendiculars drawn to the sides $\beta=0$, $\gamma=0$, respectively, from the point where the circle meets the straight line $\alpha=0$. Let r be the radius of the circle; then we know from geometry that the perpendicular from this point on $\beta=0$ is $r \cot\frac{C}{2} \sin C$ or $2r\cos^2\frac{C}{2}$; a similar expression holds for the perpendicular on $\gamma=0$.

Hence $$\frac{n}{m}=\frac{\cos^2\frac{C}{2}}{\cos^2\frac{B}{2}}. \quad \text{Similarly} \quad \frac{l}{n}=\frac{\cos^2\frac{A}{2}}{\cos^2\frac{C}{2}}.$$

Therefore the required equation is

$$\cos\frac{A}{2}\sqrt{\alpha} + \cos\frac{B}{2}\sqrt{\beta} + \cos\frac{C}{2}\sqrt{\gamma} = 0.$$

325. *To find the equation to the circle which touches one side of a triangle and the other two sides produced.*

Let the circle be required to touch the side opposite to the angle A and the other two sides produced. Suppose the origin *within* the triangle; then for all points comprised between the side $\alpha = 0$ and the other sides produced, α is positive and β and γ are negative. Hence by Art. 318, the form of the equation to the circle must be

$$\sqrt{(-l\alpha)} + \sqrt{(m\beta)} + \sqrt{(n\gamma)} = 0.$$

Hence, as before, by considering the point where the circle meets the straight line $\alpha = 0$, we have

$$\frac{n}{m} = \frac{\cos^2 \frac{\pi - C}{2}}{\cos^2 \frac{\pi - B}{2}} = \frac{\sin^2 \frac{C}{2}}{\sin^2 \frac{B}{2}}, \text{ and } \frac{l}{n} = \frac{\cos^2 \frac{A}{2}}{\cos^2 \frac{\pi - C}{2}} = \frac{\cos^2 \frac{A}{2}}{\sin^2 \frac{C}{2}}.$$

Hence the required equation is

$$\cos \frac{A}{2} \sqrt{(-\alpha)} + \sin \frac{B}{2} \sqrt{\beta} + \sin \frac{C}{2} \sqrt{\gamma} = 0.$$

Similarly the equations to the other two circles may be written down.

326. The results in Arts. 312 and 321 which hold for any conic section, will of course hold for a circle inscribed in, or described about, a triangle respectively. We have only to use the values of l, m, n, found in Arts. 323...325.

327. Many applications have been made of the method of abridged notation to express the equations to circles determined by various conditions. We will give some of these applications as specimens, and the student will have no difficulty in applying the same methods to other examples.

328. If the equation to one circle, expressed in a rational form, be denoted by $S = 0$, the equation to *any other* circle can be expressed in the form $S + \lambda\alpha + \mu\beta + \nu\gamma = 0$, by properly choosing the constants λ, μ, and ν. This result follows from the known form of the equation to a circle in the common co-ordinates; see Arts. 88, 104, 110. Thus, to take the

most general supposition, let the equations to two circles be in common oblique co-ordinates

$$K(x^2 + 2xy\cos\omega + y^2) + Lx + My + N = 0,$$
$$k(x^2 + 2xy\cos\omega + y^2) + lx + my + n = 0.$$

Denote the first equation by $S = 0$; then the second equation is equivalent to

$$S + \frac{K}{k}(lx + my + n) - Lx - My - N = 0,$$

which we may denote by $S + u = 0$. Here u is an expression of the first degree in x and y, and so will be identical with $\lambda\alpha + \mu\beta + \nu\gamma$, if we determine λ, μ, and ν suitably.

If $S = 0$ and $S + \lambda\alpha + \mu\beta + \nu\gamma = 0$ be the equations to two circles, $\lambda\alpha + \mu\beta + \nu\gamma = 0$ will be the equation to the radical axis of the two circles; see Art. 110.

Since $a\alpha + b\beta + c\gamma$ is a constant, by Art. 73, we may instead of $S + (\lambda\alpha + \mu\beta + \nu\gamma)$ use

$$S + (a\alpha + b\beta + c\gamma)(l\alpha + m\beta + n\gamma)$$

or $$S + (\alpha\sin A + \beta\sin B + \gamma\sin C)(l\alpha + m\beta + n\gamma),$$

provided we properly determine the constants in each case.

329. *To express the equation to the circle which passes through the middle points of the sides of the triangle of reference.*

Let $\alpha = 0$, $\beta = 0$, $\gamma = 0$ be the equations to the straight lines which form the triangle of reference; see Art. 78. Assume for the required equation

$$\beta\gamma\sin A + \gamma\alpha\sin B + \alpha\beta\sin C$$
$$+ (\alpha\sin A + \beta\sin B + \gamma\sin C)(l\alpha + m\beta + n\gamma) = 0;$$

see Arts. 323 and 328.

At the middle point of the side BC we have $\alpha = 0$, and

$$\frac{\beta}{\gamma} = \frac{\sin C}{\sin B};$$

substituting in the assumed equation we obtain

$$\frac{\sin A \sin C}{\sin B} + 2 \sin C \left(\frac{m \sin C}{\sin B} + n\right) = 0,$$

or

$$\sin A + 2 (m \sin C + n \sin B) = 0.$$

But $\sin A = \sin B \cos C + \cos B \sin C$; thus

$$\sin C (2m + \cos B) + \sin B (2n + \cos C) = 0.$$

In a similar manner we obtain two analogous equations; and from the three equations we deduce

$$l = -\frac{1}{2} \cos A, \quad m = -\frac{1}{2} \cos B, \quad n = -\frac{1}{2} \cos C.$$

Hence the required equation is

$$\beta\gamma \sin A + \gamma\alpha \sin B + \alpha\beta \sin C$$
$$-\frac{1}{2} (\alpha \sin A + \beta \sin B + \gamma \sin C)(\alpha \cos A + \beta \cos B + \gamma \cos C) = 0.$$

The radical axis of this circle and the circle described round the triangle of reference is therefore determined by

$$\alpha \cos A + \beta \cos B + \gamma \cos C = 0.$$

330. *To express the equation to the circle which passes through the feet of the perpendiculars from the angles of the triangle of reference on the opposite sides.*

Assume for the required equation

$$\beta\gamma \sin A + \gamma\alpha \sin B + \alpha\beta \sin C$$
$$+ (\alpha \sin A + \beta \sin B + \gamma \sin C)(l\alpha + m\beta + n\gamma) = 0$$

At the foot of the perpendicular from A on BC we have $\alpha = 0$, and $\frac{\beta}{\gamma} = \frac{\cos C}{\cos B}$; substituting in the assumed equation we obtain

$$\frac{\sin A \cos C}{\cos B} + \left(\frac{\cos C}{\cos B} \sin B + \sin C\right)\left(\frac{m \cos C}{\cos B} + n\right) = 0,$$

or

$$\sin A \cos B \cos C + \sin A (m \cos C + n \cos B) = 0;$$

therefore $\left(m + \frac{1}{2} \cos B\right) \cos C + \left(n + \frac{1}{2} \cos C\right) \cos B = 0.$

In a similar manner we obtain two analogous equations; and from the three equations we deduce

$$l = -\frac{1}{2}\cos A, \quad m = -\frac{1}{2}\cos B, \quad n = -\frac{1}{2}\cos C.$$

Hence the required equation is

$$\beta\gamma \sin A + \gamma\alpha \sin B + \alpha\beta \sin C$$

$$-\frac{1}{2}(\alpha \sin A + \beta \sin B + \gamma \sin C)(\alpha \cos A + \beta \cos B + \gamma \cos C) = 0.$$

Thus the circle is the same as that considered in the preceding Article.

331. Let O denote the intersection of the perpendiculars from the angles of a triangle on the opposite sides. Then it is known that the circle which passes through the six points specified in the preceding two Articles also passes through the middle points of OA, OB, and OC. The circle is called the *nine-points circle*. See *Appendix to Euclid*.

It is easy to shew that the circle which passes through the six points specified in the preceding two Articles also passes through the middle points of OA, OB, and OC. For consider the triangle OBC. The perpendiculars from the angular points on the opposite sides meet these sides respectively at points which coincide with the feet of the perpendiculars from the angles A, B, C on the opposite sides; thus we know that the circle considered in the preceding two Articles passes through these points: hence it also passes through the middle points of OB and OC, as well as through the middle point of BC. Similarly the circle also passes through the middle point of OA.

O is a *centre of similitude* of the circle described round the triangle ABC and the nine-points circle of the triangle; see Art. 119. For, as we have just seen, the three radii vectores drawn from O to the circumference of the former circle, namely OA, OB, OC, are respectively double the radii vectores drawn in the same direction to the latter circle; and it is easy to shew that the same ratio will hold for any corresponding radii vectores. See Example 58 of Chapter XIV.

332. *To investigate the conditions which must hold in order that the general equation of the second degree may represent a circle.*

Let the equation be

$$L\alpha^2 + M\beta^2 + N\gamma^2 + 2L'\beta\gamma + 2M'\gamma\alpha + 2N'\alpha\beta = 0.$$

Let Δ denote the area of the triangle of reference; then, by Art. 73,

$$a\alpha + b\beta + c\gamma = -2\Delta;$$

therefore $\quad a\alpha^2 = -2\Delta\alpha - (b\beta + c\gamma)\alpha.$

Similarly $\quad b\beta^2 = -2\Delta\beta - (c\gamma + a\alpha)\beta;$

and $\quad c\gamma^2 = -2\Delta\gamma - (a\alpha + b\beta)\gamma.$

Substitute for α^2, β^2, and γ^2 in the general equation, and it becomes

$$\left(2L' - \frac{Mc}{b} - \frac{Nb}{c}\right)\beta\gamma + \left(2M' - \frac{Na}{c} - \frac{Lc}{a}\right)\gamma\alpha$$
$$+ \left(2N' - \frac{Lb}{a} - \frac{Ma}{b}\right)\alpha\beta$$
$$-2\Delta\left(\frac{L\alpha}{a} + \frac{M\beta}{b} + \frac{N\gamma}{c}\right) = 0.$$

Then, by Arts. 323 and 328, we see that the necessary and sufficient conditions in order that this equation may represent a circle are

$$\frac{2L' - \frac{Mc}{b} - \frac{Nb}{c}}{a} = \frac{2M' - \frac{Na}{c} - \frac{Lc}{a}}{b} = \frac{2N' - \frac{Lb}{a} - \frac{Ma}{b}}{c};$$

that is,

$$2L'bc - Mc^2 - Nb^2 = 2M'ca - Na^2 - Lc^2 = 2N'ab - Lb^2 - Ma^2.$$

333. *To determine the radical axis of two circles represented by general equations.*

Let the equations be

$$L\alpha^2 + M\beta^2 + N\gamma^2 + 2L'\beta\gamma + 2M'\gamma\alpha + 2N'\alpha\beta = 0,$$
$$l\alpha^2 + m\beta^2 + n\gamma^2 + 2l'\beta\gamma + 2m'\gamma\alpha + 2n'\alpha\beta = 0.$$

Since, by supposition, these equations represent circles they may, by the preceding Article, be put in the form

$$\frac{1}{a}\left(2L' - \frac{Mc}{b} - \frac{Nb}{c}\right)(a\beta\gamma + b\gamma\alpha + c\alpha\beta)$$
$$- 2\Delta\left(\frac{L\alpha}{a} + \frac{M\beta}{b} + \frac{N\gamma}{c}\right) = 0,$$

$$\frac{1}{a}\left(2l' - \frac{mc}{b} - \frac{nb}{c}\right)(a\beta\gamma + b\gamma\alpha + c\alpha\beta)$$
$$- 2\Delta\left(\frac{l\alpha}{a} + \frac{m\beta}{b} + \frac{n\gamma}{c}\right) = 0.$$

Hence the equation to the radical axis is

$$\frac{\frac{L\alpha}{a} + \frac{M\beta}{b} + \frac{N\gamma}{c}}{2L'bc - Mc^2 - Nb^2} = \frac{\frac{l\alpha}{a} + \frac{m\beta}{b} + \frac{n\gamma}{c}}{2l'bc - mc^2 - nb^2},$$

or
$$\frac{L\alpha \sin B \sin C + M\beta \sin C \sin A + N\gamma \sin A \sin B}{2L' \sin B \sin C - M \sin^2 C - N \sin^2 B}$$
$$= \frac{l\alpha \sin B \sin C + m\beta \sin C \sin A + n\gamma \sin A \sin B}{2l' \sin B \sin C - m \sin^2 C - n \sin^2 B}.$$

334. *The nine-points circle of a triangle touches the inscribed and escribed circles of that triangle.*

The equation to the nine-points circle may be put in the form

$$\alpha^2 \sin A \cos A + \beta^2 \sin B \cos B + \gamma^2 \sin C \cos C$$
$$- \beta\gamma \sin A - \gamma\alpha \sin B - \alpha\beta \sin C = 0.$$

The equation to the inscribed circle is

$$\cos\frac{A}{2}\sqrt{\alpha} + \cos\frac{B}{2}\sqrt{\beta} + \cos\frac{C}{2}\sqrt{\gamma} = 0;$$

putting this in a rational form we obtain

$$\alpha^2 \cos^4\frac{A}{2} + \beta^2 \cos^4\frac{B}{2} + \gamma^2 \cos^4\frac{C}{2}$$
$$- 2\beta\gamma \cos^2\frac{B}{2}\cos^2\frac{C}{2} - 2\gamma\alpha \cos^2\frac{C}{2}\cos^2\frac{A}{2} - 2\alpha\beta \cos^2\frac{A}{2}\cos^2\frac{B}{2} = 0.$$

The equation to the radical axis of these circles by Art. 333 is

$$\frac{\sin A \sin B \sin C(\alpha \cos A + \beta \cos B + \gamma \cos C)}{\sin A \sin B \sin C + \sin B \cos B \sin^2 C + \sin C \cos C \sin^2 B}$$

$$= \frac{\alpha \cos^4 \frac{A}{2} \sin B \sin C + \beta \cos^4 \frac{B}{2} \sin C \sin A + \gamma \cos^4 \frac{C}{2} \sin A \sin B}{2 \cos^2 \frac{B}{2} \cos^2 \frac{C}{2} \sin B \sin C + \cos^4 \frac{B}{2} \sin^2 C + \cos^4 \frac{C}{2} \sin^2 B};$$

that is,

$$\alpha \cos A + \beta \cos B + \gamma \cos C$$

$$= \frac{\sin A \sin B \sin C \left(\frac{\alpha \cos^4 \frac{A}{2}}{\sin A} + \frac{\beta \cos^4 \frac{B}{2}}{\sin B} + \frac{\gamma \cos^4 \frac{C}{2}}{\sin C} \right)}{2 \cos^2 \frac{A}{2} \cos^2 \frac{B}{2} \cos^2 \frac{C}{2}},$$

or $$(\alpha \cos A + \beta \cos B + \gamma \cos C) \cos \frac{A}{2} \cos \frac{B}{2} \cos \frac{C}{2}$$

$$= 4 \sin \frac{A}{2} \sin \frac{B}{2} \sin \frac{C}{2} \left(\frac{\alpha \cos^4 \frac{A}{2}}{\sin A} + \frac{\beta \cos^4 \frac{B}{2}}{\sin B} + \frac{\gamma \cos^4 \frac{C}{2}}{\sin C} \right).$$

This equation may be simplified by substituting for the trigonometrical functions their known values in terms of the sides; let s denote the half sum of the sides, then we obtain

$$(\alpha \cos A + \beta \cos B + \gamma \cos C) \frac{s^2}{4\Delta}$$

$$= \frac{\alpha \cos^4 \frac{A}{2}}{\sin A} + \frac{\beta \cos^4 \frac{B}{2}}{\sin B} + \frac{\gamma \cos^4 \frac{C}{2}}{\sin C};$$

therefore

$$\alpha \left\{ \cos A - \frac{2(s-a)^2}{bc} \right\} + \beta \left\{ \cos B - \frac{2(s-b)^2}{ca} \right\}$$

$$+ \gamma \left\{ \cos C - \frac{2(s-c)^2}{ab} \right\} = 0,$$

or $$\frac{\alpha(c-a)(a-b)}{bc}+\frac{\beta(a-b)(b-c)}{ca}+\frac{\gamma(b-c)(c-a)}{ab}=0,$$

or $$\frac{a\alpha}{b-c}+\frac{\beta b}{c-a}+\frac{\gamma c}{a-b}=0:$$

this may also be written thus

$$\frac{\alpha\cos\frac{1}{2}A}{\sin\frac{1}{2}(B-C)}+\frac{\beta\cos\frac{1}{2}B}{\sin\frac{1}{2}(C-A)}+\frac{\gamma\cos\frac{1}{2}C}{\sin\frac{1}{2}(A-B)}=0.$$

Now the radical axis touches the inscribed circle; for it may be shewn that the condition of tangency investigated in Art. 322 is satisfied; and as the radical axis touches one of the circles the circles must touch, and the radical axis is the common tangent at the point of contact.

Similarly we may shew that the nine-points circle touches the escribed circles.

335. If $S=0$ be the equation to a circle in a rational form, the equation to any concentric circle will be of the form $S-k=0$, where k is some constant. Or, as $a\alpha+b\beta+c\gamma$ is a constant, we may put the equation in the form

$$S-l\,(a\alpha+b\beta+c\gamma)^2=0,$$

where l is some constant.

For example, required the equation to a circle which is concentric with the circle described round the triangle of reference, and which touches the side a. Assume for the equation

$$a\beta\gamma+b\gamma\alpha+c\alpha\beta-l\,(a\alpha+b\beta+c\gamma)^2=0.$$

At the point of contact with the side a we have $\alpha=0$; thus $a\beta\gamma-l(b\beta+c\gamma)^2=0$. This quadratic in $\frac{\beta}{\gamma}$ must then have *equal roots*, so that $4l^2b^2c^2=(a-2lbc)^2$: thus $l=\frac{a}{4bc}$.

336. Let there be any quadrilateral, and let its sides be represented by the equations

$$t=0,\quad u=0,\quad v=0,\quad w=0,$$

then the equation

$$tu+kvw=0,$$

where k is a constant, represents a conic section circumscribing the quadrilateral. For the equation represents a conic section passing through the four points determined respectively by

$$t=0 \text{ and } v=0, \qquad t=0 \text{ and } w=0,$$

$$u=0 \text{ and } v=0, \qquad u=0 \text{ and } w=0.$$

Also by giving a suitable value to k, the equation may be made to represent *any* conic section passing through these four points.

The above equation has the following geometrical interpretation. If any quadrilateral figure be inscribed in a conic section, the product of the perpendiculars drawn from any point of the curve on two opposite sides bears a constant ratio to the product of the perpendiculars on the other two sides.

We may observe that the term *quadrilateral* is often used in analytical geometry in a wider sense than in ordinary synthetical geometry. Thus, if we have four given points, we may obtain three different quadrilaterals by connecting these points in different ways. Take, for example, the figure in Art. 75; and let A, B, C, D be the given points. The three different quadrilaterals are (1) the figure bounded by AB, BC, CD, DA; (2) the figure bounded by AC, CD, DB, BA; which in fact consists of the two triangles GAB and GCD; (3) the figure bounded by AC, CB, BD, DA, which in fact consists of the two triangles GBC and GDA.

Similarly, four given straight lines may be considered to form three different quadrilaterals by their intersections. Take, for example, the figure in Art. 75, and let the given straight lines be EDC, EAB, AGC, BGD. The three different quadrilaterals are (1) the figure bounded by GC, CE, EB, BG; (2) the figure bounded by GD, DE, EA, AG; (3) the figure bounded by AC, CD, DB, BA.

If four straight lines have for their equations

$$t=0, \quad u=0, \quad v=0, \quad w=0,$$

the conic sections passing through the angular points of the three different quadrilaterals which these straight lines form, may be denoted by the equations

$$tu+k_1vw=0, \quad tv+k_2uw=0, \quad tw+k_3uv=0.$$

337. We shall next consider the equation $uv - w^2 = 0$. This represents a conic section which passes through the point determined by $u = 0$ and $w = 0$, and also through the point determined by $v = 0$ and $w = 0$. Also each of the straight lines $u = 0$ and $v = 0$ *touches* the conic section where it meets it; for if we combine $u = 0$ with the above equation, we see that $w = 0$ also, that is, the straight line $u = 0$, meets the curve at only *one* point, namely, that point at which $u = 0$ and $w = 0$ intersect. Similarly the straight line $v = 0$ touches the curve. Thus $u = 0$ and $v = 0$ represent two tangents to the conic section, and $w = 0$ represents the corresponding chord of contact.

We may also shew in the following way that the straight line $u = 0$ cannot *cut* the curve: for points on one side of the straight line $u = 0$, the expression u is positive, and for points on the other side of the straight line, negative; but w^2 is of invariable sign; thus the straight line $u = 0$ cannot cut the curve.

The geometrical interpretation of the above equation is as follows. The product of the perpendiculars from any point of a conic section on a pair of tangents bears a constant ratio to the square of the perpendicular from the same point on the chord of contact.

338. We will now consider the equations to a secant and a tangent to the curve denoted by $uv = w^2$; the results for this particular case are included in the general results of Art. 315, but the investigation may be put in a different form.

Let (u', v', w') denote one point on the curve, and (u'', v'', w'') another. The equation to any straight line passing through the first point may be denoted by

$$uv' - ww' = \lambda(vu' - ww'),$$

where λ is a constant. For this equation is of the first degree in u, v, w, and therefore represents some straight line; and the equation is obviously satisfied at the point (u', v', w'). Suppose the straight line to pass also through the point (u'', v'', w''); then we have

$$u''v' - w''w' = \lambda(v''u' - w''w').$$

Hence, by division,

$$\frac{uv' - ww'}{u''v' - w''w'} = \frac{vu' - ww'}{v''u' - w''w'},$$

that is,

$$\frac{uv' - ww'}{\frac{w''^2}{v''}v' - w''w'} = \frac{vu' - ww'}{\frac{v''w'^2}{v'} - w''w'},$$

or

$$\frac{v''}{w''}(uv' - ww') + \frac{v'}{w'}(vu' - ww') = 0.$$

This equation then represents the secant passing through the two given points. Hence the equation to the tangent at the point (u', v', w') is

$$uv' + vu' - 2ww' = 0.$$

Suppose $\frac{u'}{w'} = \mu'$, then from the equation to the curve $\frac{v'}{w'} = \frac{1}{\mu'}$; similarly if $\frac{u''}{w''} = \mu''$, then $\frac{v''}{w''} = \frac{1}{\mu''}$. Thus the equation to the secant may be written

$$\frac{1}{\mu''}\left(\frac{u}{\mu'} - w\right) + \frac{1}{\mu'}(v\mu' - w) = 0,$$

or

$$u + \mu'\mu''v - (\mu' + \mu'')w = 0;$$

and the equation to the tangent may be written

$$u + \mu'^2 v - 2\mu' w = 0.$$

339. Next take the equation $l^2u^2 + m^2v^2 = n^2w^2$. This may be written $(nw + mv)(nw - mv) = l^2u^2$. Hence by Art. 337 $nw + mv = 0$ and $nw - mv = 0$ are tangents to the conic section represented by the equation, and $u = 0$ is the equation to the corresponding chord of contact. Since these two tangents meet at the point of intersection of $v = 0$ and $w = 0$, it follows that this point is the *pole* of $u = 0$.

Similarly we may write the equation in the form

$$(nw + lu)(nw - lu) = m^2v^2,$$

and infer that the point of intersection of $u = 0$ and $w = 0$ is the pole of $v = 0$.

Hence it follows that the point of intersection of $u=0$ and $v=0$ is the pole of $w=0$. See Art. 291.

340. The following is a particular case of the preceding Article, $\alpha^2+\beta^2=n^2\gamma^2$. (See Art. 71.) Suppose the straight lines $\alpha=0$, $\beta=0$, at right angles; then $\alpha^2+\beta^2$ is the square of the distance of the point (x, y) from the intersection of $\alpha=0$ and $\beta=0$. Hence the above equation represents a conic section which has $\gamma=0$ for its directrix, and the intersection of $\alpha=0$ and $\beta=0$ for its focus. The straight lines $n\gamma-\alpha=0$ and $n\gamma+\alpha=0$ are tangents to the conic section, touching it at the extremities of the focal chord $\beta=0$; also these tangents meet on the straight line $\gamma=0$; hence, *the tangents at the extremities of any focal chord meet on the corresponding directrix.* Also the above tangents meet on the straight line $\alpha=0$, which by supposition is at right angles to $\beta=0$; hence, *the straight line which joins the focus to the intersection of tangents at the extremities of a focal chord is at right angles to that focal chord.*

341. If $u=0$ and $v=0$ be the equations to two conic sections which meet at four points, then $u+lv=0$ will represent any conic section which passes through the four points of intersection. This will be obvious after the proofs given of similar propositions.

Also if $w=0$ and $w'=0$ be the equations to two straight lines, $u+lww'=0$ will represent any conic section passing through the four points at which the lines $w=0$ and $w'=0$ meet the conic section $u=0$.

Also $u+lw^2=0$ will represent a conic section passing through the points of intersection of the conic section $u=0$, and the straight line $w=0$. This conic section will have the same tangent as $u=0$ at the points where $u=0$ and $w=0$ intersect; we might anticipate this would be the case from observing the interpretation of the equation $u+lww'=0$, and supposing the straight line $w'=0$ to approach the straight line $w=0$, and ultimately to coincide with it. We may prove it strictly by taking one of the points where $u=0$ meets $w=0$ for the origin, and the straight line $w=0$ for the axis of x; thus u becomes of the form

$$Ax^2+Bxy+Cy^2+Dx+Ey,$$

and we can see, by Art. 283, that

$$Ax^2 + Bxy + Cy^2 + Dx + Ey = 0$$

and $$Ax^2 + Bxy + Cy^2 + Dx + Ey + ly^2 = 0$$

have the same tangent at the origin.

Also by giving a suitable value to l the equation $u + lw^2 = 0$ may be made to represent the two straight lines which touch the conic section $u = 0$ at the points where it intersects the straight line $w = 0$. This may be inferred from Art. 293; the equation $w = 0$ is equivalent to $\frac{x}{h} + \frac{y}{k} - 1 = 0$, and the equation $u = 0$ is equivalent to $\left(\frac{x}{h} + \frac{y}{k} - 1\right)^2 + \mu xy = 0$. Thus by taking $l = -1$ we have $u + lw^2 = \mu xy$; and the equation $xy = 0$ denotes the two tangents to the conic section $u = 0$ at its points of intersection with the straight line $w = 0$.

342. Pascal's Theorem. *The three intersections of the opposite sides of any hexagon inscribed in a conic section are on one straight line.*

Let $r = 0$, $s = 0$, $t = 0$, $u = 0$, $v = 0$, $w = 0$, be the equations to the sides of a hexagon which is inscribed in the conic section $S = 0$. Let the hexagon be divided by a new straight line $\phi = 0$ into two quadrilaterals, one of which has for its sides the straight lines obtained by equating to zero successively, r, s, t, ϕ, and the other the straight lines obtained by equating to zero successively, u, v, w, ϕ. Now we know that if a, b, l, m are appropriate constants, the equation to the conic section may be written in the forms $as\phi + brt = 0$ and $lv\phi + muw = 0$; therefore $as\phi + brt$ and $lv\phi + muw$ must each be identical with S; therefore $as\phi + brt = lv\phi + muw$; therefore $(as - lv)\,\phi = muw - brt$.

The right-hand member of this equation vanishes when u and r simultaneously vanish, and when u and t simultaneously vanish; also when w and r simultaneously vanish, and when w and t simultaneously vanish. Since the left-hand member is identically equal to the right-hand, the left-hand member must also vanish in these four cases; that is, one of its two factors ϕ and $as - lv$ must vanish in each of these four cases. By construction, $\phi = 0$ represents the straight line joining the point determined by $r = 0$ and $w = 0$, with the point determined by $t = 0$ and $u = 0$; and thus we see that $as - lv = 0$ is the straight line joining the intersection of $u = 0$ and $r = 0$ with that of $t = 0$ and $w = 0$. But the straight line $as - lv = 0$ obviously passes through the intersection of $s = 0$ and $v = 0$; therefore the three points determined respectively by $u = 0$ and $r = 0$, $t = 0$ and $w = 0$, $s = 0$ and $v = 0$, lie on a straight line.

It is to be observed that if six points be connected by straight lines in different ways, as many as sixty figures can be formed which may be called *hexagons* in an extended sense of that word. Thus for six given points on a conic section there will be sixty applications of Pascal's Theorem.

343. Let $s = 0$ be the equation to a conic section, and $u = 0$, $v = 0$, $w = 0$, equations to three straight lines; then $s - l^2u^2 = 0$, $s - m^2v^2 = 0$, $s - n^2w^2 = 0$, represent curves of the second degree touching the proposed conic section. By properly choosing u, v, w, l, m, n, we may make each of the last three equations represent a pair of straight lines touching $s = 0$. (See Art. 341.) Thus, if there be a hexagon circumscribed round the conic section $s = 0$, the equations

$$s - l^2u^2 = 0 \ldots (1), \quad s - m^2v^2 = 0 \ldots (2), \quad s - n^2w^2 = 0 \ldots (3),$$

may be taken to represent the six sides of the hexagon.

By combining (1) and (2) we obtain

$$s - l^2u^2 - (s - m^2v^2) = 0, \quad \text{or } (mv - lu)(mv + lu) = 0 \ldots (4),$$

for the equation to a pair of straight lines which pass through the intersections of (1) and (2).

Similarly $$(nw - mv)(nw + mv) = 0 \ldots\ldots\ldots\ldots\ldots (5)$$

represents a pair of straight lines which pass through the intersections of (2) and (3). And

$$(lu - nw)(lu + nw) = 0 \ldots\ldots\ldots\ldots\ldots (6)$$

represents a pair of straight lines which pass through the intersections of (3) and (1).

The six straight lines which we have obtained may be

arranged in four groups, each containing three straight lines which meet at a point, namely,

$$mv - lu = 0, \qquad nw - mv = 0, \qquad lu - nw = 0,$$
$$mv + lu = 0, \qquad nw + mv = 0, \qquad lu - nw = 0,$$
$$mv + lu = 0, \qquad nw - mv = 0, \qquad lu + nw = 0,$$
$$mv - lu = 0, \qquad nw + mv = 0, \qquad lu + nw = 0.$$

This result is consistent with Brianchon's theorem; *if a hexagon be described about a conic section the three diagonals which join opposite angles meet at a point.*

For suppose that a hexagon is described round a conic section, and let its angular points be denoted by A, B, C, D, E, F. By properly choosing u, v, w, l, m, n, we may make equation (1) denote the straight lines AB and DE, equation (2) denote the straight lines BC and EF, and equation (3) denote the straight lines CD and FA. We will now examine what straight lines are determined by equations (4), (5), and (6). Equation (4) determines the two straight lines which pass through the intersections of the straight lines determined by (1) and (2); and as the signs of l and m are at present in our power we may take them so that $mv - lu = 0$ shall represent the straight line BE, and then $mv + lu = 0$ will represent the straight line joining the point which is common to AB and EF with the point which is common to BC and DE. Similarly as the sign of n is still in our power, we may take it so that $nw - mv = 0$ shall represent the straight line CF, and then $nw + mv = 0$ will represent the straight line joining the point which is common to BC and FA with the point which is common to CD and EF. One of the two straight lines represented by (6) is AD, and the other is the straight line joining the point which is common to DE and FA with the point which is common to CD and AB; it is however not obvious how we are in general to discriminate between these two straight lines. Thus the proof of Brianchon's theorem is not perfectly satisfactory, and accordingly we shall give another proof by which the theorem is deduced from that of Pascal.

Let the angular points of the hexagon be denoted as before by the letters A, B, C, D, E, F. Let the straight line be drawn which passes through the points of contact of the conic section and the tangents AB, BC; also let the straight line be drawn which passes through the points of contact of the conic section and the tangents DE, EF; and let P denote the point which is common to these two straight lines. *Then P is the pole of BE*; see Arts. 103, 120, 289. In the same way we may determine the pole of CF which we shall denote by Q, and the pole of AD which we shall denote by R. By Pascal's theorem P, Q, and R lie on a straight line; hence CF, BE, and AD meet at a point, namely, at the pole of the straight line PQR; see Art. 291.

For further information on the subject of this Chapter the student is referred to Salmon's *Conic Sections*.

EXAMPLES.

1. Shew that if $a-c : a'-c' :: b : b'$, a circle may be described through the intersections of the two conic sections

$$ax^2 + bxy + cy^2 + dx + ey + f = 0,$$
$$a'x^2 + b'xy + c'y^2 + d'x + e'y + f' = 0.$$

Find also the condition that a parabola may be described passing through the origin and the points of intersection of these curves.

2. Two conic sections have their principal axes at right angles: shew that a circle will pass through their points of intersection.

3. The equations to two conic sections are

$$Ay^2 + 2Bxy + Cx^2 + 2A'x = 0, \quad ay^2 + 2bxy + cx^2 + 2a'x = 0.$$

Shew that the straight lines joining the origin with their points of intersection will be at right angles to each other if

$$a'(A + C) = A'(a + c).$$

4. An ellipse is described so as to touch the asymptotes of an hyperbola: shew that two of the chords joining the points of intersection of the ellipse and hyperbola are parallel.

5. If $\alpha\beta = c^2$ be the equation to an hyperbola (Art. 71), then $\alpha\beta = 0$, $\alpha^2 - \beta^2 = 0$, $\alpha^2 - n^2\beta^2 = 0$, are the respective equations to the asymptotes, the axes, and a pair of conjugate diameters, n being any constant.

6. The straight lines which bisect the angles of a triangle, meet the opposite sides at the points P, Q, R, respectively: find the equation to an ellipse described so as to touch the sides of the triangle in these points.

7. From any point two straight lines are drawn, one inclined at an angle α, the other at an angle $\frac{\pi}{2} + \alpha$, to the axis of a parabola: shew that another parabola may be described which shall pass through the four points of intersection, whose axis is inclined at an angle 2α to that of the given parabola.

8. Prove that the equation to the conic section which passes through the point (h, k), and touches the parabola $y^2 = lx$ at the vertex and at an extremity of the latus rectum, is $(y^2 - lx)(k - 2h)^2 = (y - 2x)^2(k^2 - lh)$.

Shew that it is an ellipse or hyperbola according as the point (h, k) is within or without the parabola.

9. A conic section touches the sides of a triangle ABC at the points a, b, c; and the straight lines Aa, Bb, Cc, intersect the conic section at a', b', c': shew that

(1) the straight lines Aa, Bb, Cc pass respectively through the intersections of Bc' and Cb', Ca' and Ac', Ab' and Ba',

(2) the intersections of the straight lines ab and $a'b'$, bc and $b'c'$, ac and $a'c'$, lie respectively on AB, BC, CA.

10. A conic section is described round a triangle ABC; straight lines bisecting the angles of this triangle meet the conic section at the points A', B', C', respectively: express the equations to $A'B$, $A'C$, $A'B'$.

11. If a conic section be described about any triangle, and the points where the straight lines bisecting the angles of the triangle meet the conic section be joined, the intersections of

the sides of the triangle so formed with the corresponding sides of the original triangle lie on a straight line.

12. Interpret the equation

$$\left(\frac{x}{a}+\frac{y}{b}-1\right)\left(\frac{x}{a'}+\frac{y}{b'}-1\right)+\mu xy=0:$$

find how many parabolas can be drawn through four given points.

13. If $u=0$, $v=0$, $w=0$ represent the sides of a triangle, shew that the sides of any triangle which has one angle on each side of the former may be represented by

$$u+nv+\frac{w}{m}=0,\quad \frac{u}{n}+v+lw=0,\quad mu+\frac{v}{l}+w=0,$$

where l, m, n are constants.

Find also the relation which must hold between l, m, n, in order that the straight lines joining corresponding angles of the two triangles may meet at a point.

14. A circle and a rectangular hyperbola intersect at four points, and one of their common chords is a diameter of the hyperbola: shew that another of them is a diameter of the circle.

15. ACA' is the major axis of an ellipse; P is any point on the circle described on the major axis; AP, $A'P$ meet the ellipse at Q, Q': shew that the equation to QQ' is

$$(a^2+b^2)\,y\sin\theta+2b^2x\cos\theta-2ab^2=0,$$

the ellipse being referred to its axes, and θ being the angle ACP.

If an ordinate to P meet QQ' at R, the locus of R is an ellipse.

16. The locus of a point such that the sum of the squares of the perpendiculars drawn from it to the sides of a given triangle shall be constant, is an ellipse; and if the constant be so chosen that the ellipse may touch the side opposite to the angle A at D, then $CD : BD :: b^2 : c^2$.

17. With the notation of Art. 323, shew that the equation to the straight line through C and the centre of the circle is

$$\alpha \cos B = \beta \cos A.$$

18. Suppose in Art. 323 that D is the middle point of the arc AB; then the equations to BD and AD are respectively

$$\alpha \sin C + \gamma (\sin A + \sin B) = 0,$$
$$\beta \sin C + \gamma (\sin A + \sin B) = 0.$$

19. In Art. 318, equation (4), if A', B', C' be the points of contact of the triangle and conic section, shew that the equation to $A'B'$ is $lu + mv - nw = 0$.

20. In the figure of Art. 292, suppose $u = 0$ the equation to AC, $v = 0$ the equation to BD, and $w = 0$ the equation to EF, and that $l^2u^2 + m^2v^2 - n^2w^2 = 0$ represents a conic section passing through A, B, C, D: then express the equations to the tangents at A, B, C, D, and also to the straight lines AB, BC, CD, DA. Shew also that the straight line FG passes through the intersection of the tangents at A and B, and of those at C and D.

21. Express by the aid of Art. 323 the equation to the circle described round the triangle formed by the straight lines

$$y = m_1 x + \frac{a}{m_1}, \quad y = m_2 x + \frac{a}{m_2}, \quad y = m_3 x + \frac{a}{m_3}.$$

Hence deduce the last proposition of Art. 146.

22. Give a geometrical interpretation of equation (1) in Art. 310, and shew that it is a particular case of the theorem in Art. 336.

23. Interpret the last equation in Art. 323: deduce the following theorem; if from any point of the circle which circumscribes a triangle, perpendiculars are drawn on the sides of the triangle, the feet of the perpendiculars lie on one straight line.

24. If ellipses be inscribed in a triangle each with one focus on a fixed straight line, the locus of the other focus is a conic section passing through the angular points of the triangle.

25. Three conic sections are drawn touching respectively each pair of the sides of a triangle at the angular points where they meet the third side, and each passing through the centre of the inscribed circle: shew that the three tangents at their common point meet the sides of the triangle which intersect their respective conics at three points lying on a straight line. Shew also that the common tangents to each pair of conics intersect the sides of the triangle which touch the several pairs of conics at the above three points.

26. With the angular points of a triangle ABC as centres, and the sides as asymptotes, three hyperbolas are described, having A', B', C' as their vertices respectively: prove that if $AA' \sin \frac{A}{2} = BB' \sin \frac{B}{2} = CC' \sin \frac{C}{2}$, the intersections of each pair of hyperbolas lie on the axis of the third.

27. The necessary and sufficient condition in order that the equation $l\alpha^2 + m\beta^2 + n\gamma^2 = 0$ may represent a rectangular hyperbola is $l + m + n = 0$.

The necessary and sufficient condition in order that $l\beta\gamma + m\gamma\alpha + n\alpha\beta = 0$ may represent a rectangular hyperbola is $l \cos A + m \cos B + n \cos C = 0$.

28. Shew that $\sqrt{(l\alpha)} + \sqrt{(m\beta)} + \sqrt{(n\gamma)} = 0$ represents in general an ellipse, parabola, or hyperbola according as $lmn\left(\frac{l}{a} + \frac{m}{b} + \frac{n}{c}\right)$ is positive, zero, or negative; where a, b, c denote the lengths of the sides of the triangle formed by $\alpha = 0$, $\beta = 0$, $\gamma = 0$.

29. Shew that $l\beta\gamma + m\gamma\alpha + n\alpha\beta = 0$ represents in general an ellipse, parabola, or hyperbola according as

$$l^2a^2 + m^2b^2 + n^2c^2 - 2lmab - 2mnbc - 2nlca$$

is negative, zero, or positive.

30. Express the equation to the circle which is concentric with the inscribed circle of the triangle of reference, and passes through the angular point A.

31. Find the *fourth* point of intersection of the conic sections $lvw + mwu + nuv = 0$, and $l'vw + m'wu + n'uv = 0$.

32. Shew that the equation to the radical axis of the circles inscribed in a triangle and circumscribed about it is

$$\alpha \operatorname{cosec} A \cos^4 \frac{A}{2} + \beta \operatorname{cosec} B \cos^4 \frac{B}{2} + \gamma \operatorname{cosec} C \cos^4 \frac{C}{2} = 0.$$

33. Find the equation to the diameter of the curve $l\beta\gamma + m\gamma\alpha + n\alpha\beta = 0$ which passes through the point of intersection of the straight lines $\beta = 0$ and $\gamma = 0$.

34. Find the equation to the tangent to the curve $\sqrt{(l\alpha)} + \sqrt{(m\beta)} + \sqrt{(n\gamma)} = 0$, which is parallel to the straight line $\gamma = 0$; and thence shew that the centre of the curve is determined by

$$\frac{\alpha}{mc + nb} = \frac{\beta}{na + lc} = \frac{\gamma}{lb + ma}.$$

35. Employ the method of Art. 332, and the result given in Example 29 to find the condition which determines whether the general equation

$$L\alpha^2 + M\beta^2 + N\gamma^2 + 2L'\beta\gamma + 2M'\gamma\alpha + 2N'\alpha\beta = 0$$

represents an ellipse, parabola, or hyperbola.

36. A conic section passes round a triangle, and the tangent to the curve at each angular point is parallel to the opposite side of the triangle: shew that the curve is an ellipse.

37. OP, OQ are tangents to an ellipse at P, Q, and asymptotes of an hyperbola; RS is a common chord parallel to PQ: shew that if PR touches the hyperbola at R, QS touches it at S; also if PS, QR intersect at U, then OU bisects PQ.

38. If t, u, v, w be linear functions of x and y, shew that the equation to the tangent at the point (t', u', v', w') to the conic section given by $tu = vw$ is $tu' + ut' = vw' + wv'$.

39. If $\alpha=0$, $\beta=0$, $\gamma=0$, $\frac{\alpha}{a_1}+\frac{\beta}{b_1}+\frac{\gamma}{c_1}=0$, $\frac{\alpha}{a_2}+\frac{\beta}{b_2}+\frac{\gamma}{c_2}=0$, $\frac{\alpha}{a_3}+\frac{\beta}{b_3}+\frac{\gamma}{c_3}=0$, be the equations to the sides of a hexagon which circumscribes a conic section, shew that

$$a_1(b_2c_3-b_3c_2)+a_2(b_3c_1-b_1c_3)+a_3(b_1c_2-b_2c_1)=0.$$

40. *ABC* is the triangle of reference; *D*, *E*, *F* are the middle points of the sides: express the equations to the straight lines which bisect the angles of the triangle *DEF*.

41. Express by means of abridged notation the equation to the ellipse which touches the sides of a triangle at the middle points of the sides.

42. From a point *P* two tangents are drawn to a conic section meeting it at the points *M* and *N* respectively; the straight line through *P* which bisects the angle *MPN* meets the chord *MN* at *Q*; any chord of the conic section is drawn through *Q*: shew that the segments into which the chord is divided by the point *Q* subtend equal angles at *P*.

CHAPTER XVI.

SECTIONS OF A CONE. ANHARMONIC RATIO AND HARMONIC PENCIL.

Sections of a Cone.

344. We shall now shew that the curves which are included under the name *conic sections*, can be obtained by the intersection of a cone and a plane.

Definition. A cone is a solid figure described by the revolution of a right-angled triangle about one of the sides containing the right angle, which remains fixed. The fixed side is called the axis of the cone.

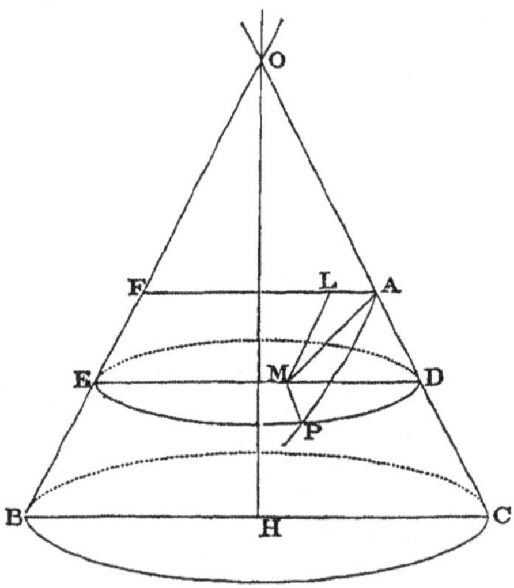

Let OH be the fixed side, and OHC the right-angled triangle which revolves round OH. In order to obtain a cone such as is considered in ordinary synthetical geometry, we should take only a *finite* straight line OC; but in analytical geometry it is usual to suppose OC *indefinitely produced both ways*. A section of the cone made by a plane through OH and OC will meet the cone in a straight line OB, which is the position OC would occupy after revolving half way round. Let a section of the cone be made by a plane perpendicular to the plane BOC; let AP be the section, A being the point where the cutting plane meets OC; we have to find the nature of this curve AP. Let a plane pass through any point P of the curve, and be perpendicular to the axis OH; this plane will obviously meet the cone in a circle DPE, having its diameter DE in the plane BOC. Let MP be the straight line in which the plane of this circle meets the plane section we are considering, M being in the straight line DE. Since each of the planes which intersect in the straight line MP is perpendicular to the plane BOC, the straight line MP is perpendicular to that plane, and therefore to every straight line in that plane.

Draw AF parallel to ED, and ML parallel to OB; join AM. Let $AM = x$, $MP = y$, $OA = c$, $HOC = \alpha$, $OAM = \theta$; the angle AML will be equal to the inclination of AM to OB, that is, to $\pi - \theta - 2\alpha$.

$$\text{Now } \frac{MD}{MA} = \frac{\sin MAD}{\sin MDA} = \frac{\sin\theta}{\cos\alpha}; \text{ therefore } MD = \frac{x\sin\theta}{\cos\alpha}.$$

$$EM = FL = FA - AL = 2c\sin\alpha - AL;$$

$$\frac{AL}{AM} = \frac{\sin AML}{\sin ALM} = \frac{\sin(\pi - \theta - 2\alpha)}{\sin\left(\frac{\pi}{2} + \alpha\right)};$$

therefore $$AL = \frac{x\sin(\theta + 2\alpha)}{\cos\alpha},$$

therefore $$EM = 2c\sin\alpha - \frac{x\sin(\theta + 2\alpha)}{\cos\alpha}.$$

But, from a property of the circle, $MP^2 = EM \,.\, MD$;

therefore $$y^2 = \frac{x\sin\theta}{\cos\alpha}\left\{2c\sin\alpha - \frac{x\sin(\theta + 2\alpha)}{\cos\alpha}\right\}.$$

If we compare this equation with that in Art. 282, we see that the section is an ellipse, hyperbola, or parabola, according as $-\frac{\sin\theta\sin(\theta+2\alpha)}{\cos^2\alpha}$ is negative, positive, or zero, that is, according as $\theta+2\alpha$ is less than π, greater than π, or equal to π.

Hence if AM is parallel to OB the section is a parabola, if AM produced through M meets OB the section is an ellipse, if AM produced through A meets OB produced through O the section is an hyperbola.

If $c=0$ the section is a point if $\theta+2\alpha$ is less than π, two straight lines if $\theta+2\alpha$ is greater than π, and one straight line if $\theta+2\alpha=\pi$. The section is also a straight line whatever c may be, if $\theta=0$ or π.

The equation above obtained may be written

$$y^2=\frac{\sin\theta\sin(\theta+2\alpha)}{\cos^2\alpha}\left\{\frac{2c\sin\alpha\cos\alpha}{\sin(\theta+2\alpha)}x-x^2\right\};$$

suppose $\theta+2\alpha$ to be less than π, so that the curve is an ellipse; then by comparing this equation with the equation $y^2=\frac{b^2}{a^2}(2ax-x^2)$, we have

$$2a=\frac{2c\sin\alpha\cos\alpha}{\sin(\theta+2\alpha)},\quad \frac{b^2}{a^2}=\frac{\sin\theta\sin(\theta+2\alpha)}{\cos^2\alpha}.$$

Thus
$$2a=\frac{c\sin 2\alpha}{\sin(\theta+2\alpha)},\qquad b^2=\frac{c^2\sin^2\alpha\sin\theta}{\sin(\theta+2\alpha)}.$$

Also
$$e^2=1-\frac{b^2}{a^2}=\frac{\cos^2\alpha-\{\sin^2(\theta+\alpha)-\sin^2\alpha\}}{\cos^2\alpha}=\frac{\cos^2(\theta+\alpha)}{\cos^2\alpha}.$$

If we suppose in the figure on page 308 that AM is produced to meet the cone again at A', then $2a=AA'$, as might have been anticipated; also b may be shewn to be a mean proportional between the perpendiculars from A and A' on the axis OH. Similar results may be obtained when the curve is an hyperbola.

345. An ellipse of given excentricity can always be obtained from a given cone by properly choosing the cutting

plane. For we have the equation $\cos^2(\theta + \alpha) = e^2 \cos^2\alpha$, in which α and e are given, e being less than unity. Now it is manifest that there must exist a value of θ between 0 and $\frac{\pi}{2} - \alpha$ which satisfies this equation, and also a value of θ between $\frac{\pi}{2} - \alpha$ and $\pi - 2\alpha$.

From a given cone we cannot obtain an hyperbola of given excentricity unless the given quantities are such that $e^2 \cos^2\alpha$ is not greater than unity.

346. Art. 344 admits of great extension.

We may first give a more general definition of a cone. If a straight line move so as always to pass through a fixed point and a fixed curve the surface generated is called a cone. The fixed point is called the vertex, and the fixed curve the directrix.

If a cone be formed with any conic section as directrix any plane section of the cone will be a conic section.

The demonstration will be similar to that in Art. 344. Let O be the vertex, and instead of the circle with BC as a diameter let there be any conic section for directrix. The plane EPD is to be taken parallel to the plane of the directrix, so that the curve EPD will be a similar conic section. The plane OBC may be any fixed plane passing through the vertex, so that it will not be necessarily perpendicular to the plane EPD. Now an equation of the second degree will hold between MP and MD, because the curve EPD is a conic section; and MD bears a constant ratio to AM; therefore an equation of the second degree holds between MP and AM. And MP is always parallel to a fixed direction. Therefore the curve AP is a conic section.

347. In consequence of the extension of the definition of a cone it is necessary to have a special name for the particular cone considered in Art. 344; and accordingly it is called a *right circular cone.* The word *circular* indicates that the directrix is a circle; and the word *right* indicates that the straight line drawn from the vertex to the centre of the directrix is at right angles to the plane of the directrix.

An *oblique circular cone* is a cone in which the directrix is a circle, but the straight line drawn from the vertex to the centre of the directrix is not at right angles to the plane of the directrix.

When the word *cone* occurs in mathematics the student will often have to determine from the context whether the word is used in the general sense of Art. 346, or is used as an abbreviation for *right circular cone.*

348. The case of an oblique circular cone deserves separate consideration.

Let O be the vertex of the cone; EPD a section parallel to the plane of the directrix, which is therefore a circle.

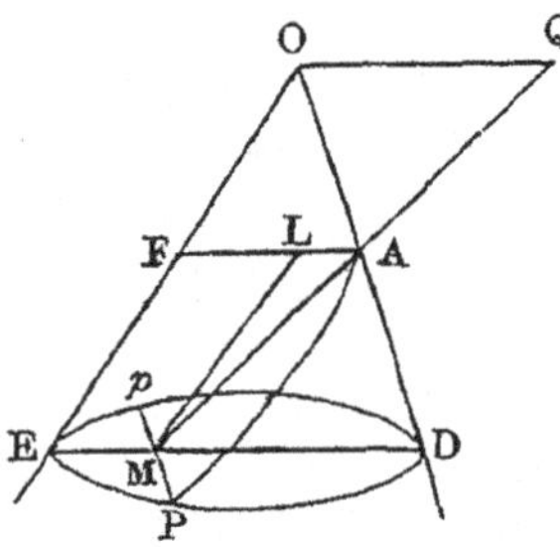

Let AP be a section made by any plane. Let Pp be the intersection of these two planes; and ED that diameter of the base which bisects Pp. Let M be the point of bisection, and MA the intersection of the plane PAp and the plane EOD. Then MP is always parallel to a fixed direction, but is not necessarily at right angles to AM.

Proceeding as in Art. 344 we have $MP^2 = EM . MD$. Now $EM = FA - AL$. Also the ratio of AL to AM is constant, and so is that of MD to AM. Thus finally we obtain $MP^2 = \lambda . AM - \mu . AM^2$, where λ and μ are constants, which involve FA and the sines and cosines of the angles MAD, MDA, AML.

It is easy to shew that in a certain special case the section is a circle. Suppose the plane OED perpendicular to the plane of the directrix; and suppose the plane MAP perpendicular to the plane OED: then MP is at right angles to

AM. Let AM produced meet OE at A'; then the section will be a circle provided $MP^2 = AM . MA'$, that is provided $AM . MA' = EM . MD$. This requires the triangles AMD and $A'ME$ to be similar; thus the angle MAD must be equal to the angle MEA', and the angle MDA equal to the angle $MA'E$. Such a section of an oblique circular cone is called a *sub-contrary section*.

349. Conversely, suppose we have a given conic section, and we require to form an oblique circular cone which shall contain the conic section.

Refer the conic section to axes consisting of a diameter and the tangent at its extremity. The angle between these axes will determine the angle AMP of the preceding figure. Then λ and μ will have known values, so that we have two equations for finding four unknown quantities, namely, FA and the angles MAD, MDA, AML. Hence the problem is indeterminate; and will remain indeterminate even if one condition is introduced.

Such a condition, for example, might be the following: let MA produced meet at Q the plane through O parallel to the plane of the directrix; and let AQ be required to have a given value.

Suppose AM produced to meet OE at A'; then

$$\frac{OQ}{AQ} = \frac{MD}{MA}, \text{ and } \frac{OQ}{A'Q} = \frac{LA}{MA}:$$

therefore
$$\frac{OQ^2}{AQ . A'Q} = \frac{MD . AL}{MA^2}.$$

The right-hand expression is what we have denoted by μ; thus when the conic section is given, and also AQ, it follows that OQ is known.

Anharmonic Ratio and Harmonic Pencil.

350. We will now give a short account of anharmonic ratios and harmonic pencils, which are often used in investigating and enunciating properties of the conic sections.

Let there be four straight lines meeting at a point; then if *any* straight line $ADCB$ be drawn across the system,

$$\frac{AB}{AC} \div \frac{DB}{DC} \text{ will be a constant ratio.}$$

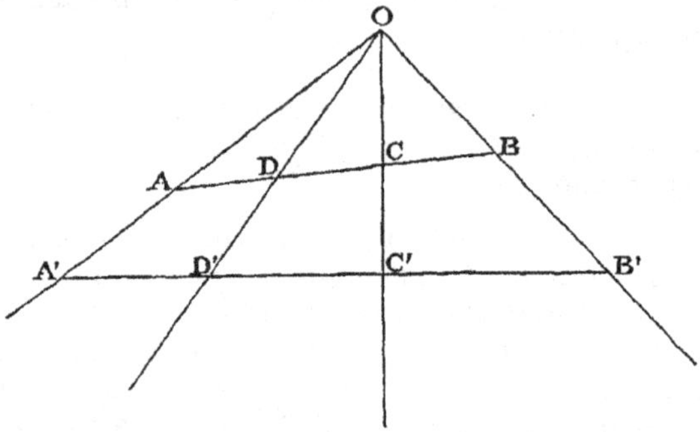

Suppose O the point where the straight lines meet; then

$$\frac{AB}{AO} = \frac{\sin AOB}{\sin ABO}, \qquad \frac{AC}{AO} = \frac{\sin AOC}{\sin ACO};$$

therefore $$\frac{AB}{AC} = \frac{\sin AOB}{\sin AOC} \cdot \frac{\sin ACO}{\sin ABO}.$$

Similarly $$\frac{DB}{DC} = \frac{\sin DOB}{\sin DOC} \cdot \frac{\sin DCO}{\sin DBO};$$

therefore $$\frac{AB}{AC} \div \frac{DB}{DC} = \frac{\sin AOB}{\sin AOC} \div \frac{\sin DOB}{\sin DOC}.$$

Now suppose any other straight line $A'D'C'B'$ drawn across the system, then since AOB and $A'OB'$ are the same angle, and so on for the other angles, we have

$$\frac{AB}{AC} \div \frac{DB}{DC} = \frac{A'B'}{A'C'} \div \frac{D'B'}{D'C'},$$

which proves the proposition.

Similarly we can prove that each of the following is a constant ratio

$$\frac{AB}{AD} \div \frac{CB}{CD} \text{ and } \frac{AC}{AD} \div \frac{BC}{BD}.$$

351. Definitions. Any four straight lines meeting at a point form a *pencil.*

A straight line drawn across a pencil is called a *transversal.*

The four points at which the straight line meets the pencil form a *range.*

Any one of the constant ratios $\frac{AB}{AC} \div \frac{DB}{DC}$, $\frac{AB}{AD} \div \frac{CB}{CD}$, $\frac{AC}{AD} \div \frac{BC}{BD}$ is called *an anharmonic ratio* of the pencil.

The pencil is called *harmonic* if $AB . DC = AD . BC$, that is, if the rectangle formed by the *whole* straight line (AB) and the middle part (DC) is equal to the rectangle of the other two parts (AD), (BC).

352. The *harmonic* pencil is so called because it divides any transversal harmonically. For since $AB . DC = AD . BC$, $\frac{AB}{AD} = \frac{BC}{DC}$, that is, if we call AB, AC, AD, the first, second, and third quantities respectively, the first is to the third as the difference of the first and second is to the difference of the second and third.

When the pencil is harmonic *one* of the three constant ratios of the pencil is equal to unity.

We shall sometimes select one of the anharmonic ratios of a pencil, and confine our attention to it, and shall then speak of the selected ratio as *the* anharmonic ratio of the pencil.

353. Suppose OA, OB, OC, OD form an harmonic pencil; if we take any new origin O', and join $O'A$, $O'B$, $O'C$, $O'D$, these four straight lines form a new harmonic pencil; for the transversal $ABCD$ is cut harmonically.

354. *The anharmonic ratio of a pencil is not altered if the transversal meet the straight lines of the pencils produced, instead of the straight lines themselves.*

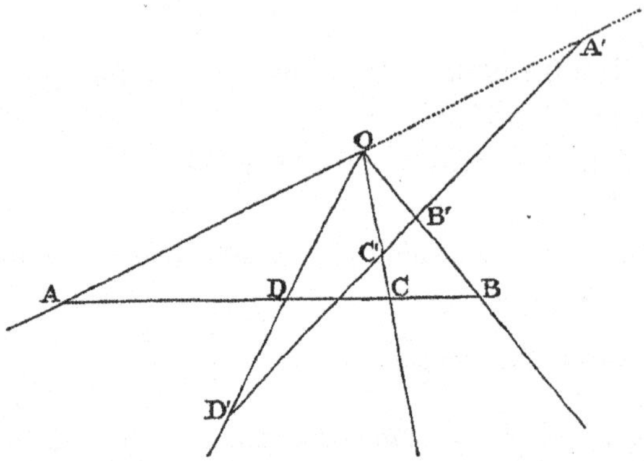

Suppose OA, OB, OC, OD to be a pencil, and let a transversal $A'B'C'D'$ meet three straight lines of the pencil, and the fourth AO produced at A'. The angles $A'OB'$, AOB are supplemental; and so are AOD, $A'OD'$; and so on. Hence any anharmonic ratio formed on $ABCD$ is equal to the corresponding ratio formed on $A'B'C'D'$.

355. Suppose $AB \,.\, CD = AD \,.\, BC$, so that OA, OB, OC, OD form an harmonic pencil. By the last proposition

$$\frac{A'B'}{A'D'} \div \frac{C'B'}{C'D'} = \frac{AB}{AD} \div \frac{CB}{CD} = 1;$$

therefore OA', OB', OC', OD' form an harmonic pencil.

Similarly OC', OB', OA', and DO produced through O will form an harmonic pencil. Thus from one harmonic pencil by producing the straight lines through the vertex, we can derive four other harmonic pencils.

356. The straight lines whose equations are $\alpha = 0$, $\beta = 0$, $\alpha - k\beta = 0$, $\alpha + k\beta = 0$ form an *harmonic* pencil.

Let OM be the straight line $\alpha = 0$, ON the straight line $\beta = 0$, OP the straight line $\alpha - k\beta = 0$, OQ the straight line $\alpha + k\beta = 0$.

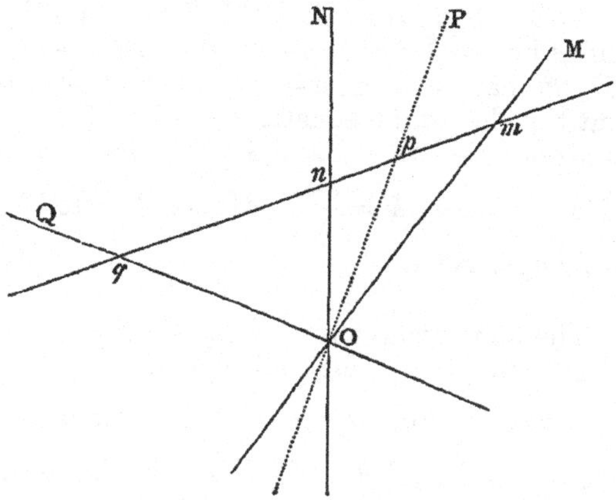

Let a transversal meet the pencil at *mpnq*; then (Art. 70)

$$\frac{\sin POM}{\sin PON} = k = \frac{\sin QOM}{\sin QON};$$

therefore $$\frac{\sin POM}{\sin PON} \cdot \frac{\sin QON}{\sin QOM} = 1;$$

therefore (as in Art. 350) $\frac{pm}{pn} \cdot \frac{qn}{qm} = 1;$

therefore $$pm \,.\, qn = pn \,.\, qm.$$

The same result will follow if we draw the transversal in a different position. The harmonic pencil is so formed that its outside straight lines are always one of the two $\alpha = 0$ and $\beta = 0$, and one of the two $\alpha - k\beta = 0$ and $\alpha + k\beta = 0$.

357. The anharmonic ratio of the four straight lines $\alpha = 0$, $\beta = 0$, $\alpha - k\beta = 0$, $\alpha + k'\beta = 0$, is $\frac{k}{k'}$.

For as in the preceding Article we have

$$\frac{\sin POM}{\sin PON} = k, \quad \frac{\sin QOM}{\sin QON} = k';$$

therefore, by Art. 351, $\frac{k}{k'}$ expresses the anharmonic ratio.

358. Article 356 will also hold if the equations to the straight lines be $u=0$, $v=0$, $u-kv=0$, and $u+kv=0$. For, by Art. 57, we have $u=\lambda\alpha$, $v=\mu\beta$, where λ and μ are constant quantities; hence the equations $u-kv=0$ and $u+kv=0$ may be written $\lambda\alpha-k\mu\beta=0$ and $\lambda\alpha+k\mu\beta=0$, or $\alpha-k'\beta=0$ and $\alpha+k'\beta=0$, where $k'=\frac{k\mu}{\lambda}$. Hence Article 356 becomes immediately applicable.

359. The four straight lines *EB, EC, EG, EF*, in Art. 75, form an harmonic pencil; for their equations are

$$u=0,\ w=0,\ lu-nw=0,\ lu+nw=0.$$

By symmetry *FB, FA, FG, FE*, will also form an harmonic pencil.

Also *GD, GC, GF, GE* form an harmonic pencil, for their equations are respectively

$$lu-mv=0,\quad mv-nw=0,\quad lu-mv-(mv-nw)=0,$$
$$lu-mv+mv-nw=0.$$

360. *A straight line drawn through the intersection of two tangents to a conic section is divided harmonically by the curve and the chord of contact.*

Refer the curve to the tangents as axes; its equation will be of the form (Art. 293)

$$\left(\frac{x}{h}+\frac{y}{k}-1\right)^2+\mu xy=0 \qquad (1).$$

Suppose a straight line drawn through the origin, and let its equation be (Art. 27)

$$\frac{x}{l}=\frac{y}{m}=r \qquad (2).$$

Thus the distances from the origin of the points of intersection of (1) and (2) will be the values of r found from the equation

$$\left(\frac{lr}{h}+\frac{mr}{k}-1\right)^2+\mu lmr^2=0,$$

or $$\left(\frac{l}{h}+\frac{m}{k}-\frac{1}{r}\right)^2+\mu lm=0 \qquad\text{.....................(3).}$$

If r' and r'' be the roots of the equation, we have

$$\frac{1}{r'}+\frac{1}{r''}=2\left(\frac{l}{h}+\frac{m}{k}\right) \qquad\text{.....................(4).}$$

Also the equation to the chord of contact is

$$\frac{x}{h}+\frac{y}{k}-1=0 \qquad\text{.....................(5).}$$

Hence for the distance (r_1) of the point of intersection of (2) and (5) from the origin, we have the equation

$$\frac{lr_1}{h}+\frac{mr_1}{k}=1, \quad\text{or } \frac{1}{r_1}=\frac{l}{h}+\frac{m}{k} \qquad\text{..............(6).}$$

From (4) and (6) we have $\frac{2}{r_1}=\frac{1}{r'}+\frac{1}{r''}$, thus r_1 is an harmonic mean between r' and r''.

Since *LMNO* is divided harmonically, if from any point in *AB* we draw straight lines to *L*, *N*, and *O*, these straight lines

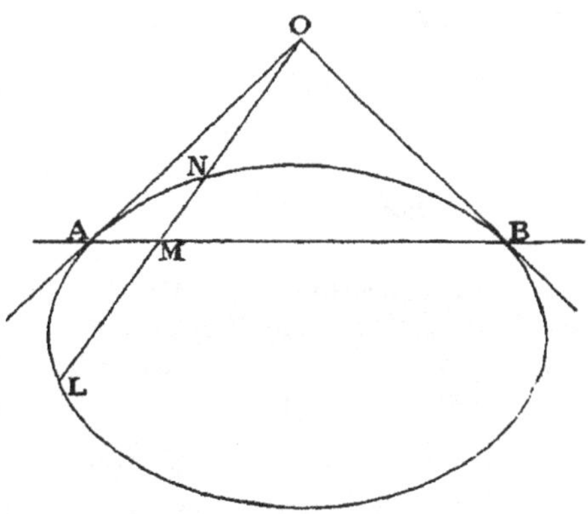

with *AB* form an harmonic pencil. A particular case is that in which the point in *AB* is the intersection of the tangents at *N* and *L*, which we know will meet on *AB* produced. (See Arts. 103, 185.)

361. Let A, B, C, D be four points on a conic section, and P any fifth point. Let α denote the perpendicular from P on AB, β the perpendicular from the same point on BC, γ on CD, δ on DA. Then by Art. 336 we know that wherever P may be, $\alpha\gamma$ bears a constant ratio to $\beta\delta$. Now $AB \,.\, \alpha =$ twice the area of the triangle PAB

$$= PA \,.\, PB \,.\, \sin APB;$$

$$\text{therefore } \alpha = \frac{PA \,.\, PB \,.\, \sin APB}{AB}.$$

Similar values may be found for β, γ, δ. Thus

$$\frac{PA \,.\, PB \,.\, PC \,.\, PD}{AB \,.\, CD} \sin APB \,.\, \sin CPD$$

bears a constant ratio to

$$\frac{PA \,.\, PB \,.\, PC \,.\, PD}{BC \,.\, AD} \sin BPC \,.\, \sin DPA,$$

therefore $\dfrac{\sin APB \,.\, \sin CPD}{\sin BPC \,.\, \sin DPA}$ is constant, that is, the pencil drawn from any point P to the four points A, B, C, D, has a constant anharmonic ratio.

EXAMPLES.

1. Different elliptic sections of a right cone are taken all perpendicular to one plane which contains the axis of the cone: if the elliptic sections have equal major axes, shew that the locus of the centres is an ellipse.

2. If two spheres be inscribed in a right cone so as to touch the plane of any section, the points of contact of the plane with the spheres will be the foci of the conic section, and the intersections of this plane with the planes of contact of the spheres and the cone will be the directrices of the conic section.

3. Find the locus of the foci of all the parabolas which can be cut from a given cone.

4. Shew that a given hyperbola cannot be cut from a given cone unless the vertical angle of the cone is greater than the angle between the asymptotes of the hyperbola.

5. Shew that the latus rectum of any section of a given cone varies as the perpendicular from the vertex of the cone on the plane of section.

6. A conic section circumscribes a triangle, and at each angular point the tangent, the two sides of the triangle, and the perpendicular on the opposite side form an harmonic pencil: determine the equation to the conic section.

7. If the equations to the three diagonals of a quadrilateral be $u=0$, $v=0$, $w=0$, shew that the equations to the four sides may be put in the form $lu+mv+nw=0$, $-lu+mv+nw=0$, $lu-mv+nw=0$, $lu+mv-nw=0$.

8. In the diagram of Art. 360 suppose a straight line Onl to be drawn meeting the curve at n and l: then shew that the straight lines Nl and Ln intersect on AB.

CHAPTER XVII.

PROJECTIONS.

362. IN the preceding Chapters we have carried on our investigations chiefly by the aid of co-ordinates; there are various other methods by which we may discover and demonstrate theorems relating to the Conic Sections: we shall now explain one of these, which is called the *method of projections.*

363. There are two kinds of projection which may be called respectively *orthogonal projection* and *conical projection:* we proceed to consider the former.

364. DEFINITIONS. From any point let a perpendicular be drawn on a fixed plane; the intersection of the perpendicular with the plane is called the *orthogonal projection* of the point on the plane. The plane on which the perpendicular is drawn is called the *plane of projection.*

The orthogonal projection of any line straight or curved is the locus of the orthogonal projection of every point in that line.

365. We shall use the word *projection* as equivalent to the term *orthogonal projection,* until the contrary is specified.

366. *The projection of a straight line is in general a straight line.*

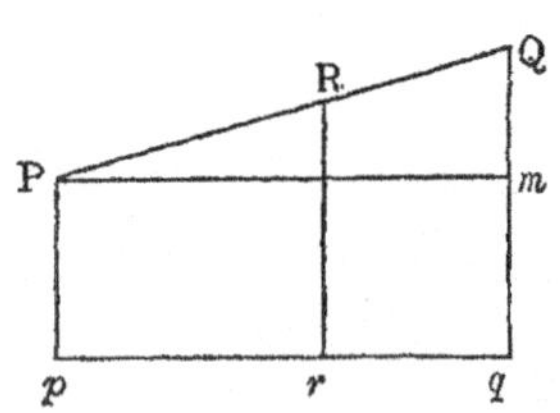

Let PQ be a straight line. From any point P in the straight line draw Pp perpendicular to the plane of projection, meeting it at p. Let a plane pass through PQ and Pp, and let its intersection with the plane of projection be pq, and draw Qq in that plane parallel to Pp.

Then pq is a straight line, by Euclid, XI. 3: and we shall shew that pq is the projection of PQ.

Take any point R in PQ; and in the plane QPp draw Rr parallel to Pp, meeting pq at r: then Rr is perpendicular to the plane of projection, by Euclid, XI. 8, so that r is the projection of R.

If the given straight line be perpendicular to the plane of projection, its projection is a point.

367. *The length of the projection of a straight line is equal to the length of the original straight line multiplied by the cosine of the angle between the straight line and its projection.*

Let PQ be a straight line, pq its projection; draw Pm parallel to pq meeting Qq at m. Then

$$pq = Pm = PQ \cos QPm.$$

And by the angle between PQ and pq is meant the angle between one of these straight lines as PQ, and any straight line Pm parallel to the other. Thus QPm is the angle between PQ and pq.

368. *The projections of parallel straight lines are themselves parallel straight lines.*

Let there be two parallel straight lines: denote them by PQ and RS. Let p denote the projection of P, and r the projection of R.

The plane QPp is parallel to the plane SRr, by Euclid, XI. 15; the intersections of these planes with the plane of projection are parallel by Euclid, XI. 16; and these intersections are the projections of PQ and RS by Art. 366.

The angle between PQ and its projection is equal to the angle between RS and its projection by Euclid, XI. 10.

369. *Let the boundary of any plane figure be projected; then the area of the projected figure is equal to the area of the*

original figure multiplied by the cosine of the angle between the two planes.

First, suppose the figure to be a triangle having one side in the plane of projection.

Let ABC be the triangle, having the side AB in the

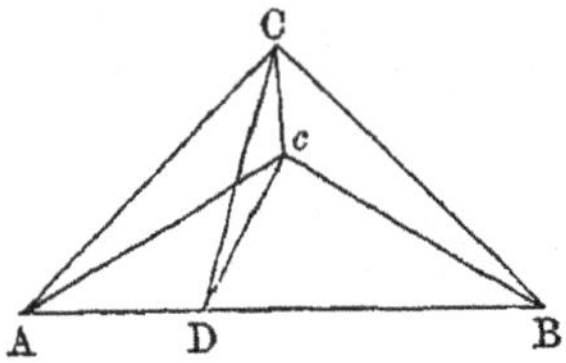

plane of projection. Let c be the projection of C. Draw CD perpendicular to AB, and join cD.

Cc is perpendicular both to Ac and Dc; thus

$$AD^2 = AC^2 - CD^2 = Ac^2 + Cc^2 - (Dc^2 + Cc^2) = Ac^2 - Dc^2;$$

therefore the angle ADc is a right angle.

Now the area of $ABc = \frac{1}{2} AB \,.\, Dc$; and the area of $ABC = \frac{1}{2} AB \,.\, DC$: therefore

$$\frac{\text{area of } ABc}{\text{area of } ABC} = \frac{Dc}{DC} = \cos CDc;$$

and CDc is the angle of inclination of the planes.

Next, suppose the figure to be any triangle.

Let ABC be any triangle. Let the plane of ABC meet

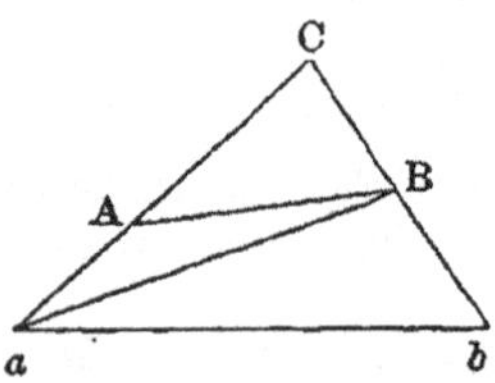

the plane of projection in the straight line ab. Let γ denote the angle between the planes. Join aB. Then, by the former case,

$$\text{area of projection of } aCb = \text{area of } aCb \times \cos\gamma,$$

$$\text{area of projection of } aBb = \text{area of } aBb \times \cos\gamma;$$

therefore, by subtraction,

$$\text{area of projection of } aCB = \text{area of } aCB \times \cos\gamma.$$

This shews that the proposition is true for any triangle which has one angular point in the plane of projection. Hence it is also true for the triangle aAB. And therefore, by subtraction, it is true for the triangle ABC.

Next, suppose that the area is any plane rectilinear figure. Then the figure may be decomposed into triangles, and as the proposition is true for each triangle, it is true for the whole figure.

Lastly, suppose that the area is bounded by curved lines. We may inscribe any rectilinear polygon in this figure, and the proposition will be true of the polygon; and by sufficiently increasing the number of sides of the polygon, and diminishing the length of each side, the area of the polygon can be made to differ as little as we please from the area of the figure. Thus we may admit that the proposition is also true for the area bounded by the curved lines.

370. *The projection of the tangent at any point of a curve is the tangent at the corresponding point of the projection of the curve.*

Let P and Q be two points on a curve; let p and q be their projections. Then the straight line through p and q is the projection of the straight line through P and Q. Let Q move along the curve to P; then the limiting position of the secant through P and Q is the tangent at P to the curve: and as Q moves to P along the curve, q moves to p along the projection of the curve, and the limiting position of the secant through p and q is the tangent at p to the projection of the curve.

371. *The projection of a circle is an ellipse.*

Let C be the centre of a circle; let BCB' be that diameter which is perpendicular to the intersection of the plane of the

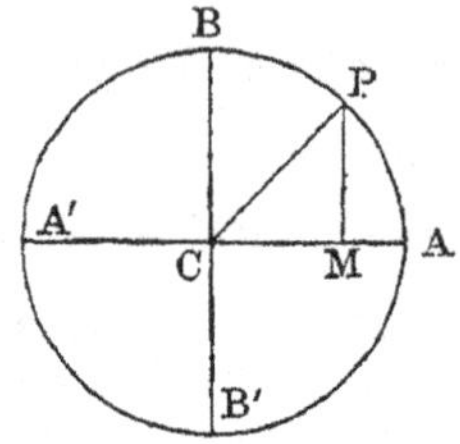

circle and the plane of projection. Let AA' be the diameter at right angles to BB'.

Take any point P on the circumference of the circle, and draw PM perpendicular to AA'. Then

$$CM^2 + PM^2 = CP^2 = CA^2.$$

Now, suppose the projections of C, P, M to be denoted by c, p, m respectively. Then cm is parallel and equal to CM, and $pm = PM \cos\gamma$, where γ is the angle of inclination of PM to its projection. Thus

$$(cm)^2 + \frac{(pm)^2}{\cos^2\gamma} = CA^2.$$

Let $cm = x$, $pm = y$, $CA = r$: then

$$x^2 + \frac{y^2}{\cos^2\gamma} = r^2.$$

Now γ is the same for every ordinate; see Art. 368; and cm and mp are at right angles by the reasoning in the first part of Art. 369. Thus the above equation represents an ellipse having r for the major semi-axis, and $r\cos\gamma$ for the minor semi-axis.

The straight line ACA' is either the line of intersection of the plane of the circle and the plane of projection, or is parallel to this line. In the former case m and M coincide, and γ is the angle of inclination of the two planes; in the latter case γ is equal to the angle of inclination of the two planes by Euclid, XI. 10.

It is obvious that the centre of the circle is projected into the centre of the ellipse.

372. *Conjugate diameters of an ellipse are the projections of diameters of a circle which are at right angles to each other.*

For if diameters of a circle are at right angles to each other, each diameter is parallel to the tangent at the extremity of the other diameter.

Hence, by Arts. 368 and 370, the projection of each diameter is parallel to the tangent to the projection of the circle at the extremity of the other diameter. Therefore, by Art. 191, the projections of diameters of the circle which are at right angles to each other are conjugate diameters of the ellipse.

373. The area bounded by two radii of a circle which are at right angles to each other, and the corresponding arc of the circle, is a quarter of the area of the circle. Therefore, by Arts. 369 and 372, *the area bounded by two conjugate semi-diameters of an ellipse and the corresponding arc of the ellipse is one quarter of the area of the ellipse.*

374. *To find the area of an ellipse.*

Let a and b be the major and minor semi-axes of the ellipse; therefore, by Art. 371, the ellipse can be obtained by projection from a circle of radius a; and the cosine of the angle between the plane of the circle and the plane of the ellipse is $\frac{b}{a}$. The area of the circle is known to be πa^2; see *Trigonometry.* Hence, by Art. 369, the area of the ellipse is $\frac{b}{a} \times \pi a^2$, that is πab.

375. It is now easy to see that certain properties may be immediately inferred to belong to the ellipse from the fact that they belong to the circle; for example, the results of Arts. 182, 184, 185, and 194 may be thus obtained. Such properties are called *projective properties.*

Also Art. 203 may be thus obtained; for it is obvious by Euclid, III. 31, that if a chord and a diameter of a circle are parallel, the supplemental chord is parallel to the diameter of the circle which is at right angles to the first.

376. Again, Art. 208 may be obtained by projection. For by Euclid, III. 35, if two chords of a circle intersect, the rectangles contained by their segments are equal. Denote the chords by POp and QOq, so that $PO \times Op = QO \times Oq$. Let CD be a radius parallel to OP, and CE a radius parallel to OQ. Then as $CD = CE$, we have

$$\frac{PO \times Op}{CD^2} = \frac{QO \times Oq}{CE^2}.$$

Now when the circle is projected into an ellipse, since PO is parallel to CD, the projection of PO bears to the projection of CD the same ratio as PO bears to CD; and in like manner the projection of Op bears to the projection of CD the same ratio as Op bears to CD. A similar remark applies to the projections of QO and CE, and to the projections of Oq and CE. Hence the property of the ellipse follows from that of the circle by projection.

377. It will be instructive for the student to apply the method of projections to the following examples: 20, 38, 42, 50 of the Examples attached to Chapter IX, and 2, 3, 9, 19, 23, 24, 25, 31, 32, 43, 52 of the Examples attached to Chapter X.

378. We proceed to consider the other kind of projection which is called *conical projection*, and sometimes *perspective projection*.

379. DEFINITIONS. The conical projection of any point on a given plane is the intersection of the plane by a straight line drawn from a fixed origin through the point. The conical projection of any line, straight or curved, is the locus of the conical projection of every point in that line. Or we may put the definition thus: if every point in a line, straight or curved, be joined with a fixed origin, the assemblage of these joining straight lines will constitute a cone, and the intersection of the cone with any given plane is called the conical projection of the line on the plane.

The fixed origin is called the vertex, the given plane is called the plane of projection; when the projected line lies in a plane, that plane is called the original plane.

380. It is obvious from our definition that the *shadow* formed on any plane by a figure when light falls on it from a point, is the conical projection of the figure corresponding to the point as vertex.

By the single word *projection* in the remainder of this Chapter is to be understood *conical projection.*

381. *The projection of a straight line is in general a straight line.*

For the projection of a straight line is the intersection of two planes, namely, the plane of projection and the plane passing through the given straight line and the vertex.

If the given straight line passes through the vertex, its projection on any plane not passing through the vertex is a point.

382. *The projection of the tangent to a curve at any point is the tangent to the projection of the curve at the corresponding point.*

This may be established in the manner of Art. 370.

383. *Projections on parallel planes of the same figure with the same vertex are similar.*

Let O denote the vertex. Let P, Q, R be the projections of three points of a figure on any plane; p, q, r the projections of the same points respectively on a parallel plane.

Join PQ, QR, pq, qr. Then by Euclid, VI. 4,

$$\frac{PQ}{pq} = \frac{OP}{Op} = \frac{OQ}{Oq} = \frac{QR}{qr}.$$

Also the angle PQR = the angle pqr by Euclid, XI. 10.

In this way we can shew when the projections are *rectilinear* figures that they are similar; and the proposition may be extended to *curvilinear* figures in the manner exemplified in Art. 369.

384. It follows from our definitions that if two plane sections be made of a cone, each curve thus obtained may be considered as the projection of the other. Now it appears from Art. 349, that any conic section may be projected into a circle by a suitable adjustment of the cone and the plane of projection. And thus it will follow that certain properties may be inferred to be true for the conic sections when they have been shewn to be true for the circle. Such properties of a figure as may be inferred to be true for the projection of the figure are called *projective properties.* It is not possible to give a brief definition of these properties; it will be seen that they relate to the *positions* of points and straight lines, and not to the *magnitudes* of lines.

385. For an example of projective properties we may take the theory of poles and polars. Thus the properties of Arts. 101 and 102 being demonstrated for a circle, may be inferred to be true of any conic section by Arts. 382 and 384.

Again, Pascal's Theorem and Brianchon's Theorem might be demonstrated for the circle, and then inferred to hold for any conic section. Of course the method would be advantageous only in the case in which it would be easier to demonstrate the property for the circle than for a conic section generally.

386. But the great advantage of the method of projection arises from the fact that by properly choosing the projecting cone and the plane of projection, we are able to simplify the theorem we wish to establish by substituting some particular case instead of the general enunciation: this we shall now proceed to explain.

387. It is obvious from our definition that every point has its projection unless it lies in a plane through the vertex parallel to the plane of projection; and then the straight line from the vertex through the point never meets the plane of projection. Hence we may say that *points in the plane through the vertex parallel to the plane of projection have no projections;* this is usually expressed by saying that such points *are projected to infinity.*

The plane through the vertex parallel to the plane of projection may be called the *vertex plane.*

388. *If two straight lines intersect in the vertex plane their projections are parallel straight lines.*

The projections cannot meet because the point at which the original lines meet is projected to infinity by Art. 387.

389. Conversely, if the projection of a point is at an infinite distance, that point must be in the vertex plane. If the projections of two or more points are at an infinite distance, those points must be in the vertex plane; if the points are known to be also in another plane, they must be in the intersection of this plane and the vertex plane, so that they must be in a straight line.

390. *Any angle may be projected into an angle of assigned magnitude.*

Let A denote the angular point; let a plane be drawn parallel to the plane of projection meeting the straight lines which form the angle at B and C. On BC in the plane parallel to the plane of projection describe a segment of a circle containing an angle equal to the given angle. Join A with any point on this segment and take any point on the joining straight line for the vertex. Then the angle BAC will be projected into an angle of the assigned magnitude.

391. In the preceding investigation, the plane of projection may be any plane which does not coincide with the plane of the angle, and is not parallel to it: we may, therefore, take the plane of projection such that the corresponding vertex plane shall pass through an assigned straight line, that is, so that an assigned straight line shall be projected to infinity.

If we require a second angle to be also projected into an angle of assigned magnitude, we must determine in the manner employed a second segment of a circle; and when these segments intersect, the point of intersection may be taken as the vertex.

392. *Any quadrilateral may be projected into a parallelogram having a given angle.*

Draw the *third* diagonal of the quadrilateral; see Art. 75. Take the plane of projection such that this straight line is projected to infinity: then the projection of the quadrilateral is a parallelogram, for the opposite sides of the projection do not meet. Then apply Art. 390.

393. As an example of the application of projections we will take the following theorem: a quadrilateral is inscribed in a conic section; tangents are drawn at the angular points, thus forming a second quadrilateral: the diagonals of both quadrilaterals intersect at a point.

Project the figure so that the inscribed quadrilateral may be a rectangle; see Art. 392. The sides of a rectangle inscribed in a conic section are parallel to the axes. Hence, by symmetry, the diagonals of this figure and of that formed by the tangents at the angular points will intersect at a point.

394. *Any conic section may be projected into a circle, and a given straight line in the plane of the conic which does not intersect the conic section may be at the same time projected to infinity.*

This has been shewn in Art. 349.

395. The following demonstration of Pascal's Theorem has been given by the method of projections: Project the conic into a circle, so that the quadrilateral formed by two pairs of opposite sides may become a parallelogram; see Arts. 392 and 394: the theorem then reduces to a simple property of the circle, namely, if a hexagon be inscribed in a circle, and two pairs of opposite sides be parallel, so is the third pair. This simple property may be established by elementary geometry.

This demonstration is however not complete. For in Art. 394, there is the limitation that the straight line which is projected to infinity does not intersect the conic section, and thus Pascal's Theorem is only established for such figures as conform to this restriction. Writers who use the method of projections are accustomed to *assume* that theorems which are demonstrated under certain limitations will hold when those limitations are removed. For example, a straight line which really does not meet a curve, is called an *ideal* secant, and treated in effect as if it did meet the curve. But it would be beyond the scope of an elementary work like the present to discuss the grounds on which the assumption is based.

EXAMPLES.

The following Examples may be treated by the method of projections:

1. CP and CD are any two conjugate semi-diameters of a given ellipse; tangents to the ellipse at P and D meet at T: shew that the triangle CPT is equal to the triangle CDT.

2. CP and CD are any conjugate semi-diameters of a given ellipse; K is any point in PD, and CL is the semi-diameter parallel to PD: shew that the triangle KCL is of constant area.

3. CP and CD are any conjugate semi-diameters of a given ellipse; PQ is a chord drawn parallel to a fixed straight line: shew that DQ will be parallel to a fixed straight line.

4. If from the extremities of the axes of an ellipse any four parallel straight lines be drawn, the points at which they cut the curve will be the extremities of conjugate diameters.

5. CP and CQ are any two semi-diameters of an ellipse; from P a straight line is drawn parallel to the conjugate to CP, meeting CQ at M; from Q a straight line is drawn parallel to the conjugate to CQ, meeting CP at N: shew that the triangles CPM and CQN are equal.

6. PQ is any diameter of an ellipse; R, S any two points on the curve; let PR and QS, or these straight lines

produced, meet at M, and PS and QR at T: shew that TM is parallel to the diameter conjugate to PQ.

7. If parallelograms which circumscribe an ellipse have their areas constantly equal to n times that on the major and minor axes, all the angular points of the parallelograms lie on two ellipses similar to the given one, and having their axes to those of the given ellipse as $\sqrt{(n^2+n)} \pm \sqrt{(n^2-n)}$ to unity.

8. If a parallelogram be inscribed in the inner of two similar, concentric, and similarly situated ellipses, and its sides be produced to meet the outer, and the adjacent points of intersection belonging to each pair of parallel lines be joined, shew that the quadrilateral figure formed by producing these joining straight lines will be a parallelogram, having its corners situated on a third ellipse, similar to the two former, and independent of the original parallelogram.

ANSWERS TO THE EXAMPLES.

CHAPTER I.

8. The co-ordinates of D are $\frac{1}{2}(x_1 + x_2)$ and $\frac{1}{2}(y_1 + y_2)$. The co-ordinates of G are $\frac{1}{3}(x_1 + x_2 + x_3)$ and $\frac{1}{3}(y_1 + y_2 + y_3)$.

10. Let r and θ be the polar co-ordinates of C. Then the angle AOC = the angle BOC; or $\theta - \theta_1 = \theta_2 - \theta$; thus $\theta = \frac{1}{2}(\theta_1 + \theta_2)$.

Again, from the known expression for the area of a triangle (see *Trigonometry*, Chapter XVI.), triangle $AOB = \frac{1}{2}r_1r_2 \sin(\theta_2 - \theta_1)$, triangle $AOC = \frac{1}{2}r_1 r \sin(\theta - \theta_1)$, triangle $BOC = \frac{1}{2}r_2 r \sin(\theta_2 - \theta)$.

$$\text{Thus } r_1r_2 \sin(\theta_2 - \theta_1) = r_1 r \sin(\theta - \theta_1) + r_2 r \sin(\theta_2 - \theta)$$
$$= r(r_1 + r_2) \sin \tfrac{1}{2}(\theta_2 - \theta_1);$$

therefore $r(r_1 + r_2) = 2r_1r_2 \cos \frac{1}{2}(\theta_2 - \theta_1)$.

CHAPTER III.

1. (1) $y + 2x = 1$. (2) $x = 2$. (3) $y = x$. (4) $x = 0$.
2. $y - 4 = -3(x - 4)$, $y - 4 = \frac{1}{3}(x - 4)$.
3. $y - 1 = (\sqrt{3} - 2)x$, $y - 1 = -(\sqrt{3} + 2)x$.
4. $y = x$, $y = -x$. 5. $y = \dfrac{1}{\sqrt{3}}x$, $x = 0$.
6. $90°$, $x = -\frac{1}{2}$, $y = \frac{3}{2}$. 7. $60°$. 8. $45°$.
9. $y = \pm(x - a)$. 10. $y = \mp x$. 11. $2\sqrt{2}$.
12. $\dfrac{ab}{\sqrt{(a^2 + b^2)}}$. 13. $x = y = \dfrac{ab}{a + b}$. 14. $\dfrac{x}{a^2} - \dfrac{y}{b^2} = \dfrac{1}{a} - \dfrac{1}{b}$.

15. (1) The origin. (2) Two straight lines, $y = x$ and $y = -x$. (3) Two straight lines, $x = 0$ and $x + y = 0$. (4) The axes. (5) Impossible. (6) Two straight lines, $x = 0$ and $y = a$. 16. (1) Two

straight lines, $x = a$ and $y = b$. (2) The point (a, b). (3) The point $(0, a)$. 17. The straight lines $y = x$ and $y = 3x$. 19. $4y = 5x$, and $3y + 2x - 20 = 0$. 20. Let a be the length of the side of the hexagon; the equations are, to AB, $y = 0$; AC, $y\sqrt{3} = x$; AD, $y = x\sqrt{3}$; AE, $x = 0$; AF, $y + x\sqrt{3} = 0$; BC, $y = \sqrt{3}(x - a)$; BD, $x = a$; BE, $y + \sqrt{3}(x - a) = 0$; BF, $y\sqrt{3} + x - a = 0$; CD, $y + x\sqrt{3} = 2a\sqrt{3}$; CE, $y\sqrt{3} + x = 3a$; CF, $2y = a\sqrt{3}$; DE, $y = a\sqrt{3}$; DF, $y\sqrt{3} - x = 2a$; EF, $y - x\sqrt{3} = a\sqrt{3}$. 21. If (x_1, y_1), (x_2, y_2), (x_3, y_3) be the angular points, the co-ordinates of the point midway between the first and second are $\frac{x_1 + x_2}{2}$, $\frac{y_1 + y_2}{2}$; similarly the co-ordinates of the point midway between the second and third points are known; and then the required equation can be found by Art. 35. 22. $\frac{m^2 + 1}{m^2 - 1} \tan \omega$. 24. $\frac{x}{a} + \frac{y}{b} = 1$, $\frac{x}{a} = \frac{y}{b}$; tangent of the angle between them $\frac{2ab \sin \omega}{a^2 - b^2}$. 29. The points whose abscissæ are $a + \frac{a}{b}\sqrt{(a^2 + b^2)}$ and $a - \frac{a}{b}\sqrt{(a^2 + b^2)}$.

31. $\frac{\sqrt{(B^2 - 4AC)}}{A + C}$. 35. 90°. 36. $F(\theta) = 0$ gives a system of straight lines through the origin; $\sin 3\theta = 0$ gives the three straight lines $y = 0$, $y = x\sqrt{3}$, $y = -x\sqrt{3}$. 40. The second pair of straight lines bisect the angles included by the first pair. 44. Let ABC be the triangle; take A for the origin and straight lines through A parallel to the two given straight lines as axes; let x_1, y_1 be the co-ordinates of B, and x_2, y_2 those of C. Then it may be shewn that the equations to the three diagonals are

$$y - y_2 = \frac{y_1 - y_2}{x_2 - x_1}(x - x_1), \quad y - y_2 = -\frac{y_2}{x_2}x, \quad y - y_1 = -\frac{y_1}{x_1}x;$$

from these equations it may be shewn that the three diagonals meet at a point. 45. Take O as origin and use polar equations to the given fixed straight lines. 46. Let x_1 be the abscissa of the point of intersection of the two straight lines; then the area of the triangle is $\frac{1}{2}(c_2 - c_1)x_1$. 47. This may be solved by Art. 11.

Or we may use the result of the preceding Example; for by drawing a figure we shall obtain *three* triangles to which the preceding Example applies, and the required area is the difference between two of these triangles and the third. The result is

$$\pm\left\{\frac{(c_3-c_2)^2}{2(m_3-m_2)}+\frac{(c_1-c_3)^2}{2(m_1-m_3)}+\frac{(c_2-c_1)^2}{2(m_2-m_1)}\right\},$$

which may also be written thus

$$\mp\frac{\{c_1(m_3-m_2)+c_2(m_1-m_3)+c_3(m_2-m_1)\}^2}{2(m_3-m_2)(m_1-m_3)(m_2-m_1)}.$$

That sign should be taken which gives a positive result. It will be seen that the numerator vanishes if the three points of intersection of the straight lines lie on a straight line; the denominator vanishes if any two of the straight lines are parallel.

CHAPTER IV.

1. $\frac{x}{a}+\frac{y}{b}=\frac{x}{a'}+\frac{y}{b'}$. 7. Since the required straight line is parallel to that considered in Example 5, we may assume for its equation $\alpha\cos A-\beta\cos B+k=0$, where k is some *constant* to be determined. Now at the middle point of AB, we have $\alpha=\frac{c}{2}\sin B$, $\beta=\frac{c}{2}\sin A$; therefore $\frac{c}{2}\sin B\cos A-\frac{c}{2}\sin A\cos B+k=0$; thus k is determined. 13. Assume for the equation $\lambda u+\mu v+\nu w=0$; then since the straight line passes through the first point, $\lambda l+\mu m+\nu n=0$, and since it passes through the second point, $\lambda l'+\mu m'+\nu n'=0$. From these two equations find the ratios of λ, μ, ν; thus we obtain for the required equation

$$(mn'-m'n)u+(nl'-n'l)v+(lm'-l'm)w=0.$$

14. $ab(u-v)+c(b+a)w=0$.

15. Assume for the required equation $l\alpha+m\beta+n\gamma=0$; at the centre of the inscribed circle $\alpha=\beta=\gamma$; thus $l+m+n=0$; at the centre of the circumscribed circle α, β, γ are proportional respectively to $\cos A$, $\cos B$, $\cos C$; thus $l\cos A+m\cos B+n\cos C=0$. Hence the required result may be obtained.

18. To CP, $2mv - nw = 0$; to DP, $2lu - 2mv + nw = 0$;

to AQ, $lu - 2mv + 2nw = 0$; to BQ, $lu - 2mv = 0$.

26. Take $\alpha = 0$, $\beta = 0$, $\gamma = 0$ to represent the sides of the triangle $A'B'C'$; then the equations to BC, CA, AB will be respectively $\beta + \gamma = 0$, $\gamma + \alpha = 0$, $\alpha + \beta = 0$. Then the equation to AA' will be $\beta - \gamma = 0$, so that AA' is perpendicular to BC.

27. The equation to OO' is $\beta - \gamma = 0$; take $\beta - \gamma - \lambda\alpha = 0$ for the equation to the straight line drawn through D. Then it will be found that the equation to OF is $\beta - \gamma - \lambda(\alpha - \gamma) = 0$, and that the equation to $O'E$ is $\beta - \gamma - \lambda(\alpha + \beta) = 0$. Thus at the point P we have $\beta = -\gamma$. The same relation holds at the point Q.

28. $$\frac{l\alpha + m\beta + n\gamma}{\sqrt{(l^2 + m^2 + n^2 - 2mn\cos A - 2nl\cos B - 2lm\cos C)}}$$
$$= \pm \frac{l'\alpha + m'\beta + n'\gamma}{\sqrt{(l'^2 + m'^2 + n'^2 - 2m'n'\cos A - 2n'l'\cos B - 2l'm'\cos C)}}.$$

30. See Ex. 29 and page 72. 31. Denote the triangle by PQR; the area is $\dfrac{p^2 \sin P}{2 \sin Q \sin R}$ where p is the perpendicular from P on QR. The length of this perpendicular is known from Example 30; and $\sin P$, $\sin Q$, and $\sin R$ are known from page 70.

32. $a\lambda + b\mu + c\nu = 0$. 33. $l\lambda + m\mu + n\nu = 0$.

34. We must have $\dfrac{\alpha - \alpha'}{\lambda} = \dfrac{\beta - \beta'}{\mu}$ identical with

$$\frac{\alpha - \alpha'}{\lambda} = -\frac{a\alpha + b\beta - (a\alpha' + b\beta')}{c\nu}:$$

this gives $a\lambda + b\mu + c\nu = 0$. 35. A point. 36. We shall find that $\dfrac{AE}{AC} = \dfrac{lc}{lc + na}$; and $\dfrac{AF}{AB} = \dfrac{lb}{lb + ma}$;

hence $$\frac{\text{triangle } AEF}{\text{triangle } ABC} = \frac{l^2bc}{(lc + na)(lb + ma)}.$$

And $$\frac{\text{triangle } DEF}{\text{triangle } ABC} = \frac{2abc\, lmn}{(lc + na)(lb + ma)(nb + mc)}.$$

38. Divide by γ^2; thus we have a quadratic in $\frac{\beta}{\gamma}$: then as in Art. 60 we obtain $(F^2 - AB)(D^2 - BC) = (FD - BE)^2$, that is

$$AD^2 + BE^2 + CF^2 - ABC - 2DEF = 0.$$

39. See Art. 9 and XII. of Art. 78: thus we get the first form. Also $(\alpha_1 - \alpha_2)\sin A + (\beta_1 - \beta_2)\sin B + (\gamma_1 - \gamma_2)\sin C = 0$; transpose the last term and square; thus we express $(\alpha_1 - \alpha_2)(\beta_1 - \beta_2)$ in terms of $(\alpha_1 - \alpha_2)^2$, $(\beta_1 - \beta_2)^2$, and $(\gamma_1 - \gamma_2)^2$, and so obtain the second form. To obtain the third form from the first we put $-\frac{(\alpha_1 - \alpha_2)}{\sin A}\left\{(\beta_1 - \beta_2)\sin B + (\gamma_1 - \gamma_2)\sin C\right\}$ for $(\alpha_1 - \alpha_2)^2$, and make a similar substitution for $(\beta_1 - \beta_2)^2$.

CHAPTER V.

1. $(x^2 + y^2)^2 = a^2(x^2 - y^2)$.

3. $y'^2 = cx'\ 2 - \frac{c^2}{2}$.

4. $y'^2 \sin^2 \alpha = 4ax'$.

6. By Art. 83, we have

$$m = \frac{\sin(\omega - \alpha)}{\sin \omega},\quad n = \frac{\sin(\omega - \beta)}{\sin \omega},\quad m' = \frac{\sin \alpha}{\sin \omega},\quad n' = \frac{\sin \beta}{\sin \omega}.$$

CHAPTER VI.

1. (1) Co-ordinates of the centre 2 and -2, radius 3.
   (2) Co-ordinates of the centre -3 and $\frac{3}{2}$, radius $\frac{7}{2}$.

2. The first straight line meets the circle at the points $(-4, 3)$ and $(3, -4)$; the second at the points $(0, -5)$ and $(-5, 0)$; the third *touches* it at the point $(-4, -3)$.

3. $x^2 + y^2 = hx + ky$.

5. $x^2 - x(x' + x'') + y^2 - y(y' + y'') + x'x'' + y'y'' = 0$.

7. Let O and O' denote the centres of the circles; r and r' their radii; P any point on the locus. Then

$$\frac{r}{OP} = \frac{r'}{O'P}.$$

8. For determining the abscissæ of the points of intersection we have $x^2\left(1 + \frac{k^2}{h^2}\right) + \frac{2k}{h}(b-k)x - 2ax + k^2 - 2bk = 0$; if the straight line *touches* the circle we must have $(kb - ha)^2 + 2kh(ka + hb) = h^2k^2$.

9. $2y + 3x = 0$. 14. $x^2 + y^2 - xy - hx - ky = 0$.
15. Inclination of axes $120°$; co-ordinates of the centre each $= h$; radius $= h$. 16. Inclination of axes $60°$; co-ordinates of the centre each $= \frac{h}{3}$; radius $= \frac{h}{\sqrt{3}}$. 17. $x^2 + y^2 + xy\sqrt{2} - 9 = 0$.

18. $x^2 + y^2 + xy + x + y - 1 = 0$. 19. $\frac{\sqrt{(h^2 + k^2 - 2hk\cos\omega)}}{2\sin\omega}$.

23. $x^2 + y^2 = a\left(x + \frac{y}{\sqrt{3}}\right)$; $r = \frac{2a}{\sqrt{3}}\cos\left(\theta - \frac{\pi}{6}\right)$. 27. A circle.

28. Use the equation in Example 26. 29. Using polar co-ordinates, we have

$$r + \sqrt{(r^2 + l^2 - 2rl\cos\theta)} = \sqrt{\left\{r^2 + l^2 - 2rl\cos\left(\frac{\pi}{3} - \theta\right)\right\}},$$

where l is the length of a side. Reduce and we get

$$\left\{\sqrt{3}r - 2l\cos\left(\theta - \frac{\pi}{6}\right)\right\}^2 = 0;$$

thus the locus is the circle circumscribing the triangle.

30. $\sin^2\alpha + \sin^2\beta + \sin^2\gamma + \ldots = \cos^2\alpha + \cos^2\beta + \cos^2\gamma + \ldots$
and $\sin 2\alpha + \sin 2\beta + \sin 2\gamma + \ldots = 0$.

32. If the perpendiculars are both on the same side of the straight line the locus is a circle; if on different sides the locus consists of two straight lines. 33. A circle. 34. A circle.
36. Solve the quadratic in r; it will be found that $r = 2a\cos\theta$ or $-a\sec\theta$; thus the locus consists of a straight line and a circle.

38. Take the extremity of the diameter as the pole; it will follow from Example 37, that the tangent at P is represented by the equation $2c\cos^2\alpha = r\cos(2\alpha - \theta)$, and the tangent at Q by the equation $2c\cos^2\beta = r\cos(2\beta - \theta)$. These tangents meet at T, so at that point we have $\frac{\cos(2\alpha - \theta)}{\cos^2\alpha} = \frac{\cos(2\beta - \theta)}{\cos^2\beta}$; from this we shall find $\tan\theta = \frac{\sin(\beta + \alpha)}{2\cos\beta\cos\alpha}$, so that if C be the centre of the circle $Ct = \frac{c\sin(\beta + \alpha)}{2\cos\beta\cos\alpha}$. Hence we can shew that $Cq - Ct = Ct - Cp$.

39 and 40. Assume $x^2 + y^2 + Ax + By + C = 0$ for the required equation, and substitute successively the co-ordinates of the given

points (x_1, y_1), (x_2, y_2) and (x_3, y_3). It will be found that the values of A, B, and C have for their common denominator $x_2y_1 - x_1y_2 + x_3y_2 - x_2y_3 + x_1y_3 - x_3y_1$. Then see Art. 36.

CHAPTER VII.

4. $x = y$, and $x + y = \dfrac{3ab}{a+b}$.

5. Let $y = mx$ be the equation to one straight line; then

$$\delta = \frac{y_1 - mx_1}{\pm\sqrt{(1+m^2)}}\text{; therefore } \delta^2(1+m^2) = (y_1 - mx_1)^2;$$

this is a quadratic for finding m, and we may replace m by $\dfrac{y}{x}$.

6. $A^2c^2 + C^2a^2 + B^2ac + ACb^2 - 2ACac - Bb(Ac + Ca) = 0.$

10. Let the given ratio be that of q to p; then the required equations are $\dfrac{p}{P}(l\alpha + m\beta + n\gamma) = \pm\dfrac{q}{Q}(l'\alpha + m'\beta + n'\gamma)$, where

$$P^2 = l^2 + m^2 + n^2 - 2mn\cos A - 2nl\cos B - 2lm\cos C,$$
$$Q^2 = l'^2 + m'^2 + n'^2 - 2m'n'\cos A - 2n'l'\cos B - 2l'm'\cos C.$$

11. Let α and β be the inclinations to the axis of x of the straight lines represented by the given equation; and let θ be the inclination of one of the bisecting straight lines. Then $\theta = \frac{1}{2}(\alpha+\beta)$, or $\frac{\pi}{2} + \frac{1}{2}(\alpha+\beta)$, so that $2\theta = \alpha + \beta$, or $\pi + \alpha + \beta$. In both cases $\tan 2\theta = \tan(\alpha+\beta)$, so that

$$\frac{2\tan\theta}{1-\tan^2\theta} = \frac{\tan\alpha + \tan\beta}{1 - \tan\alpha\tan\beta}.$$

Now by the theory of quadratic equations $\tan\alpha + \tan\beta = -\dfrac{B}{A}$ and $\tan\alpha\tan\beta = \dfrac{C}{A}$. And as the equation to one of the required straight lines is $y = x\tan\theta$, we have finally $\dfrac{2yx}{y^2 - x^2} = \dfrac{B}{A-C}$.

12. $\dfrac{B}{A-C} = \dfrac{b}{a-c}$.

CHAPTER VIII.

1. $y = 2x$. 2. $y^2 = 5ax - x^2$. 3. The locus consists of two parabolas of which the centre of the circle is the common focus, and the directrices are the two tangents to the circle which are parallel to the fixed diameter. 4. The second curve is a parabola having its axis coinciding with the negative part of the axis of y; the curves intersect at the origin and at the point $x = 4a$, $y = -4a$. 5. $y = x + a$. 6. $\tan^{-1}\frac{1}{3}$. 7. $y + x = 3a$. 8. At the point $(9a, -6a)$; length $8a\sqrt{2}$. 9. $y = 2a\sqrt{3}$, $x = 3a$. 11. The abscissa of the required point is 0 or $3a$. 13. The curve is a parabola having its axis parallel to that of y, and its vertex at the point $x = \frac{1}{2}$, $y = \frac{1}{4}$. The straight line is a tangent at the point $x = 1$, $y = 0$. 20. Abscissa of required point is $\frac{1}{4a}\left(\frac{8a^2}{y'} + y'\right)^2$, ordinate $-\left(\frac{8a^2}{y'} + y'\right)$; length of chord $\frac{2}{y'^2}(4a^2 + y'^2)^{\frac{3}{2}}$. 22. Locus of Q, $x = -2a$. Locus of Q', $x^3 = ay^2$. 23. Refer the parabola to PT and the diameter at P as axes. See Art. 151. 25. See Art. 155. 27. Transform equation (1) of Art. 125 to polar co-ordinates, and we shall deduce $r = 2a\dfrac{\cos\theta \pm \sqrt{(\cos 2\theta)}}{\sin^2\theta}$.

28. Use the result of the preceding Example.

29. $r = 2a\dfrac{\sin\theta \pm \sqrt{(-\cos 2\theta)}}{\cos^2\theta}$. 30. The locus is a parabola; see Art. 147. 32. $\sqrt{x} + \sqrt{y} = \sqrt{(a2\sqrt{2})}$. 33. $(y' - x')^2 - 8ax'\sqrt{2} = 0$. 34. $x^2 + y^2 - x(a + x') - yy' + ax' = 0$. 37. Use the result of Example 5, Chap. VI. 41. The equation to one tangent can be written $y = m(x + a) + \dfrac{a}{m}$, (see Example 40), and that to the other $y = -\dfrac{1}{m}(x + a') - a'm$. By eliminating m we have for the required locus $x + a + a' = 0$. This supposes the parabolas both to extend along the *positive* direction of the axis of x; if the second parabola extends along the *negative* direction the final result will be $x + a - a' = 0$. 42. Take for the equation to the chord $y = mx + n$; then to find the abscissa of the middle point of the chord we must take *half the sum of the roots* of the

equation $(mx+n)^2 = 4ax$; so that the abscissa is $\frac{2a-mn}{m^2}$. Now since the chord *touches* the parabola $y^2 = 8a(x-c)$ the equation $(mx+n)^2 = 8a(x-c)$ must have *equal* roots; by means of this condition it can be shewn that $\frac{2a-mn}{m^2} = c$. 44. The equation to the normal at a point (x', y') is $y - y' = -\frac{y'}{2a}(x-x')$. If the normal is to pass through a given point (h, k) we have $k - y' = -\frac{y'}{2a}(h-x')$; also $x' = \frac{y'^2}{4a}$. Thus we obtain a cubic equation for determining y', namely $y'^3 + 4a(2a-h)y' - 8a^2k = 0$. By Chapter III. of the *Theory of Equations* the sum of the roots of this cubic equation is zero. The points of intersection of the parabola with a circle $(x-b)^2 + (y-c)^2 = r^2$ are found by combining the equations to the two curves. Thus we obtain

$$\left(\frac{y^2}{4a} - b\right)^2 + (y-c)^2 = r^2,$$

which is an equation of the fourth degree in y. By the *Theory of Equations* the sum of the roots is zero. If then three of the roots coincide with those already shewn to have a sum equal to zero, the fourth root is zero; and the corresponding point is therefore the vertex of the parabola. 45. The tangents of the inclinations to the axis of x of the three normals that can be drawn through a point (x, y) are determined by the equation $m^3 + m\left(2 - \frac{x}{a}\right) + \frac{y}{a} = 0$. See Art. 135. Suppose m_1, m_2, m_3 the roots of this cubic, then by Chapter III. of the *Theory of Equations*

$$m_1 + m_2 + m_3 = 0, \quad m_2m_3 + m_3m_1 + m_1m_2 = 2 - \frac{x}{a}, \quad m_1m_2m_3 = -\frac{y}{a};$$

if two of the normals are at right angles we may put $m_2m_3 = -1$; from these equations by eliminating m_1, m_2, and m_3, we find $y^2 = a(x-3a)$. 46. By the length is meant the length of the common chord; by the breadth is meant the distance between the two tangents which are parallel to the common chord.

47. $\frac{k^2 - 4ah}{\sqrt{(k^2 + 4a^2)}}$. 55. The equation $y = mx + \frac{a}{m}$ repre-

sents a tangent to the parabola; if this passes through the point (h, k) we have $k = mh + \frac{a}{m}$; also $m = \frac{y-k}{x-h}$, where (x, y) is any point on the tangent; thus $k - \frac{y-k}{x-h}h = \frac{a(x-h)}{y-k}$; this will give the first form of the equation. The second form may be deduced from the first; the student will see hereafter what suggested the second form; see Arts. 341 and 343. 56. The equation $y^2 = 4ax$ represents the parabola; and the equation $ky - 2ax = 2ah$ represents the chord of contact; hence it follows that the equation $4ax(ky - 2ax) = 2ahy^2$ represents *some locus* passing through the intersection of the parabola and chord; then see Art. 61.

57. $x = \frac{a}{m_1 m_2}$, $y = a\left(\frac{1}{m_1} + \frac{1}{m_2}\right)$. If the equation to the third tangent is $y = m_3 x + \frac{a}{m_3}$ the required ordinate is

$$a\left(\frac{1}{m_1} + \frac{1}{m_2} + \frac{1}{m_3} + \frac{1}{m_1 m_2 m_3}\right).$$

CHAPTER IX.

1. $\frac{1}{\sqrt{3}}$. 2. $y + ex = a$; the intercept on the axis of $x = \frac{a}{e}$; and the intercept on the axis of $y = a$. 3. $y + ae^2 = \frac{x}{e}$.

4. The excentricity is determined by $e^4 + e^2 = 1$. 5. $y = \frac{b}{a}(x + a)$; $y = \frac{b^2 x}{a^2 e}$; the straight lines are parallel if $2e^2 = 1$. 6. $y = \frac{b}{ae}(x - ae)$; the abscissa of the point of intersection is $\frac{2ae}{1+e^2}$.

7. $y = -(1+e)(x-a)$; $\tan^{-1}\frac{1}{1+e+e^2}$. 8. $\frac{2a^2 e - ax'(1+e^2)}{a(1+e^2) - 2ex'}$.

9. The co-ordinates of the point are $x = \frac{a^2}{\sqrt{(a^2+b^2)}}$, $y = \frac{b^2}{\sqrt{(a^2+b^2)}}$.

10. The co-ordinates of the point are $x = \frac{a}{\sqrt{2}}$, $y = \frac{b}{\sqrt{2}}$.

19. It will be found that the circle falls entirely without the ellipse if the inclination of the two parallel straight lines to the major axis be greater than $\tan^{-1}\frac{ae}{b}$. 22. $\frac{x}{a}\cos\phi+\frac{y}{b}\sin\phi=1$.
25. The co-ordinates of the required point are $x=\frac{ae(b-a)}{b-ae^2}$, $y=\frac{ab(1-e^2)}{b-ae^2}$; the straight lines are parallel when $e^4+e^2=1$.

28. $x^2+y^2-x(ae+x')-yy'+aex'=0$. 30. If the point (h, k) be *between* the directrices, the sum of the perpendiculars is $\frac{2a^2b^2}{\sqrt{(a^4k^2+b^4h^2)}}$; if the point (h, k) be *not* between the directrices, the sum of the perpendiculars is $\pm\frac{2ab^2he}{\sqrt{(a^4k^2+b^4h^2)}}$, the upper or lower sign being taken according as h is positive or negative. 31. A circle having its centre at the centre of the ellipse and radius $=a+b$.

32. $y=\pm x\pm\sqrt{(a^2+b^2)}$. See Art. 171. 34. Locus is the circle $x^2+y^2=a^2+b^2$; this may be deduced from the second part of Example 33. 35. See remark on Example 55 of Chap. VIII.
42. The first part of this Example may be solved by finding the equation to the straight line passing through the points of intersection of the two ellipses. 45. $x^2+y^2=(a^2+b^2)^{\frac{1}{2}}(x+y)$.
46. Let h, k be the co-ordinates of an external point; the equation to the corresponding chord of contact is $a^2ky+b^2hx=a^2b^2$; the equation to the straight line through (h, k) perpendicular to the chord is $(y-k)\,b^2h=a^2k\,(x-h)$. We require that the latter straight line shall be a tangent to the ellipse; the necessary condition may be found by comparing this equation with the equation $y=mx+\sqrt{(m^2a^2+b^2)}$; thus we shall obtain for the condition $k^2a^6+h^2b^6=h^2k^2(a^2-b^2)^2$. 48. $a^2(y^2+2yk)+b^2(x^2+2xh)=0$.
51. Transferring the origin to the vertex of the ellipse the equation becomes

$$y=m(x-a)+\sqrt{(m^2a^2+b^2)}=mx-ma+ma\left(1+\frac{1-e^2}{m^2}\right)^{\frac{1}{2}}$$

$$=mx-ma+ma\left\{1+\frac{(1+e)\,c}{m^2a}\right\}^{\frac{1}{2}},\text{ where } c=(1-e)\,a.$$

Expand the square root by the Binomial Theorem; then ultimately when $e=1$ and a is infinite, we have $y=mx+\frac{c}{m}$.

52. An ellipse. 53. The locus is an ellipse; if A be the origin, AB the axis of x, each of the co-ordinates of the focus is equal to half the radius of the circle. 54. $\frac{e^2xy}{b^2}$. 55. Put $a\cos\phi$ for x and $b\sin\phi$ for y in the preceding result (Art. 168); then the greatest value is $\frac{e^2a}{2b}$. 57. Let P denote a point on the ellipse, and Q the centre of the circle inscribed in the triangle SPH; then if y' be the ordinate of P it may be shewn that the radius of the circle which $=\frac{\text{area of triangle } SPH}{\text{semiperimeter of triangle}}=\frac{ey'}{1+e}$; this is the ordinate of Q. Let x' be the abscissa of P, then it may be shewn that the abscissa of Q is ex'; thus it will be found that the required locus is an ellipse. 58. Find the point at which SZ meets the normal at P; also find the point at which HZ' meets the normal at P; it will then appear that the points coincide. 60. See Example 12 of Chapter VII.

CHAPTER X.

1. $xb(bx'-ay')+ya(ay'+bx')=a^2b^2$. 2. Refer the ellipse to the diameter and its conjugate as axes. 3. See Art. 11. 8. $r(a^2\sin^2\theta+b^2\cos^2\theta)=2ab^2\cos\theta$. 9 and 10. Use the result of 8. 12. Result the same as that in Example 11. 13. They intersect when $\theta=0$ and when $\theta=\frac{\pi}{2}$. 14. The equations to the tangents at the ends of the latera recta are (Art. 205)

$$r(e\cos\theta+\sin\theta)=a(1-e^2); \qquad r(\sin\theta-e\cos\theta)=a(1+e^2);$$
$$r(e\cos\theta-\sin\theta)=a(1-e^2); \qquad r(\sin\theta+e\cos\theta)=-a(1+e^2).$$

The equations to the tangents at the ends of the minor axis are $r\sin\theta=b$; $r\sin\theta=-b$. 15. A straight line through S. See Art. 205. 17. $\cos\theta=-\frac{e+e'}{1+ee'}$, $r=a(1+ee')$. 18. Between $\frac{b}{a}$ and $\frac{a}{b}$. 20. See Art. 208. 22. The sine of the angle between the radius vector from the centre and the tangent is $\frac{p}{r}$, where $p^2(a^2+b^2-r^2)=a^2b^2$ by Art. 196; then the least value of $\frac{p^2}{r^2}$ may be shewn to be when $2r^2=a^2+b^2$. 29. It may be shewn that the axis of the parabola must coincide with *one* of the axes of the ellipse, hence the latus rectum will be either $\frac{2a^2}{\sqrt{(a^2+b^2)}}$ or $\frac{2b^2}{\sqrt{(a^2+b^2)}}$. 31. An ellipse. 32. An ellipse. 35. Use the polar equations to PQ and pq; see Art. 205. 38. Two of the sides of the parallelogram are determined by the equations $\frac{x}{a}\cos\phi+\frac{y}{b}\sin\phi=\pm 1$, and the other two by the equations $\frac{x}{a}\cos\phi'+\frac{y}{b}\sin\phi'=\pm 1$; see Example 22 of Chap. IX. It may be shewn that the diagonals of the parallelogram intersect at the centre of the ellipse; then if the centre of the ellipse be joined with two adjacent corners of the parallelogram the triangle thus formed is one fourth of the parallelogram; and the area of the triangle is known by Example 7 of Chap. I. 41. The abscissa is $\frac{bx'-ay'}{b}$, and the ordinate $\frac{ay'+bx'}{a}$. 42. The co-ordinates of the intersection of the tangents are found in Example 41; call them h and k, then use the second form given in Example 35 of Chap. IX. 44. The greatest value may be found by substituting for x' and y' their values from Art. 168; it is $ab(\sqrt{2}-1)$. 48. An ellipse referred to its *equal* conjugate diameters. 51. This may be solved by means of Example 50. Or we may take the usual axes, then if x', y' be the co-ordinates of P those of M will be $\frac{a(ax'+by')}{a^2+b^2}$ and $\frac{b(ax'+by')}{a^2+b^2}$; those of N will be $\frac{a(ax'-by')}{a^2+b^2}$ and $\frac{b(by'-ax')}{a^2+b^2}$. Hence the solution can be completed. 52. See Art. 208.

CHAPTER XI.

1. $y^2 - 3x^2 = -3a^2$. 2. A straight line. 7. See Arts. 178 and 228.

CHAPTER XII.

4. Let a straight line be drawn through the focus meeting the hyperbola at P and p and the asymptotes at Q and q; then it may be shewn that $Pp = \frac{2a(e^2-1)}{1-e^2\cos^2\theta} = \frac{2a\sin^2\alpha}{\cos^2\alpha - \cos^2\theta}$, $Qq = \frac{2a\sin\alpha\sin\theta}{\cos^2\alpha - \cos^2\theta}$, and the required length is half the difference of Pp and Qq. 5. Take the centre of the circle as the origin, AB as the axis of x, and a diameter parallel to PQ as the axis of y; then the locus is given by the equation $y^2 = x^2 - a^2$, and is therefore a rectangular hyperbola referred to conjugate diameters. 11. By Example 53 of Chapter VIII. we shall obtain $\tan\alpha = \frac{\sqrt{(k^2-4ah)}}{h+a}$; thus $(h+a)^2\tan^2\alpha = k^2 - 4ah$; therefore $(h+a)^2\sec^2\alpha = k^2 + (h-a)^2$. 12. Both the diameters must *meet* the curve; it will be found that this requires the conjugate axis to be *greater* than the transverse axis.

CHAPTER XIII.

1. The equation may be written $(x-2y)(x-2y-2a) = 0$, and therefore represents two *parallel* straight lines; a straight line parallel to them, and midway between them, will be a line of centres. 2. $h = \frac{b}{3}$, $k = \frac{c}{3}$. 3. Two parallel straight lines. 4. A parabola. 5. An hyperbola if the angle A is less than $\frac{\pi}{2}$, an ellipse if it is greater than $\frac{\pi}{2}$, a straight line if it is equal to $\frac{\pi}{2}$. 6. The equation to the hyperbola is $a^2y^2 = a^2b^2 - 4ab^2x + 3b^2x^2$; the asymptotes are determined by the equations $ay = \pm\left(x - \frac{2a}{3}\right)b\sqrt{3}$.

8. The locus is then a straight line which coincides with the equal axes. 10. Use Art. 205. 11. $\frac{a\sqrt{2}}{4}$. 13. $\text{Tan}^{-1}\frac{1}{b}$.
14. $\{ay + x\sqrt{(\beta\beta')}\}^2 - 2a\beta\beta'x - a^2(\beta+\beta')\,y + a^2\beta\beta' = 0$.

17. (1) A circle about the other focus of the given ellipse as centre; (2) an ellipse about the other focus of the given ellipse as focus, and having the same excentricity as the given ellipse. 18. The equation is $(y - 3x + 1)(y - 2x + 4) = 0$, and therefore represents two straight lines. 24. Use the result given in Example 56 of Chap. VIII. 26. The equation may be written

$$(x^2 + y^2 + xy\sqrt{2} - a^2)(x^2 + y^2 - xy\sqrt{2} - a^2) = 0.$$

27. Take AB and AC as axes of x and y. Let the angle PBA and the angle PCA be each equal to α, and the angle $BAC = \omega$. Let x and y be the co-ordinates of P; then

$$PB = \frac{y\sin\omega}{\sin\alpha}, \quad PC = \frac{x\sin\omega}{\sin\alpha}.$$

And $BC^2 = BP^2 + CP^2 - 2BP.CP\cos BPC$. 28. Take the given point as the origin, the common tangent at that point as the axis of y, and the diameter through that point as the axis of x. Then the equation to the parabola will be of the form $y^2 = 4cx$, and the equation to the other tangent $y = mx + \frac{c}{m}$, where m is constant for all the parabolas. Whatever be the value of c the point of contact is on the locus $y^2 = 4xm(y - mx)$, which is obtained by eliminating c; that is on the locus $(y - 2mx)^2 = 0$. 29. We may take for the equation to the ellipse $y^2 = \frac{b^2}{a^2}(2ax - x^2)$. Let (x', y') be a point on it; then the equation to one of the straight lines is $y + 2b = \frac{y' + 2b}{x'}x$; put $y = 0$, then $x = \frac{2bx'}{y' + 2b}$: this gives the length of one segment. The length of the segment at the other end of the major axis will be $\frac{2b(2a - x')}{y' + 2b}$; and therefore the length of the third segment $\frac{2ay'}{y' + 2b}$. 30. Take one of the ellipses, and refer it to its equal conjugate diameters as axes, the axis of x being that which passes through the fixed point. Let C be the centre of the ellipse, P the point of contact of the tangent from the fixed point, PM the ordinate of P: then it may be shewn that M is a fixed point and MP a constant length.

31. $\left(\frac{TP}{TQ}\right)^2 = \frac{SP \cdot HP}{SQ \cdot HQ}$ by Arts. 208 and 193. Also $\frac{SP}{SQ} = \frac{PR}{QR}$, and $\frac{HP}{HQ} = \frac{PR'}{QR'}$, by Art. 158.

32. Refer the ellipse to rectangular axes with P as origin and PK as the axis of x. The equation will be of the form $ax^2 + bxy + cy^2 + dx + ey = 0$. Assume $y = mx + n$ for the equation to the straight line QR; then the following equation will represent the two straight lines PQ and PR,

$$ax^2 + bxy + cy^2 + \frac{(dx + ey)(y - mx)}{n} = 0.$$

In order that these two straight lines may be equally inclined to PK we must have $b + \frac{d - em}{n} = 0$, so that $n = \frac{em - d}{b}$. Thus the equation to QR becomes $y = mx + \frac{em - d}{b}$, and this is satisfied by $x = -\frac{e}{b}$ and $y = -\frac{d}{b}$, whatever m may be.

CHAPTER XIV.

2. Each locus is an ellipse.

4, 5, 6. Use the equation in Art. 294.

7. The equation to the ellipse is $\frac{x^2}{a^2} + \frac{y^2}{b^2} = 1$; the equation to the chord of contact is $\frac{xh}{a^2} + \frac{yk}{b^2} = 1$; hence the equation $\frac{x^2}{a^2} + \frac{y^2}{b^2} = \frac{xh}{a^2} + \frac{yk}{b^2}$ represents some locus passing through the points of contact.

10. The equation to the hyperbola is $(y - k)\, b^2 x = (x - h)\, a^2 y$.

12. Let y', y'' denote the two ordinates which correspond to the same abscissa x'; then

$$y' = -bx' + \sqrt{(b^2x'^2 - ax'^2 - f)}, \quad y'' = -bx' - \sqrt{(b^2x'^2 - ax'^2 - f)}.$$

The equations to the normals are, by Art. 284,

$$(y - y')(ax' + by') = (y' + bx')(x - x'), \text{ and}$$
$$(y - y'')(ax' + by'') = (y'' + bx')(x - x');$$
by addition $(a - b^2)\,x'(y + 2bx') + bf = 0 \ldots (1)$;
by subtraction $b(y + bx') - (a - b^2)\,x' = x - x'$,
therefore $x'(1 + 2b^2 - a) = x - by \ldots\ldots\ldots\ldots (2)$.

Substitute the value of x' from (2) in (1) and the required equation will be obtained. The locus is an hyperbola. 13. Locus a conic section, which passes through H and R, and through the intersection of the fixed straight lines. 18. A circle having its centre on the straight line joining the two points. 19. Two loci, an ellipse, and a parabola. 20. A circle. 23. See Art. 293. 26. Use the equation to the parabola given in Art. 294, and the equation to the circle given in Example 21 to Chap. VI. 29. $r \sin 2\theta = c$. 30. $x^{\frac{4}{3}}y^{\frac{2}{3}} + y^{\frac{4}{3}}x^{\frac{2}{3}} = a^2$. 32. See Example 30 to Chap. X. 35. An ellipse. 37. In the first case the locus is a circle; in the second case it is a straight line. 38. A circle having its centre at H. 41. The equation to the parabola may be written $y^2 = 4a(x - a) + 4a^2$; the equation to the chord of contact is $ky = 2a(x - a) + 2a(a + h)$; therefore the following equation represents some locus passing through the intersection of the parabola and the chord of contact,

$$y^2 = 4a(x - a)\frac{ky - 2a(x - a)}{2a(a + h)} + \left\{\frac{ky - 2a(x - a)}{a + h}\right\}^2.$$

44. $\dfrac{b^2x^2}{a^4} + \dfrac{y^2}{b^2} = 1$. 46. The equation is $y^2 = 4a(x - 8a)$.

50. The straight line $\dfrac{x}{a} - \dfrac{y}{b} = 0$ bisects the chord of contact, and is therefore parallel to the axis of the parabola; if through the point $(a, 0)$ a straight line be drawn making the same angle with the tangent at that point as the axis makes, the focus must be in this straight line: $y(a + 2b\cos\omega) + b(x - a) = 0$ is the equation to this straight line. Similarly we can draw a straight line through the point $(0, b)$ which will also contain the focus. 52. We may take for the equation to one normal $y = mx - am - am^3$, and for the other $x = m'y - am' - am'^3$; also $m' = -m$. Then by addition $y + x = m(x - y)$. Substitute for m in the first equation

and reduce; thus we obtain $2a(x+y)=(x-y)^2$. 53. We have to eliminate m between

$$y-mx=-\frac{m(a^2-b^2)}{\sqrt{(a^2+m^2b^2)}}, \text{ and } my+x=\frac{m(a^2-b^2)}{\sqrt{(m^2a^2+b^2)}}.$$

Square and add; we shall obtain after reduction

$$y^2+x^2=\frac{(a^2+b^2)(a^2-b^2)^2}{a^2b^2\left(m-\frac{1}{m}\right)^2+(a^2+b^2)^2} \quad\ldots\ldots\ldots\ldots\ldots\ldots(1).$$

Also $\quad (y-mx)^2(a^2+m^2b^2)=(my+x)^2(m^2a^2+b^2)$;

by reduction we obtain

$$(a^2y^2-b^2x^2)\left(m-\frac{1}{m}\right)=-2xy(a^2+b^2)\ldots\ldots\ldots\ldots(2).$$

From (1) and (2)

$$(a^2+b^2)(x^2+y^2)(a^2y^2+b^2x^2)^2=(a^2-b^2)^2(a^2y^2-b^2x^2)^2.$$

54. Suppose the figure in Art. 192 to represent the ellipse and the conjugate diameters. Take the equation in Example 23 of Chapter IX. for the equation to the normal at P, and an analogous equation for the equation to the normal at D. Let Q denote the point of intersection of these normals, and x, y its co-ordinates. Then it will be found that

$$ax=(a^2-b^2)\sin\phi\cos\phi(\sin\phi-\cos\phi),$$
$$by=(b^2-a^2)\sin\phi\cos\phi(\sin\phi+\cos\phi).$$

Similarly we can determine the co-ordinates of the point of intersection of the normals at P' and D; denote this point by R. Then express the area of the triangle CQR, which is one-fourth of the required area.

55. Take the centre of the square as the origin, and the axes parallel to the sides of the square. Then for the equation to the circle take $x^2+y^2=2a^2$, and for the equation to the conic take $y^2-a^2=\lambda(x^2-a^2)$. The equation to the tangent to the circle at the point (x_1, y_1) is $xx_1+yy_1=2a^2$. The equation to the tangent to the conic at the point (x', y') is $yy'-\lambda xx'=a^2(1-\lambda)$. These equations must represent the same straight line. Hence eliminating λ and x_1 and y_1 we shall arrive at an equation which deter-

mines the required locus. It will be found that this equation may be written $\{(x'^2+y'^2-2a^2)\}\{a^2(x'^2+y'^2)-2x'^2y'^2\}=0$.

56. This follows from Art. 288. 57. Let a perpendicular be drawn from H on the tangent TQ, and let R denote the intersection of this perpendicular with SQ produced. Then $SR=SQ+QR=2a$; and $TR=TH$. We have to find the value of the perpendicular from T on SR; denote it by r; then $r2a=$ twice the area of the triangle TSR. Let $TS=c_1$, and TR or $TH=c_2$; then by using the known expression for the area of a triangle in terms of its sides, we have

$$4ra=\sqrt{(2c_1^2c_2^2+8a^2c_1^2+8a^2c_2^2-c_1^4-c_2^4-16a^4)}.$$

This will lead to the required result. Or thus: Let ϕ denote the angle between HP and TP; then we shall have

$$r=TP\sin\phi=TP\times\frac{b}{CD},$$

where CD is conjugate to CP; see Arts. 181 and 193. And it may be shewn by Art. 208 that $\left(\frac{TP}{CD}\right)^2=\frac{x^2}{a^2}+\frac{y^2}{b^2}-1$.

59. Determine the co-ordinates of the points of intersection of the tangents by Art. 288; it will be seen that they satisfy the given equation. 60. It may be shewn that the equation

$$x(2ah+bk+d)+y(2ck+bh+e)+dh+ek+2f=0$$

represents the *chord of contact*: see Arts. 183 and 283. Thus the equation represents some locus passing through the intersections of the curve and the chord of contact. Also the equation may be put in the form

$$A(x-h)^2+B(x-h)(y-k)+C(y-k)^2=0.$$

Hence it represents two straight lines through the point (h, k). See also the remark on Example 55 of Chapter VIII.

CHAPTER XV.

6. $\sqrt{\alpha}+\sqrt{\beta}+\sqrt{\gamma}=0$. 10. The equation to the conic section being $l\beta\gamma+m\gamma\alpha+n\alpha\beta=0$, that to $A'B$ is $(m+n)\alpha+l\gamma=0$, that to $A'C$ is $(m+n)\alpha+l\beta=0$, and that to $A'B'$ is $(m+n)\alpha+(l+n)\beta-n\gamma=0$. 13. $lmn+1=0$.

21. $$\frac{y - m_1 x - \frac{a}{m_1}}{\sqrt{(1+m_1^2)}} \frac{y - m_2 x - \frac{a}{m_2}}{\sqrt{(1+m_2^2)}} \frac{m_2 - m_1}{\sqrt{(1+m_1^2)}\sqrt{(1+m_2^2)}} + \ldots = 0,$$

or $$(1+m_3^2)\left(y - m_1 x - \frac{a}{m_1}\right)\left(y - m_2 x - \frac{a}{m_2}\right)(m_2 - m_1) + \ldots = 0.$$

24. Suppose the focus S is to lie on the straight line $l\alpha + m\beta + n\gamma = 0$. Let α', β', γ' denote the values of α, β, γ respectively for the other focus H of one of the ellipses. Then, by Art. 181, $\alpha\alpha' = \beta\beta' = \gamma\gamma' =$ the square of half the minor axis. Hence, substituting in the given equation we obtain $\frac{l}{\alpha'} + \frac{m}{\beta'} + \frac{n}{\gamma'} = 0$, that is, $l\beta'\gamma' + m\gamma'\alpha' + n\alpha'\beta' = 0$. This shews that the locus of H is a conic section passing through the angular points of the triangle.

25. It will be found that the conic sections may be represented by the equations

(1) $\beta\gamma - \alpha^2 = 0$, (2) $\gamma\alpha - \beta^2 = 0$, (3) $\alpha\beta - \gamma^2 = 0$.

Now, (1) may be written $\beta(\gamma + \beta - 2\alpha) - (\alpha - \beta)^2 = 0$,

(2) may be written $\gamma(\alpha + \gamma - 2\beta) - (\beta - \gamma)^2 = 0$,

(3) may be written $\alpha(\beta + \alpha - 2\gamma) - (\gamma - \alpha)^2 = 0$;

this shews that the tangents to the conic sections at the common point are given by

$$\gamma + \beta - 2\alpha = 0, \quad \alpha + \gamma - 2\beta = 0, \quad \beta + \alpha - 2\gamma = 0;$$

these three straight lines intersect respectively the straight lines

$$\alpha = 0, \quad \beta = 0, \quad \gamma = 0,$$

at three points which all lie on the straight line $\alpha + \beta + \gamma = 0$. Again, (1) may be written $\beta(\gamma + 4\alpha + 4\beta) - (\alpha + 2\beta)^2 = 0$, and (2) may be written $\alpha(\gamma + 4\alpha + 4\beta) - (\beta + 2\alpha)^2 = 0$; and this shews that $\gamma + 4\alpha + 4\beta = 0$ is a common tangent of (1) and (2), and this common tangent meets $\gamma = 0$ at the point where $\beta + \alpha - 2\gamma = 0$ meets it. And so on.

26. The equation to the first hyperbola is $\beta\gamma = AA'^2 \sin^2 \frac{A}{2}$; similarly for the others.

27. See Art. 274.

28 and 29. These may be solved by taking oblique axes coinciding with the sides of the triangle. For instance, consider 29.

We have $a\alpha + b\beta + c\gamma = -ab \sin C$. Thus the equation may be written $cn\alpha\beta - (l\beta + m\alpha)(ab \sin C + a\alpha + b\beta) = 0$; and taking CA for the axis of x, and CB for the axis of y, we have $\alpha = x \sin C$, $\beta = y \sin C$. Substitute for α and β and then to the equation in x and y we may apply the ordinary test; see Arts. 272 and 278.

30. $S = \dfrac{\cos^4 \dfrac{A}{2}}{a^2} (a\alpha + b\beta + c\gamma)^2$, where S denotes $\alpha^2 \cos^4 \dfrac{A}{2} + \dots$; see Art. 334.

31. $u(mn' - m'n) = v(nl' - n'l) = w(lm' - l'm)$.

32. Let $S_1 = 0$ be the equation to the inscribed circle, $S_2 = 0$ the equation to the circumscribed circle, these equations not being necessarily in their simplest forms; see Art. 110. Then, if k be a suitable constant, $S_1 - kS_2 = 0$ will represent the straight line required. In this way we shall have

$$\alpha^2 \cos^4 \frac{A}{2} + \beta^2 \cos^4 \frac{B}{2} + \gamma^2 \cos^4 \frac{C}{2} - 2\beta\gamma \cos^2 \frac{B}{2} \cos^2 \frac{C}{2}$$
$$- 2\gamma\alpha \cos^2 \frac{C}{2} \cos^2 \frac{A}{2} - 2\alpha\beta \cos^2 \frac{A}{2} \cos^2 \frac{B}{2}$$
$$- k(\beta\gamma \sin A + \gamma\alpha \sin B + \alpha\beta \sin C)$$
$$= (a\alpha + b\beta + c\gamma)(l\alpha + m\beta + n\gamma),$$

where l, m, n, are to be found. Then by comparing like terms we can find l, m, n.

33. It may be shewn that the equation $\dfrac{n\beta + m\gamma}{a} = \dfrac{l\gamma + n\alpha}{b}$ represents a diameter; for this equation represents a straight line passing through the intersection of the tangents at A and B, and through the middle point of AB. Hence the centre of the conic section is determined by $\dfrac{n\beta + m\gamma}{a} = \dfrac{l\gamma + n\alpha}{b} = \dfrac{m\alpha + l\beta}{c}$; and then the required equation can be found. It is

$$\frac{\beta}{m(al - bm + cn)} = \frac{\gamma}{n(al + bm - cn)}.$$

34. Assume for the required equation $\gamma =$ constant, that is $\gamma = k(a\alpha + b\beta + c\gamma)$. Then by applying the result of Art. 322 we

shall obtain for the required equation $(lb+ma)(a\alpha+b\beta)-nab\gamma=0$. 36. It may be shewn that the equation to the conic section is $\frac{\beta\gamma}{a}+\frac{\gamma\alpha}{b}+\frac{\alpha\beta}{c}=0$: then apply the condition given in Example 29. 37. The Example depends chiefly on the fact that a straight line can be drawn through the intersection of PR and QS so as to bisect both PQ and RS. 38. It may be shewn that the equation to the secant through the points (t', u', v', w') and (t'', u'', v'', w'') is $(t-t')(u-u'')-(v-v')(w-w'')=tu-vw$; from this we can obtain the equation to the tangent. 39. Since the conic section touches $\alpha=0$, $\beta=0$, and $\gamma=0$, we may assume for its equation $\sqrt{(l\alpha)}+\sqrt{(m\beta)}+\sqrt{(n\gamma)}=0$; then apply the conditions given in Art. 322 under which the conic section touches the other sides. 40. It may be shewn that the expression for the length of the perpendicular on DE from the point (α, β, γ) is $\frac{\alpha \sin A+\beta \sin B-\gamma \sin C}{2 \sin C}$. Hence the equation to the straight line which bisects the angle EDF is

$$\frac{\alpha \sin A+\beta \sin B-\gamma \sin C}{2 \sin C}=\frac{\alpha \sin A-\beta \sin B+\gamma \sin C}{2 \sin B}.$$

41. $\sqrt{(a\alpha)}+\sqrt{(b\beta)}+\sqrt{(c\gamma)}=0$. 42. The equation to the conic section may be taken to be $\alpha\beta=k\gamma^2$; and the equation to the straight line PQ will be $\alpha-\beta=0$. The equation to the chord will be $\alpha-\beta=k'\gamma$. Thus $k(\alpha-\beta)^2=k'^2\alpha\beta$ will represent the straight lines joining P with the points of intersection of the chord and the conic section. From the symmetrical form of the last equation we infer that one straight line makes the same angle with the straight line $\alpha=0$ which the other makes with the straight line $\beta=0$.

CHAPTER XVI.

1. See Example 35 of Chapter XIV. 2. Suppose the conic section to be an ellipse. Let S denote the point of contact of the plane with the sphere which is between the plane and the vertex of the cone; and let H denote the point of contact of the plane

with the sphere which is on the other side of the plane. Join any point P of the conic section with S and H and with the vertex O: then we shall shew that $SP+PH$ is constant. Since all tangents to a sphere from a given point are of equal length, PS is equal to that portion of PO which is between P and that point of PO which is common to the smaller sphere and the cone. Similarly PH is equal to that portion of OP produced which is between P and that point of OP produced which is common to the larger sphere and the cone. Thus $SP+PH$ is equal to the part of a generating line of the cone which is terminated by the two spheres, and is therefore constant.

Next, let A be that vertex of the ellipse which is the nearer to S. Let T' be the point in OA where the cone is touched by the smaller sphere, X the intersection of SA produced with the plane of contact of the smaller sphere and cone. Then AS and AT are equal, being tangents from A to the sphere. And with the notation of Art. 344 we have $\frac{AX}{AT}=\frac{\cos\alpha}{\cos(\alpha+\theta)}=\frac{1}{e}$. Therefore $AX=\frac{AS}{e}$. Thus X is the intersection of the axis of the conic section with the directrix corresponding to S. In a similar manner the other directrix is determined.

If the conic section is an hyperbola the demonstration remains substantially the same. For the history of this theorem see Hutton's *Course of Mathematics* by T. S. Davies, Vol. II. page 208.

3. In Art. 344 if the section be a parabola, it will be found that the latus rectum varies as OA. Hence so long as we keep to sections perpendicular to the same plane OBC, the required locus consists of two straight lines passing through O. Thus on the whole the locus is the surface of a certain right cone which has the same axis and vertex as the given cone.

6. $\beta\gamma\cos A+\gamma\alpha\cos B+\alpha\beta\cos C=0$. 7. Take the figure of Art. 292. Let $u=0$ denote AC, $v=0$ denote BD, $w=0$ denote EF: then we may assume $lu+mv=0$ as the equation to FG, and $lu+mv+nw=0$ as the equation to FA. Then by Arts. 358 and 359, the equation to FB is $lu+mv-nw=0$. It may now be

shewn that $lu - mv + nw = 0$ denotes EC, and that $mv + nw - lu = 0$ denotes EB. 8. Join Nl cutting AB at G; join Gn and GL. Let Ol cut AB at m. Then GO, GN, GM, GL form an harmonic pencil; and so also do GO, Gn, Gm, Gl: see Art. 360. Therefore, by the aid of Art. 354, it follows that LGn is a straight line. Or the Example may be deduced immediately from Art. 292.

CHAPTER XVII.

2. It is easily seen that the triangle KCL is the projection of a triangle of constant area in a circle. Since the area of a triangle is half the product of the base into the perpendicular from the vertex on the base, the result may be put in this form: the length of the perpendicular from C on PD varies inversely as the semidiameter parallel to PD. 8. This is to be considered in the first place with respect to *concentric circles* and *rectangles*. Let C denote the centre of the circles, L a corner of the inscribed rectangle, so that L is on the circumference of the inner circle. Let r be the radius of this circle, and R the radius of the outer circle; let x and y be the co-ordinates of L. Draw through L a straight line parallel to the axis of x meeting the outer circumference at M, and a straight line parallel to the axis of y meeting the outer circumference at N. Complete the rectangle of which LM and LN are adjacent sides; and let P denote the other corner of this rectangle. Then the abscissa of M is $\sqrt{(R^2 - y^2)}$, and the ordinate of N is $\sqrt{(R^2 - x^2)}$; and these are the co-ordinates of P. Thus $CP^2 = R^2 - x^2 + R^2 - y^2 = 2R^2 - r^2$; so that the locus of P is a concentric circle, the radius of which is independent of the original rectangle.

THE END.

CAMBRIDGE: PRINTED BY C. J. CLAY, M.A. AND SONS, AT THE UNIVERSITY PRESS.

WORKS

BY

I. TODHUNTER, Sc.D., F.R.S.

Natural Philosophy for Beginners. In two parts. Part I. The Properties of Solid and Fluid Bodies. With numerous Examples. 18mo. 3s. 6d. Part II. Sound, Light, and Heat. 3s. 6d.

Euclid for Colleges and Schools. New Edition. 18mo. cloth. 3s. 6d.

Key to Exercises in Euclid. Crown 8vo. cloth. 6s. 6d.

Mensuration for Beginners. With numerous Examples. New Edition. 18mo. 2s. 6d.

Key to the Mensuration for Beginners. By Rev. Fr. Lawrence McCarthy, Professor of Mathematics in St Peter's College, Agra. Crown 8vo. 7s. 6d.

Algebra for Beginners. With numerous Examples. New Edition. 18mo. cloth. 2s. 6d.

Key to the Algebra for Beginners. New Edition. Crown 8vo. cloth. 6s. 6d.

Trigonometry for Beginners. With numerous Examples. New Edition. 18mo. cloth. 2s. 6d.

Key to the Trigonometry for Beginners. Crown 8vo. cloth. 8s. 6d.

Mechanics for Beginners. With numerous Examples. New Edition. 18mo. cloth. 4s. 6d.

Key to the Mechanics for Beginners. Crown 8vo. cloth. 6s. 6d.

Algebra for the use of Colleges and Schools. With numerous Examples. New Edition. Crown 8vo. cloth. 7s. 6d.

Key to the Algebra for the use of Colleges and Schools. New Edition. Crown 8vo. cloth. 10s. 6d.

A Treatise on Analytical Statics. With numerous Examples. Fifth Edition. Edited by J. D. Everett, M.A., F.R.S. Crown 8vo. 10s. 6d.